Advances in Marine Gas Hydrate Exploration and Discovery

Advances in Marine Gas Hydrate Exploration and Discovery

Guest Editors

Wei Zhang
Pibo Su
Jiliang Wang
Qianyong Liang

MDPI

Basel • Beijing • Wuhan • Barcelona • Belgrade • Novi Sad • Cluj • Manchester

Guest Editors

Wei Zhang
Sanya Institute of South China Sea Geology
Guangzhou Marine Geological Survey
Sanya
China

Pibo Su
Sanya Institute of South China Sea Geology
Guangzhou Marine Geological Survey
Sanya
China

Jiliang Wang
Institute of Deep-sea Science and Engineering
Chinese Academy of Sciences
Sanya
China

Qianyong Liang
National Engineering Research Center of Gas Hydrate Exploration and Development
Guangzhou Marine Geological Survey
Guangzhou
China

Editorial Office

MDPI AG
Grosspeteranlage 5
4052 Basel, Switzerland

This is a reprint of the Special Issue, published open access by the journal *Journal of Marine Science and Engineering* (ISSN 2077-1312), freely accessible at: https://www.mdpi.com/journal/jmse/special_issues/WC1883YBSP.

For citation purposes, cite each article independently as indicated on the article page online and as indicated below:

Lastname, A.A.; Lastname, B.B. Article Title. *Journal Name* **Year**, *Volume Number*, Page Range.

ISBN 978-3-7258-5649-7 (Hbk)
ISBN 978-3-7258-5650-3 (PDF)
https://doi.org/10.3390/books978-3-7258-5650-3

Cover image courtesy of Wei Zhang

Contents

Journal of
Marine Science and Engineering

MDPI

Editorial

Advances in Marine Gas Hydrate Exploration and Discovery

Wei Zhang [1,2,*], Pibo Su [1,2], Jiliang Wang [3] and Qianyong Liang [2]

1 Sanya Institute of South China Sea Geology, Guangzhou Marine Geological Survey, China Geological Survey, Sanya 572024, China; spb_525@sina.com
2 National Engineering Research Center of Gas Hydrate Exploration and Development, Guangzhou 511458, China; tomlqy@163.com
3 Institute of Deep Sea Science and Engineering, Chinese Academy of Sciences, Sanya 572000, China; wangjl@idsse.ac.cn
* Correspondence: zwgmgs@foxmail.com; Tel.: +86-134-2218-5987

Natural gas hydrates are extensively distributed across terrestrial permafrost zones and continental margins worldwide [1]. Recognized as a clean and efficient energy resource, they have attracted significant exploration interest from major global economies [2–5]. In recent decades, substantial progress has been achieved in hydrate exploration and discovery, driven by advancements in multidisciplinary approaches, including geological, geophysical, and geochemical investigations, as well as numerical simulation techniques [6]. Current research on gas hydrate focuses on several critical aspects: (1) the origin of hydrate gas and its genetic relationship with deep conventional oil and gas reservoirs [7–9]; (2) the mechanisms governing hydrate accumulation and reservoir formation [10]; (3) the spatial heterogeneity in hydrate deposits, including their characteristics, evolution, and controlling factors [11]); and (4) innovations in drilling and production testing technologies [12–14]. This research domain represents a highly interdisciplinary field, integrating knowledge from tectonics, sedimentology, geophysics, geochemistry, petrology, mineralogy, and numerical modeling [15]. The expansion of this field has been facilitated by advancements in field exploration methodologies, laboratory analytical techniques, and computational simulation capabilities. Substantial progress in gas hydrate research has been driven by scientific and technological breakthroughs, including field exploration and discoveries in the South China Sea (SCS) [16] and other continental margins worldwide [17], as well as laboratory-based numerical simulations [18,19]. This Special Issue aims to advance research on gas hydrate exploration and discovery in continental margins as well as gas hydrate test mining simulation, with an emphasis on the mechanism of accumulation and the assessment of marine hydrate's environmental and resource benefits. By addressing these key aspects, we seek to provide theoretical insights and methodological frameworks to support future exploration, drilling, and test production of marine hydrates.

Understanding the controlling factors of submarine gas seepage and bottom current activity on pockmark formation and development is crucial for interpreting seafloor geomorphological evolution. Li et al. (contribution 1) revealed that in the northwestern Xisha Uplift region, deep-sourced gas migrates along fault systems, fractures, and gas chimneys before escaping at the seafloor, where it forms characteristic pockmark features. Their study further demonstrates that established pockmarks undergo significant modification under the influence of bottom current activity.

Precise identification of mass transport deposits (MTDs) in gas hydrate exploration significantly improves bottom simulating reflector (BSR) detection accuracy and facilitates refined resource assessment. Gong et al. (contribution 2) demonstrated that MTDs with low-permeability, high-density properties effectively cap shallow deep-sea fluids.

These deposits inhibit upward free gas migration from underlying strata, promoting gas accumulation for hydrate formation below.

Identifying hydrate gas sources is crucial for predicting their distribution. Zhong et al. (contribution 3) employed acid extraction in deepwater depressions to detect hydrocarbons, proposing a "lower-generation, upper-accumulation, and micro-fracture leakage" model to explain hydrocarbon anomalies in the Chaoshan Depression, SCS. Here, underlying reservoirs control hydrocarbon migration via micro-fractures to seabed surfaces. Su et al. (contribution 4) suggested thermogenic gas, associated with mud diapirism and fault migration, as the primary hydrate source.

Shallow gas, with substantial resource potential, has been verified through 3D seismic data and drilling in deepwater areas. Yang et al. (contribution 5) employed seismic data and geophysical attributes to examine how sedimentary facies interfaces control shallow gas distribution. Their study revealed that vertically stacked channelized submarine fans, which onlap these facies interfaces, serve as primary shallow gas reservoirs. The researchers found sand-rich sediments aligned along the southwest-northeast trending facies boundary, effectively constraining shallow gas accumulation within the basin.

The three-phase coexistence zone (hydrate-gas-water) represents a critical research focus internationally due to its high exploitation potential. Yu et al. (contribution 6) established an advanced characterization model by applying deterministic complex geological modeling techniques combined with seismic and logging data. Their model accurately captures geological structure interactions while using them as auxiliary constraints, significantly reducing accuracy uncertainties caused by complex geological conditions.

Understanding how mineral composition, grain size, and sedimentary processes govern gas hydrate accumulation is crucial for predicting reservoir distribution and guiding exploration. Bai et al. (contribution 7) demonstrated that while grain size shows limited influence, clay mineral content (particularly smectite) in fine-grained sediments significantly inhibits hydrate formation, highlighting mineralogy's dominant role in accumulation efficiency.

Understanding the accumulation and evolutionary processes of gas hydrates is crucial for elucidating their formation mechanisms and assessing resource potential. Guan et al. (contribution 8) developed a novel flow-reaction model to investigate the spatiotemporal evolution of gas hydrate systems in the SCS. Their study systematically compared six distinct environmental scenarios and three cases of paleo-hydrate occurrence. Remarkably, the simulation results demonstrate strong concordance with observed field distributions of massive hydrate reservoirs dating back to 30 kyr BP (before present), providing valuable validation of the model's predictive capabilities.

Understanding the mechanical properties of gas hydrate-bearing sediments (GHBS) is essential for safe and sustainable hydrate production. Yuan et al. (contribution 9) revealed that elevated hydrate saturation and confining pressure substantially improve GHBS strength and stiffness, while inducing more significant dilatancy during shear. Their study further demonstrates hydrate's dual role in pore-filling and sediment-cementation, with both effects being strongly dependent on saturation and confinement conditions.

Wan et al. (contribution 10) developed a gas hydrate reservoir model using China's offshore hydrate test data, numerically analyzing production behavior and reservoir evolution during stepwise depressurization via vertical wells. Their results demonstrate that this approach sustains favorable gas-to-water ratios while enhancing production efficiency—reducing water output mitigates sanding issues and improves economic viability.

Sand production significantly challenges marine gas hydrate extraction. Liu et al. (contribution 11) investigated gas-water-sand inflow patterns in both well orientations. For screen-only completions, sand control precision should be tiered based on screen plugging heterogeneity. Gravel-packed completions require (1) increased gravel density without reservoir destabilization, (2) cement-coated gravel for enhanced stability, and (3) screen precision tailored to particle size distribution for sustained production.

Shi et al. (contribution 12) conducted physical and numerical simulations of gravel packing to evaluate reduced-density materials and their packing/sand control performance. Their numerical results demonstrated that lightweight ceramsite performs poorly in horizontal and highly deviated wells due to inadequate compaction and insufficient sand retention, particularly with viscous slurries. In contrast, high-density particles promote better gravitational settling, yielding superior packing density and sand control effectiveness.

Assessing seafloor subsidence during hydrate extraction is crucial for safe gas hydrate production. Song and Zou (contribution 13) employed a modified Mohr–Coulomb model to characterize hydrate-bearing sediments, tracking strength degradation during depressurization. Using shear strength analysis to evaluate slope stability, they demonstrated that hydrate dissociation progressively slows during extraction. Their results indicate that large-scale submarine landslides triggered by hydrate decomposition are unlikely to occur.

Funding: This research was funded by the Project of Sanya Yazhou Bay Science and Technology City (No. SCKJ-JYRC-2023-02). First batch of the "Nanhai New Star" project (NHXXRCXM202357). Sanya Science and Technology Special Fund (No. 2022KJCX14). National Engineering Research Center of Gas Hydrate Exploration and Development (No. NERC2024002). National Natural Science Foundation of China (No. 42176215). Hainan Province Science and Technology Special Fund (No. GHYF2025017).

Acknowledgments: We sincerely appreciate all contributors for their insightful research in this Special Issue, the reviewers for their rigorous evaluations and constructive feedback, and the JMSE editorial team for their exceptional support and professional dedication.

Conflicts of Interest: The authors declare no conflicts of interest.

List of Contributions:

1. Li, X.; Guo, X.; Tian, F.; Fang, X. The Effects of Controlling Gas Escape and Bottom Current Activity on the Evolution of Pockmarks in the Northwest of the Xisha Uplift, South China Sea. *J. Mar. Sci. Eng.* **2024**, *12*, 1505. https://doi.org/10.3390/jmse12091505.
2. Gong, Y.; Yang, S.; Liang, J.; Tian, D.; Lu, J.; Deng, W.; Meng, M. Identification of Mass Transport Deposits and Insights into Gas Hydrate Accumulation in the Qiongdongnan Sea Area, Northern South China Sea. *J. Mar. Sci. Eng.* **2024**, *12*, 855. https://doi.org/10.3390/jmse12060855.
3. Zhong, G.; Zhao, J.; Zhao, Z.; Zhang, K.; Yu, J.; Shang, C.; Tu, G.; Feng, C. Acid-Extracted Hydrocarbon Anomalies and Significance in the Chaoshan Depression of the Northern South China Sea. *J. Mar. Sci. Eng.* **2024**, *12*, 909. https://doi.org/10.3390/jmse12060909.
4. Su, P.; Zhao, Z.; Zhang, K. The Mesozoic Subduction Zone over the Dongsha Waters of the South China Sea and Its Significance in Gas Hydrate Accumulation. *J. Mar. Sci. Eng.* **2024**, *12*, 1432. https://doi.org/10.3390/jmse12081432.
5. Yang, T.; Li, X.; Jin, J.; Chen, J.; Gong, Z.; Zhao, L.; Wang, W.; Liu, B.; Hu, J.; Wang, W.; et al. Shallow Gas Distribution Influenced by the Interface of Sedimentary Facies in the Southwest of the Qiongdongnan Basin. *J. Mar. Sci. Eng.* **2025**, *13*, 301. https://doi.org/10.3390/jmse13020301.
6. Yu, H.; Wang, J.; Deng, W.; Kuang, Z.; Li, T.; Lei, Z. High-Resolution 3D Geological Modeling of Three-Phase Zone Coexisting Hydrate, Gas, and Brine. *J. Mar. Sci. Eng.* **2024**, *12*, 2171. https://doi.org/10.3390/jmse12122171.
7. Bai, C.; Wang, H.; Li, Q.; Zhang, Y.; Xu, X. Controls on Deep and Shallow Gas Hydrate Reservoirs in the Dongsha Area, South China Sea: Evidence from Sediment Properties. *J. Mar. Sci. Eng.* **2024**, *12*, 696. https://doi.org/10.3390/jmse12050696.
8. Guan, J.; Wang, M.; Zhang, W.; Wan, L.; Haeckel, M.; Wu, Q. Representative Dynamic Accumulation of Hydrate-Bearing Sediments in Gas Chimney System since 30 Kyr BP in the QiongDongNan Area, Northern South China Sea. *J. Mar. Sci. Eng.* **2024**, *12*, 834. https://doi.org/10.3390/jmse12050834.

9. Yuan, Q.; Kong, L.; Liang, Q.; Liang, J.; Yang, L.; Dong, Y.; Wang, Z.; Wu, X. Mechanical Characteristics of Gas Hydrate-Bearing Sediments: An Experimental Study from the South China Sea. *J. Mar. Sci. Eng.* **2024**, *12*, 301. https://doi.org/10.3390/jmse12020301.
10. Wan, T.; Li, Z.; Lu, H.; Wen, M.; Chen, Z.; Tian, L.; Li, Q.; Qu, J.; Wang, J. Numerical Simulation of Gas Production Behavior Using Stepwise Depressurization with a Vertical Well in the Shenhu Sea Area Hydrate Reservoir of the South China Sea. *J. Mar. Sci. Eng.* **2024**, *12*, 1169. https://doi.org/10.3390/jmse12071169.
11. Liu, C.; Dong, C.; Shi, H.; Yu, Y.; Yin, B. Gas–Water–Sand Inflow Patterns and Completion Optimization in Hydrate Wells with Different Sand Control Completions. *J. Mar. Sci. Eng.* **2024**, *12*, 2071. https://doi.org/10.3390/jmse12112071.
12. Shi, H.; Dong, C.; Zhan, X.; Liu, C.; Li, L.; Ji, J.; Yu, Y.; Li, Z. Selection Results of Solid Material for Horizontal and Highly-Deviated Well Completion Gravel-Packing: Experiments, Numerical Simulation and Proposal. *J. Mar. Sci. Eng.* **2024**, *12*, 1690. https://doi.org/10.3390/jmse12101690.
13. Song, B.; Zou, Q. Seafloor Subsidence Evaluation Due to Hydrate Depressurization Recovery in the Shenhu Area, South China Sea. *J. Mar. Sci. Eng.* **2024**, *12*, 1410. https://doi.org/10.3390/jmse12081410.

References

1. Makogon, Y.F. Natural gas hydrates—A promising source of energy. *J. Nat. Gas Sci. Eng.* **2010**, *2*, 49–59. [CrossRef]
2. Matsumoto, R.; Ryu, B.J.; Lee, S.R.; Lin, S.; Wu, S.; Sain, K.; Pecher, I.; Riedel, M. Occurrence and exploration of gas hydrate in the marginal seas and continental margin of the Asia and Oceania region. *Mar. Pet. Geol.* **2011**, *28*, 1751–1767. [CrossRef]
3. Bahk, J.J.; Kim, D.H.; Chun, J.H.; Son, B.K.; Kim, J.H.; Ryu, B.J.; Torres, M.E.; Riedel, M.; Schultheiss, P. Gas hydrate occurrences and their relation to host sediment properties: Results from second Ulleung Basin gas hydrate drilling expedition, East Sea. *Mar. Pet. Geol.* **2013**, *47*, 21–29. [CrossRef]
4. Oyama, A.; Masutani, S.M. A review of the methane hydrate program in Japan. *Energies* **2017**, *10*, 1447. [CrossRef]
5. Boswell, R.; Myshakin, E.; Moridis, G.; Konno, Y.; Collett, T.S.; Reagan, M.; Ajayi, T.; Seol, Y. India National Gas Hydrate Program Expedition 02 summary of scientific results: Numerical simulation of reservoir response to depressurization. *Mar. Pet. Geol.* **2019**, *108*, 154–166. [CrossRef]
6. Chong, Z.R.; Yang, S.H.B.; Babu, P.; Linga, P.; Li, X.S. Review of natural gas hydrates as an energy resource: Prospects and challenges. *Appl. Energy* **2016**, *162*, 1633–1652. [CrossRef]
7. Pohlman, J.W.; Canuel, E.A.; Chapman, N.R.; Spence, G.D.; Whiticar, M.J.; Coffin, R.B. The origin of thermogenic gas hydrates on the northern Cascadia Margin as inferred from isotopic (13C/12C and D/H) and molecular composition of hydrate and vent gas. *Org. Geochem.* **2005**, *36*, 703–716. [CrossRef]
8. Zhang, W.; Liang, J.; Wei, J.; Su, P.; Lin, L.; Huang, W. Origin of natural gases and associated gas hydrates in the Shenhu area, northern South China Sea: Results from the China gas hydrate drilling expeditions. *J. Asian Earth Sci.* **2019**, *183*, 103953. [CrossRef]
9. Dixit, G.; Ram, H.; Kumar, P. Origin of gas in gas hydrates as interpreted from geochemistry data obtained during the National Gas Hydrate Program Expedition 02, Krishna Godavari Basin, offshore India. *Mar. Pet. Geol.* **2019**, *108*, 389–396. [CrossRef]
10. Santra, M.; Flemings, P.B.; Scott, E.; Meazell, P.K. Evolution of gas hydrate–bearing deep-water channel-levee system in abyssal Gulf of Mexico: Levee growth and deformation. *AAPG Bull.* **2020**, *104*, 1921–1944. [CrossRef]
11. Tréhu, A.M.; Long, P.E.; Torres, M.E.; Bohrmann, G.R.R.F.; Rack, F.R.; Collett, T.S.; Goldberg, D.S.; Milkov, A.V.; Riedel, M.; Schultheiss, P.; et al. Three-dimensional distribution of gas hydrate beneath southern Hydrate Ridge: Constraints from ODP Leg 204. *Earth Planet. Sci. Lett.* **2004**, *222*, 845–862. [CrossRef]
12. Konno, Y.; Fujii, T.; Sato, A.; Akamine, K.; Naiki, M.; Masuda, Y.; Yamamoto, K.; Nagao, J. Key findings of the world's first offshore methane hydrate production test off the coast of Japan: Toward future commercial production. *Energy Fuels* **2017**, *31*, 2607–2616. [CrossRef]
13. Li, J.F.; Ye, J.L.; Qin, X.W.; Qiu, H.J.; Wu, N.Y.; Lu, H.L.; Xie, W.W.; Lu, J.A.; Peng, F.; Xu, Z.Q.; et al. The first offshore natural gas hydrate production test in South China Sea. *China Geol.* **2018**, *1*, 5–16. [CrossRef]
14. Ye, J.L.; Qin, X.W.; Xie, W.W.; Lu, H.L.; Mam, B.J.; Qiu, H.J.; Liang, J.Q.; Lu, J.A.; Kuang, Z.G.; Lu, C.; et al. The second natural gas hydrate production test in the South China Sea. *China Geol.* **2020**, *3*, 197–209. [CrossRef]
15. Wei, N.; Pei, J.; Li, H.; Zhou, S.; Zhao, J.; Kvamme, B.; Coffin, R.B.; Zhang, L.; Zhang, Y.; Xue, J. Classification of natural gas hydrate resources: Review, application and prospect. *Gas Sci. Eng.* **2024**, *124*, 205269. [CrossRef]
16. Liu, J.W.; Li, X.S. Recent advances on natural gas hydrate exploration and development in the South China Sea. *Energy Fuels* **2021**, *35*, 7528–7552. [CrossRef]

17. Medhi, N.; Borthakur, P.P. An Extensive Review on Gas Hydrates: Recent Patents, Properties, Formation, Detection, Production, Importance, and Challenges. *Recent Pat. Eng.* **2025**, *19*, E080524229746. [CrossRef]
18. Schicks, J.M.; Haeckel, M.; Janicki, G.; Spangenberg, E.; Thaler, J.; Giese, R.; Strauch, B.; Heeschen, K.; Priegnitz, M.; Luzi-Helbing, M.; et al. Development, test, and evaluation of exploitation technologies for the application of gas production from natural gas hydrate reservoirs and their potential application in the Danube Delta, Black Sea. *Mar. Pet. Geol.* **2020**, *120*, 104488. [CrossRef]
19. Wang, Z.; Zhang, Y.; Peng, Z.; Shan, Z.; Sun, B.; Sun, J. Recent advances in methods of gas recovery from hydrate-bearing sediments: A Review. *Energy Fuels* **2022**, *36*, 5550–5593. [CrossRef]

Journal of
Marine Science and Engineering

MDPI

Article

The Effects of Controlling Gas Escape and Bottom Current Activity on the Evolution of Pockmarks in the Northwest of the Xisha Uplift, South China Sea

Xuelin Li [1,2], Xudong Guo [1,2], Fei Tian [1,2] and Xiaochen Fang [1,2,*]

1 Sanya Institute of South China Sea Geology, Guangzhou Marine Geological Survey, 2 Yumin Road, Sanya 572025, China; lxlingeo@126.com (X.L.)
2 Guangzhou Marine Geological Survey, China Geological Survey, 1133 Haibin Road, Guangzhou 510075, China
* Correspondence: xcgeo628@163.com; Tel.: +86-0898-88033822

Abstract: Submarine pockmarks are typical indicators of submarine gas escape activity. The deep strata of the Xisha Uplift are rich in biogenic and thermogenic gas, accompanied by strong bottom current activity. Investigating the effects of controlling submarine gas escape and bottom current activity on the formation and development of pockmarks in the Xisha Uplift is significant for understanding the evolution of submarine topography and geomorphology. This study utilized high-resolution multibeam data to identify 261 submarine pockmarks in the northwest of the Xisha Uplift. These pockmarks were categorized based on their morphology into circular, elliptical, elongated, crescent-shaped, and irregular types. The diameters of pockmarks in the study area range from 0.21 to 4.96 km, with maximum depths reaching 30.88 m. Using high-resolution multi-channel seismic data, we conducted a detailed analysis of the subsurface strata characteristics of the pockmarks, identifying chaotic weak reflections, bright spots, and high-angle reflectors. We believe that deep gas in the northwest of the Xisha Uplift escapes to the seafloor through migration pathways, such as faults, fractures, and gas chimneys, resulting in the formation of submarine pockmarks. Bottom current activity has a significant impact on already-formed pockmarks. Crescent-shaped and elongated pockmarks in the Xisha Uplift are largely the result of bottom current modifications of pre-existing pockmarks.

Keywords: submarine pockmark; gas escape; ocean currents; morphology characteristics; currents; Xisha Uplift

1. Introduction

Submarine pockmarks are common seabed features along continental margins, characterized by gas leakage through migration pathways in the strata, eroding seabed surface sediments to form negative seabed topographies. They typically develop in fine-grained mud sediment areas on the seabed [1–4]. Crater-like "pockmarks" were first discovered in Canada in the 1970s [5]. With the advancement of marine exploration technology, an increasing number of submarine pockmarks have been observed in different marine areas around the world, such as the Bering Sea [6], the Norwegian Trench [7], the North Sea [8], the Baltic Sea [9], the Black Sea [10], the Santos Basin [11], the western continental margin of India [12], the Gulf of Mexico [13], and the South China Sea [14]. Most pockmarks range in diameter from 10 to 250 m. The smallest pockmarks have diameters of less than 5 m, while the largest can reach several kilometers [2]. The largest known pockmark has a diameter of 3210 m and a depth of 165.2 m [14]. Based on their planar distribution characteristics, pockmarks are classified into isolated pockmarks, pockmark groups, and pockmark strings. Morphologically, most pockmarks are circular or elliptical. Under the influence of geological processes or marine environments, they can also exhibit crescent-shaped (semi-circular) or elongated forms. Vertically, the strata beneath pockmarks typically display three main shapes, "V", "U", and "W", with asymmetrical sides in most cases [15]. Due to the destabilization of deep-seated pressure, localized gas rapidly escapes upwards, potentially causing the formation of pockmarks in seabed surface sediments. Subsequent continued gas leakage leads to the sustained existence of pockmarks [16,17]. There are also viewpoints suggesting that bottom currents contribute to the formation and maintenance of pockmarks [18].

Submarine pockmark distribution is controlled by seabed geological structures such as overpressure-related structures, diapirs, pores, and faults [18]. These geological structures typically form pathways for gas migration and accumulation, leading to the gradual formation and development of pockmarks as the gas escapes. Mud diapirs and salt diapirs are diapiric structures driven by gravity sliding, gravity spreading, and differential compaction. Mud layers and salt layers are soft plastic layers that can easily deform due to gravitational differences and compressive forces, providing gas pathways with higher permeability than the surrounding sediments [19]. Landslides, polygonal faults, and tectonic faults all belong to sediment fractures, and they are formed under the influence of gravity, compressive forces, and tectonic forces, respectively. Pockmarks typically appear at the end of linear paths above faults and fault anticlines [15]. Gas chimneys, paleo-channels, paleo-pockmarks, and sediment waves are the preferred pathways for the release of subsurface overpressure gas.

In the northern South China Sea, several typical pockmarks have been identified in the Taixinan Basin [19], Yinggehai Basin [20], Qiongdongnan Basin [21], Guangle Uplift [14], and Zhongjiannan Basin [22]. A significant number of pockmarks have also developed in the northwestern Xisha Uplift, where deep strata are rich in gas. Yet, the processes of deep gas migration and escape leading to the formation of these pockmarks have not been thoroughly discussed, and the impact of strong bottom current activity on pockmark morphology in the Xisha Uplift has not been revealed. This study reports on the pockmarks located near the Ganquan Platform in the northwestern Xisha Uplift (Figure 1). Utilizing multibeam bathymetric data and high-resolution multi-channel seismic data, we provide a comprehensive description and classification of the morphological characteristics of these pockmarks. By analyzing the stratigraphic changes beneath the pockmarks and the surrounding seabed topographic variations, we investigate the effects of gas escape and strong bottom current activity on the pockmarks.

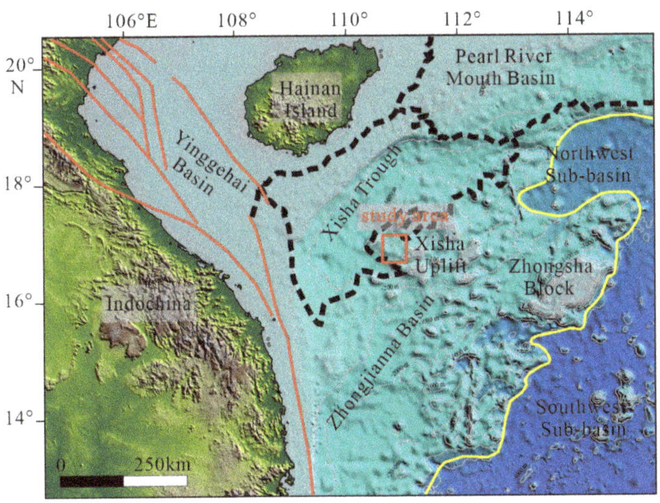

Figure 1. *Cont.*

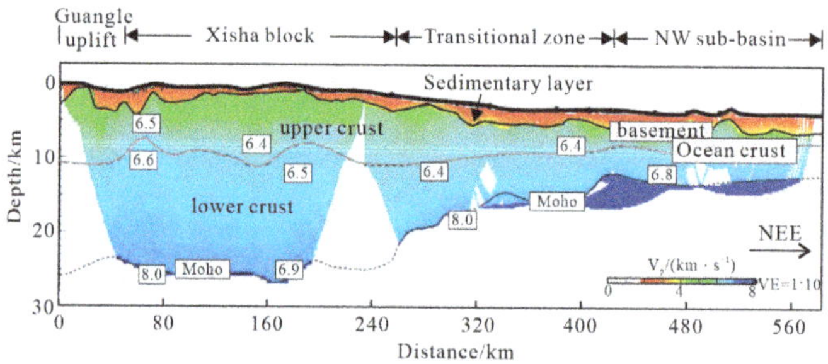

Figure 1. The location of the study area (upper image) and the crustal velocity structure of the Xisha Block and its surrounding areas (lower image) [23].

2. Geological Setting

The Xisha Uplift is one of the numerous continental blocks that formed during the evolution of the South China Sea, situated at the junction of the Indochina Block, the South China Block, and the South China Sea. To the north, it borders the Xisha Trough, to the east lies the Northwest Sub-basin, to the southeast is the Zhongsha Trough and the Zhongsha Block, to the south is the Southwest Sub-basin, and to the west, it faces the Guangle Uplift (Figure 1). Surrounding the Xisha Uplift are several Cenozoic basins, including the Qiongdongnan Basin, the Pearl River Mouth Basin, the Zhongjiannan Basin, the Beibu Gulf Basin, and the Yinggehai Basin (Figure 1). These basins were formed due to large-scale extensional and thinning processes along the northern margin of the South China Sea [24,25]. Gravity inversion data indicate that the average crustal thickness of the Xisha Uplift is approximately 23 km [26], which contrasts sharply with the significantly thin crust of the surrounding basins. For example, the crust in the Northwestern Sub-basin of the South China Sea is 6 km thick [23], and the continental crust in the Changhua Depression of the Qiongdongnan Basin has a minimum thickness of 2.8 km [27].

The sedimentary layer of the Xisha Uplift overlies a Precambrian metamorphic crystalline basement [28]. From the Triassic to the Late Cretaceous, the continental margin of the South China Block was influenced by the subduction of the Pacific Plate, resulting in large-scale magmatic structures and causing heterogeneity in the crust [23,29]. Subsequently, the extensional processes and magmatic activities during the Cenozoic disrupted the early crystalline basement structure, establishing the overall structural characteristics of the present-day Xisha Uplift [30]. The Neogene sediments of the Xisha Uplift are primarily derived from carbonate debris of biogenic reefs growing on the Xisha Uplift. Terrigenous clastics from the South China Block and the Indochina Block cannot cross the central sag of the Qiongdongnan Basin and the Guangle Uplift to reach the Xisha Uplift [31,32]. Previous studies have utilized high-resolution multi-channel seismic data to conduct sequence stratigraphic division of the Xisha Uplift. Five sequence boundaries have been identified in the strata since the Neogene, namely T60, T50, T40, T30, and T20. These sequence boundaries divide the strata into five sequences. T60 marks the base of the Early Miocene sequence, characterized by moderately continuous seismic reflections and interpreted as siliciclastic–carbonate mixed debris. T50 marks the base of the Middle Miocene sequence, characterized by strong-amplitude mound-shaped reflections and interpreted as carbonates. T40 marks the base of the Late Miocene sequence, characterized by moderately continuous seismic reflections and interpreted as carbonate debris from biogenic reefs. T30 marks the base of the Pliocene sequence, and T20 marks the base of the Quaternary sequence. Both the Pliocene and Quaternary sequences display similar moderately continuous seismic reflections and are interpreted as semi-pelagic sediments [33–36] (Figure 2).

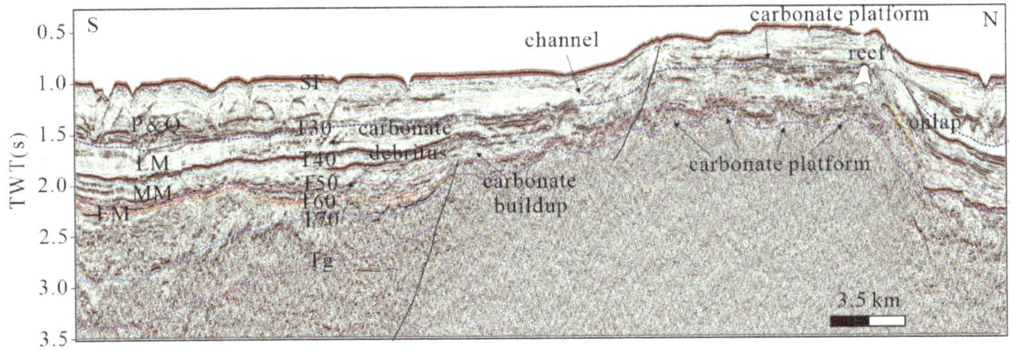

Figure 2. The sequence stratigraphy based on seismic reflection data around the study area. The dotted lines in different colors represent sequence boundaries. See the location in Figure 1.

The study area is located in the northwestern part of the Xisha Uplift, between the Ganquan Platform and the Xisha Islands, with water depths ranging from approximately 800 m to 1300 m (Figure 3). The study area lies at the junction of the Xisha Uplift and the Xisha Trough, which is significantly affected by extensional processes, resulting in well-developed underground strata porosity and faults.

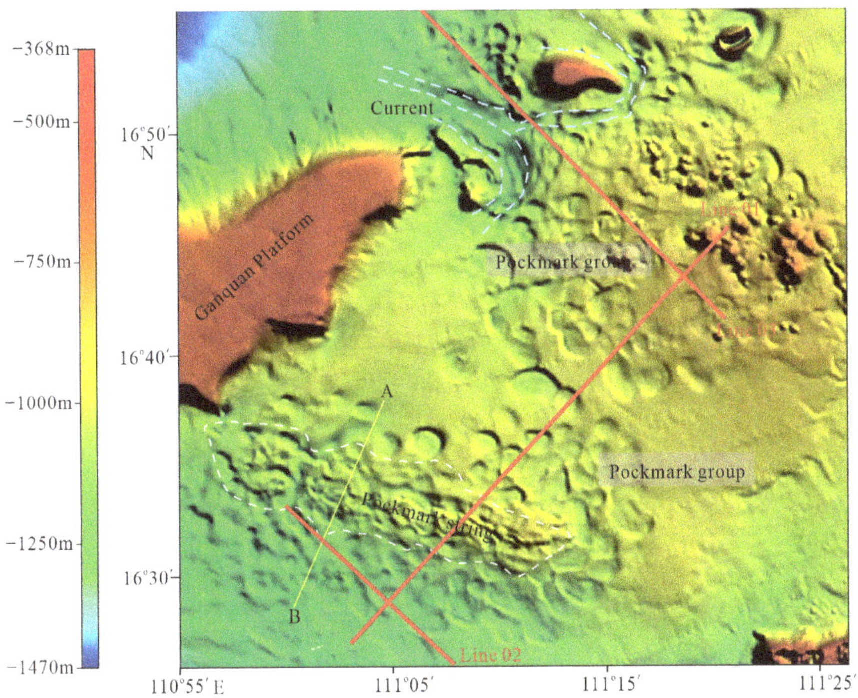

Figure 3. Pockmarks and bottom current channels are well shown on the multibeam bathymetry map. The white dashed line outlines the pockmark string in the study area, and the yellow line AB represents the profile AB passing through the pockmark string.

3. Materials and Methods

The base map for Figure 1 uses the 1 × 1 arc-minute global terrain data from GEBCO (General Bathymetric Chart of the Oceans), which is available for free download. The multibeam bathymetry data and 2D multi-channel reflection seismic data were provided by the Guangzhou Marine Geological Survey (GMGS), offering reliable information for describing the morphology, distribution, and stratigraphic characteristics of the pockmarks. The water depth in the study area ranges from approximately −300 m to −1500 m. Raster grids generated from the multibeam bathymetry data have a resolution of about 50 m × 50 m, with a vertical resolution of approximately 3‰ of the water depth. The 2D reflection multichannel seismic data were collected using a single-cable acquisition system with non-zero offset, single-side shooting, and single-side receiving. The seismic source used was an airgun array, and the streamer was 4500 m long with a channel separation of 6.25 m. The source distance is 25 m, resulting in a 175 m migration distance for the seismic system. Data processing was accomplished using the Omega processing platform. The original seismic data were interpreted using the Petrel 2024 E&P software platform (Mark of Schlumberger). All vertical scales for the seismic profiles shown in the study represent two-way travel time.

4. Results

4.1. Interpretation of Multibeam Bathymetry Data

Pockmarks can be classified by their planar shapes into circular pockmarks, elliptical pockmarks, crescent-shaped pockmarks, elongated pockmarks, annular pockmarks, and irregular pockmarks [22,37]. Multibeam bathymetric data visually represent the geomorphic features of the seafloor in the study area (Figure 3). According to the bathymetric data, the morphology of the submarine pockmarks is delineated, compared, and verified, serving as a basis for morphological statistical research. In the study area, we identified 261 pockmarks, including 69 circular, 79 elliptical, 40 irregular, 28 crescent-shaped, 27 elongated, and 18 annular pockmarks (Table 1).

Table 1. Main morphological parameters of pockmarks.

Shape	Long Axis Diameter/km	Short Axis Diameter/km	Depth/m	Number
Circular	0.28–2.59	0.23–2.43	10.334–30.882	69
Elliptical	0.34–4.62	0.21–2.98	9.228–28.166	79
Crescent-shaped	0.51–3.83	0.31–1.98	9.235–26.731	28
Elongated	0.75–4.96	0.25–2.21	8.965–26.556	27
Annular	0.66–2.62	0.51–2.33	11.564–25.776	18
Irregular	0.35–2.63	0.29–2.33	10.528–31.002	40
Shape	**Volume/10^6 m^3**	**Surface Area/km^2**	**Perimeter/km**	**Orientation/°**
Circular	0.5–73.16	0.05–4.91	0.79–7.85	6.3–351.8
Elliptical	0.62–80.43	0.05–10.88	0.91–12.69	5.7–349.5
Crescent-shaped	0.39–65.04	0.02–6.39	0.83–9.57	12.3–355.2
Elongated	0.34–38.76	0.03–6.24	0.93–11.76	20.5–324.9
Annular	0.38–54.26	0.28–5.41	1.81–8.22	4.6–358.2
Irregular	0.65–75.48	0.06–10.76	1.12–9.72	7.1–342.5

These pockmarks are distributed in pockmark groups and pockmark strings, with pockmark groups being the predominant form covering the study area. The pockmark strings are located in the southeastern part of the study area (Figure 3). The pockmarks within the string are generally smaller in area, mainly consisting of circular and irregular pockmarks distributed along the 800 m–1200 m depth transition zone (Figure 3). The pockmark strings are oriented in a northwest–southeast direction.

We calculated the long axes, short axes, perimeters, surface areas, volumes, and depths of the pockmarks and then conducted a statistical analysis. The results are shown in Figure 4. In the study area, the long axes of the pockmarks range from 0.28 to 4.96

km, the short axes range from 0.21 to 2.98 km, the perimeters range from 0.79 to 12.69 km, the areas range from 0.02 to 10.88 km^2, and the volumes range from 0.34 to 80.43 km^3 (Table 1). Hovland classifies pockmarks into small, large, and extra-large categories based on diameter. Small pockmarks have a diameter of less than 250 m, large pockmarks range from 250 to 1000 m in diameter, and extra-large pockmarks have a diameter greater than 1000 m [2,37]. In the study area, large pockmarks account for 47.2%, extra-large pockmarks account for 52.7%, while small pockmarks constitute only 0.1%. The depths of the pockmarks range between 8.965 and 31.002 m, with pockmarks less than 10 m deep accounting for 11% of the total, those with a depth of 10 to 25 m accounting for 76%, and those deeper than 25 m accounting for 13%.

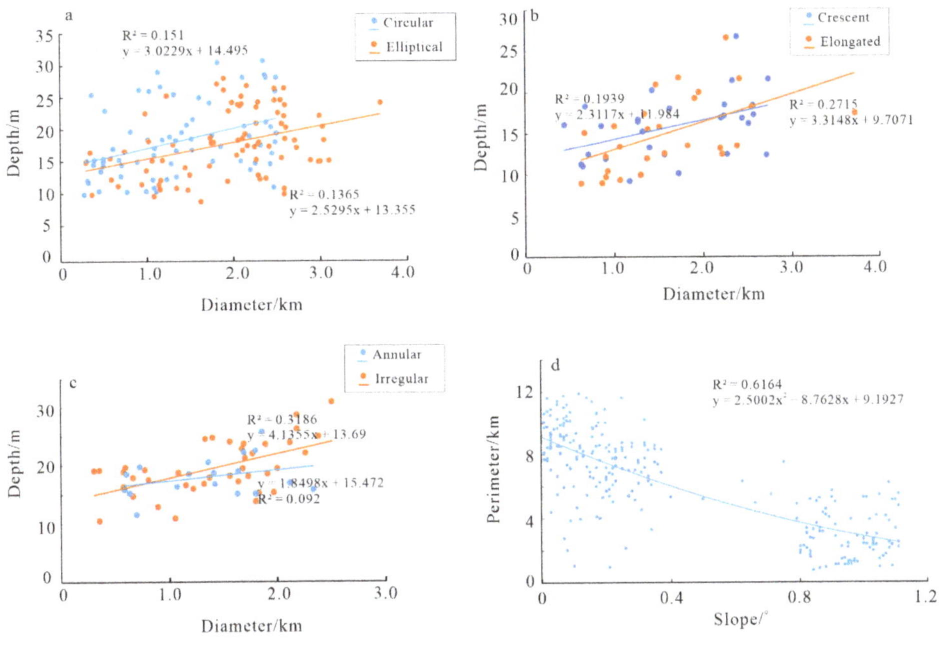

Figure 4. (**a**) Correlation diagram of diameters and depths of circular and elliptical pockmarks. (**b**) Correlation diagram of diameters and depths of crescent-shaped and elongated pockmarks. (**c**) Correlation diagram of diameters and depths of annular and irregular pockmarks. (**d**) Correlation diagram of pockmark perimeters and slopes of study area.

Statistical studies of various types of pockmarks distributed in the study area, including parameters such as the diameter and depth, show that after first-order linear fitting, the highest linear regression coefficient is observed for irregular pockmarks (R^2 = 0.3186) (Figure 4). This indicates that there is no significant linear relationship between the diameters and depths of circular, elliptical, crescent, elongated, annular, and irregular pockmarks in the study area. However, by fitting the linear relationship between pockmark perimeter and topographic slope, a good second-order correlation is found with R^2 = 0.6164, indicating that larger topographic slopes generally correspond to smaller pockmark perimeters. This correlation is mainly due to the large pockmark string south of the Ganquan Platform. In this area, the water depth varies by nearly 250 m, with maximum slopes exceeding 1°. The pockmark string is primarily composed of small-scale pockmarks with smaller diameters and shallower depths densely distributed in a banded pattern (Figure 5).

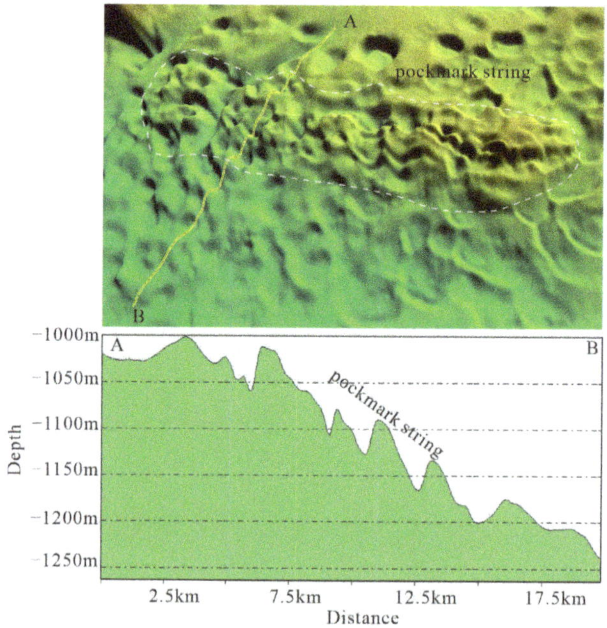

Figure 5. Three-dimensional view of pockmark string and cross section. The white dashed line outlines the pockmark string, and the yellow line AB represents the profile AB through the pockmark string shown in the lower image.

4.2. Interpretation of 2D Seismic Data

The seismic facies of lines 01, 02, and 03 in the study area were identified, revealing significant differences between the seismic facies in the southern and northern parts of the study area (Figure 6). The most notable difference is that paleo-volcanic activity is less pronounced in the southern region compared to the northern region (Figures 7 and 8). Furthermore, drift development is common in the northern region, while it is less common in the southern region (Figures 6 and 8). Detailed descriptions of the seismic facies in both the southern and northern parts are provided below.

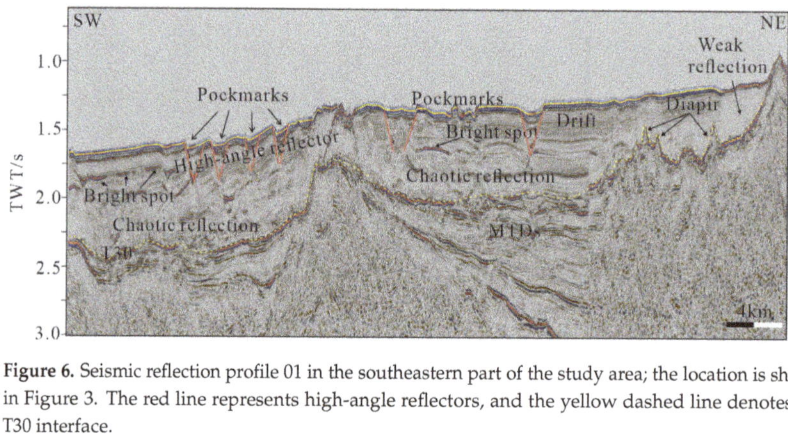

Figure 6. Seismic reflection profile 01 in the southeastern part of the study area; the location is shown in Figure 3. The red line represents high-angle reflectors, and the yellow dashed line denotes the T30 interface.

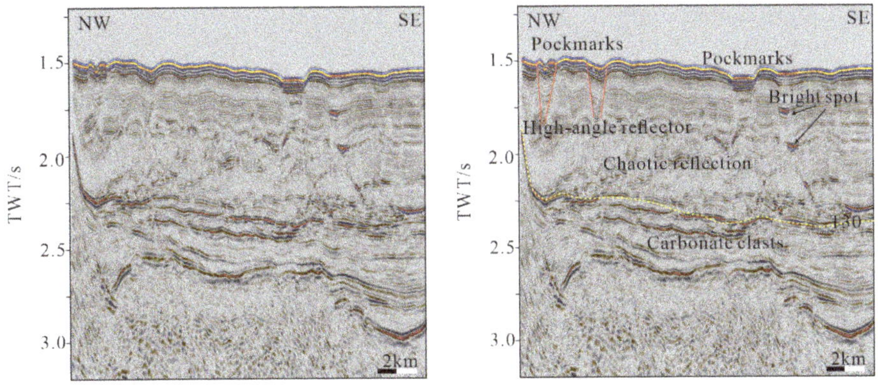

Figure 7. Seismic reflection profile 02 in the southwestern part of the study area; the location is shown in Figure 3. The red line represents high-angle reflector, and the yellow dashed line indicates the T30 interface.

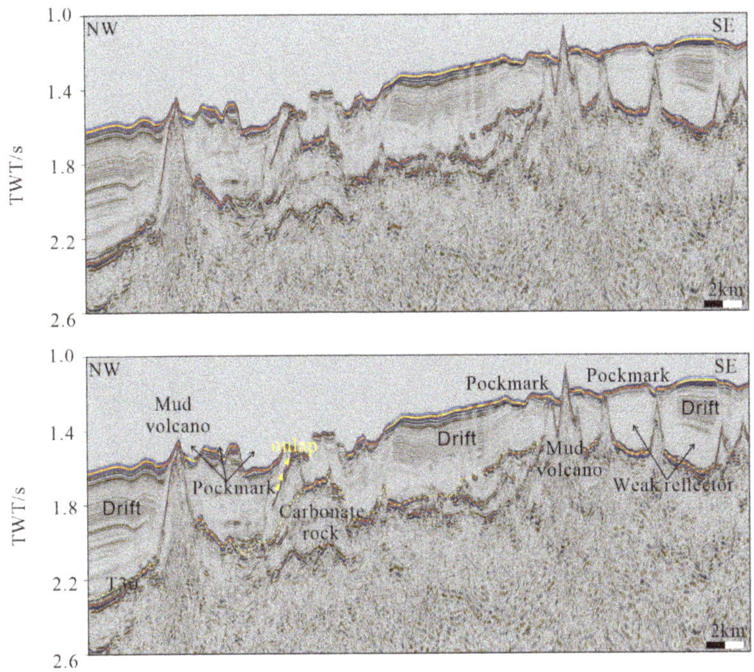

Figure 8. Seismic reflection profile 03 in the northern part of the study area; the location is shown in Figure 3. The yellow dashed line indicates the T30 interface, and the yellow triangle indicates onlap.

4.2.1. Seismic Facies in the Southern Part of the Study Area

In the southern part of the study area, Mass Transport Deposits (MTDs), diapir structures, and carbonate debris have been identified in strata below T30. Pockmarks appear as concave isochronous axes in seismic profiles. The strata beneath the pockmarks exhibit high-amplitude, high-continuity, parallel, curved, and medium-frequency reflection axes (Figures 6 and 7). Several distinctive seismic facies have also been identified in the seismic profiles, such as chaotic weak reflections, bright spots, and high-angle reflectors.

Chaotic weak reflections exhibit a low amplitude and poor continuity, primarily occurring around 2.2–2.0 s on the seismic profile, and they are prevalent throughout the southern region, overlaying T30. When the gas content in the strata is high, it increases the wavelength of seismic waves, resulting in a decrease in frequency. Chaotic weak reflections generally exhibit lower frequencies compared to the underlying strata, indicating a presence of gas within these reflections. Bright spots are characterized by punctiform or short-line, high-amplitude, and low-frequency anomalies in the seismic profiles. When gas accumulates in the strata, the acoustic impedance significantly increases, the frequency of seismic pulses correspondingly decreases, and the amplitude increases. Because gas usually accumulates only in strata with high porosity, they appear as punctiform or short-line anomalies on stacked seismic profiles [38,39]. Bright spots are commonly found near pockmarks, and the underlying strata are chaotic weak reflections, while the overlying strata show a moderate amplitude, strong continuity, and high-frequency parallel reflections. High-angle reflectors are identified in the lateral flanks of the strata beneath the pockmarks, characterized by low-amplitude, discontinuous, and low-frequency seismic facies. When gas migrates upward, it deforms the strata, resulting in curved isochronous axes in the seismic profile. These regions with higher porosity gradually connect to form pathways for gas escape, presenting as high-angle reflectors.

4.2.2. Seismic Facies in the Northern Part of the Study Area

In the northern part of the study area, seismic facies characteristics below T30 are predominantly chaotic with a moderate amplitude, representing the basement of the Xisha Block, possibly consisting of pre-Cambrian metamorphic rocks [33,40]. Small-scale carbonate platforms are also identified in the northern part of the study area, characterized by weak continuity and moderate-amplitude seismic facies. The demise of platform margins occurred later, overlain by the surrounding strata. Several mud volcanoes are identified in the seismic profiles, characterized by chaotic reflections internally and strong-amplitude isochronous axes at their boundaries. Some mud volcanoes penetrate through the seafloor, while others are buried by hemi-pelagic sediments. Pockmarks typically develop near mud volcanoes, appearing as depressions in isochronous axes. Unlike pockmarks in the southern part of the study area, those in the northern region exhibit overall higher frequencies in the underlying strata, and no chaotic weak reflections, bright spots, or high-angle reflectors are identified. This indicates that the underlying strata of pockmarks in the northern area do not contain gas. Numerous drifts are identified in the seismic profiles, characterized by moderate to weak amplitude parallel reflections, indicating strong hydrodynamic conditions in the northern part of the study area. These strong hydrodynamic conditions are primarily due to the development of strong bottom currents [14,20]. Between mud volcanoes and beneath drifting layers, weak blank reflections are identified. The seismic facies exhibit strong continuity and weak amplitude parallel reflections, indicating a weak hydrodynamic environment.

5. Discussion

5.1. Gas Migration in the Southern Part of the Study Area

Seismic profiles show that there are abundant chaotic reflections in the Pliocene strata (above T30), indicating a significant amount of gas in the Pliocene strata. During the Miocene (T60-T30), a large number of biogenic reefs grew in the Xisha Uplift area (Figure 2) [41]. These reefs were buried by hemipelagic sediments due to the rapid rise in the sea level during the Pliocene [42]. Biogenic reefs contain substantial organic matter, which decomposes gradually after burial, producing large amounts of gas [43]. The sedimentary layers formed by the hemipelagic sediments are impermeable, trapping the gas produced from the decomposition of organic matter. Therefore, it is inferred that the gas in the Pliocene strata is biogenic in origin.

Various types of escape pathways have been identified in seismic profiles, including faults, fractures, and gas chimneys, which can connect deep strata with shallow strata.

Faults are one of the main pathways for vertical gas migration, providing the path of least resistance for gas movement [44–46]. The high-angle reflectors in the underlying strata of the submarine pockmarks are most likely faults. Fractures often develop around faults, and both faults and fractures have the ability to disrupt the continuity of the strata, compromising the sealing capacity of the underlying strata of pockmarks (Figure 9(a1,a2)). Since the Piocene, gas migrates along fractures or faults to the shallower strata (Figure 9(b1,b2)).

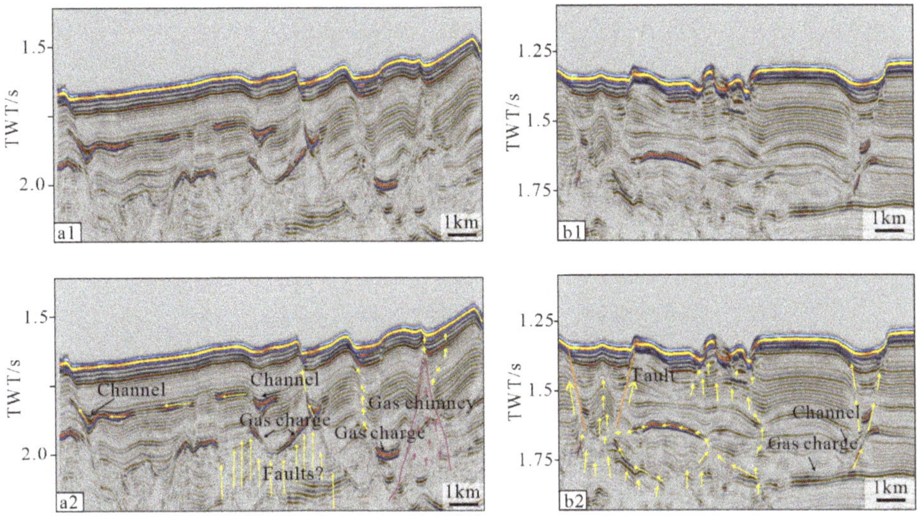

Figure 9. Gas migration in profile line 01. Faults and gas chimneys are the main pathways for gas to escape from the seabed. The yellow arrows indicate the directions of gas migration, and the orange lines indicate faults, (**a1,b1**) are seismic profiles without interpretation, while (**a2,b2**) are interpreted seismic profiles. The pink solid circle indicates the shape of the gas chimney, and the pink arrow indicates the direction of gas migration in the gas chimney.

The gas chimney structure is also a common escape pathway, and its formation requires three conditions: (1) a rich gas source in the deep strata; (2) the presence of structural weak zones in the strata, such as small fractures; and (3) a good cap rock structure [47]. The Xisha Uplift has abundant biogenic gas and deposited a good cap rock structure during the Piocene and Quaternary [48]. Biogenic gas continually accumulates, creating overpressure. Under the sealing effect of the cap rock, the pore pressure gradually increases. When the pore pressure exceeds the sealing pressure of the cap rock, episodic pressure release occurs [49]. As gas is released, the pore pressure in the underlying strata decreases, causing the weak zones to close. Pore pressure then accumulates again until the next episodic pressure release. Multiple episodic pressure releases ultimately form gas chimney structures.

In the southern part of the study area, gas primarily escapes through focused migration. The Pliocene strata contain significant amounts of biogenic gas and pore water. With the continuous deposition of overlying strata, the burial depth of gas-rich strata increases. This reduces the proportion of compressible materials, gradually increases the geotemperature, and causes gas expansion. This expansion generates overpressure within the Pliocene strata, promoting gas migration. The overlying strata contain numerous escape pathways, allowing gas to migrate upward or laterally along these pathways (Figure 9(b1,b2)). During migration, they accumulate in high-porosity areas, forming gas charges (Figures 9 and 10).

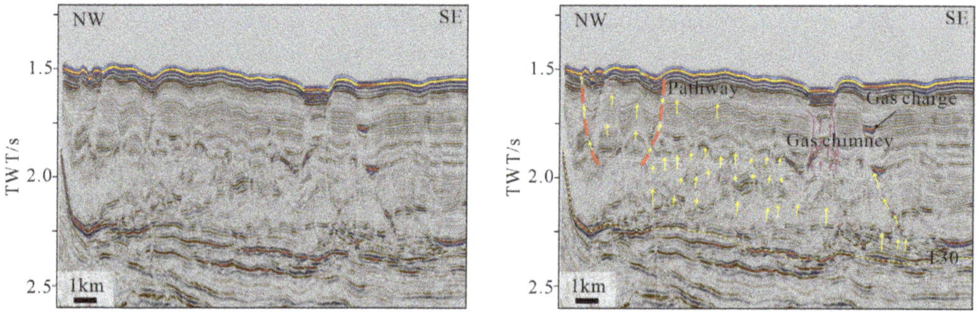

Figure 10. Gas migration in profile line 02. The red dashed line represents high-angle pathways for gas migration. The yellow arrows indicate the direction of gas migration, and the red dotted lines indicate the pathway, the pink solid circle indicates the shape of the gas chimney, and the pink arrow indicates the direction of gas migration in the gas chimney.

5.2. The Genesis of Pockmarks in the Study Area Is Primarily Attributed to Gas Escape

The genesis of pockmarks is related to gas escape. Gas migrating along conduits enriches the shallow strata, causing overpressure and deforming weak seafloor layers to form pockmarks. The formation of pockmarks requires the presence of gas, with variations in velocity, rate, and properties potentially leading to different erosion levels on the seafloor sediments and influencing the formation of different pockmark morphologies. Based on the process of pockmark formation, the mechanisms by which gas influences pockmark development can be categorized into four types: pockmarks formed by gas escape leading to strata collapse, pockmarks formed by gas escape causing strata deformation, pockmarks related to gas chimneys, and pockmarks associated with diapirs.

Pockmarks formed by gas escape leading to strata collapse. High-amplitude gas pathways are typically identified in the deeper strata beneath the pockmarks, connecting to a downwarp seismic facies above (Figure 11a). This downwarp seismic facies is interpreted as a paleo-pockmark or paleo-channel [50]. Although the paleo-pockmark or paleo-channel has been filled, it still retains relatively high porosity. Gas from deeper strata migrates vertically or laterally along the pathway and accumulates in the paleo-pockmark, appearing as high-amplitude reflections in seismic profiles. As gas accumulates, pressure gradually increases. When the pressure becomes sufficiently high, weak points in the overlying strata rupture, allowing gas to escape rapidly. The escaping gas may distort the overlying strata. The significant escape of gas causes volume loss in the paleo-pockmark, leading to the collapse of the overlying strata and the formation of new pockmarks on the seafloor. Subsequently, gas continues to escape along the faults created by the collapse, allowing the pockmark to develop further. This mechanism typically forms irregular pockmarks (Figure 11a).

Pockmarks formed by gas escape causing strata deformation. This type of pockmark is typically underlain by thick hemipelagic sediments, beneath which chaotic reflections rich in gas are present. As biogenic and thermogenic gas continuously form, the pressure in the gas-bearing layer exceeds that of the overlying strata. With the continuous thermal subsidence of the Xisha Uplift, gas also migrates upwards. The upward migration of a large amount of gas causes the strata with lower porosity to arch upwards, resulting in positive topography on the seabed, with negative topography forming pockmarks around the positive topography. The upward migration of gas deforms the overlying sedimentary layers, eventually causing the porous areas to fracture and form faults. Later, most of the gas escapes along these faults, leading to the continued existence of the pockmarks. In the Xisha Uplift, this mechanism primarily generates elliptical pockmarks (Figure 11b).

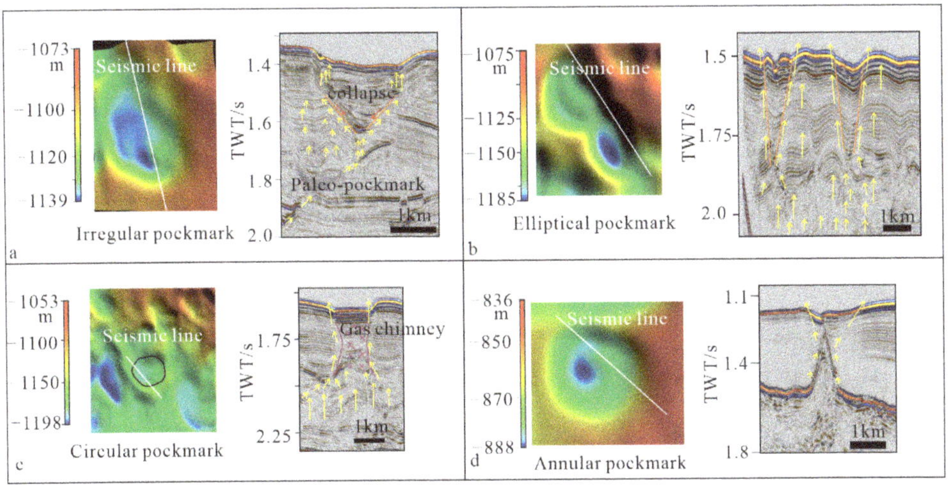

Figure 11. Formation mechanisms of various types of pockmarks. (**a**) Irregular pockmarks, (**b**) elliptical pockmarks, (**c**) circular pockmarks, and (**d**) annular pockmarks. Yellow arrows indicate direction of gas migration, orange lines indicate faults, gray lines outline gas chimneys, and red arrows indicate direction of collapse.

Pockmarks related to gas chimneys. The formation of this type of pockmark is primarily influenced by the presence of a gas chimney below. As the pressure in the Pliocene strata gradually increases, the overlying sedimentary layers remain relatively dense, with only small pores. When the accumulation of gas in the Pliocene strata reaches a certain level, the enormous pressure causes the gas to rapidly surge upwards, disrupting the overlying strata and forming a gas chimney that escapes to the seabed. The formation of a gas chimney simultaneously causes the overlying strata to collapse, leading to the creation of a pockmark. Due to the continued escape of gas through the gas chimney, the pockmark remains unfilled and persists. This mechanism typically forms circular pockmarks (Figure 11c).

Pockmarks related to diapirs. The gas associated with this type of pockmark is primarily thermogenic, with the gas source located in diapir structures. After diapir activity ceases, gas continues to escape from the diapir structures, migrating through the pores of the overlying strata and eventually reaching the seabed. The continued escape of gas leads to the stabilization of the escape pathway, forming a closed circular shape at the seabed. The strata within the pathway collapse, resulting in the formation of a pockmark. Due to the uplift of the overlying strata caused by diapir activity, the center of the pockmark is elevated relative to its boundaries, giving the pockmark an annular shape (Figure 11d).

5.3. The Impact of Bottom Currents on Pockmarks in the Xisha Uplift

The development and evolution of submarine pockmarks are not only controlled by gas seepage activities but also influenced by bottom current scouring and transport processes [51,52]. During the formation of submarine pockmarks, gas seeps into shallow subsurface layers, eroding loose seabed sediments. Fine-grained sediments are suspended in the water column and swiftly transported by intense bottom currents, while coarse-grained sediments remain on the inactive walls of the pockmarks where gas seepage is minimal [53]. Bottom currents can prevent pockmarks from being filled and buried by other sedimentary deposits, thereby maintaining their morphology. However, when bottom current action and gas seepage weaken, pockmarks may eventually be buried by subsequent sedimentation [54,55]. Crescent-shaped and elongated pockmarks typically develop in regions with strong bottom current activity, which can erode the inner walls of early-formed pockmarks, thereby enlarging them and altering their geomorphic features [54].

In the South China Sea, there exists a unique three-layer cyclonic–anticyclonic–cyclonic circulation pattern (upper layer < 750 m, middle layer 750–1500 m, deep layer > 1500 m) [56]. Within the study area, crescent-shaped pockmarks are predominantly located within the influence of the middle-layer circulation, which currents from southwest to northeast [57,58]. According to simulations by Hammer [18], when bottom currents pass over submarine pockmarks, sedimentation rates increase on the upstream side of the pockmark and decrease or even erode on the downstream side. Crescent-shaped pockmarks are widespread in the study area, and multibeam and seismic profiles indicate that fillings on the northeast side are thinner compared to the southwest side, with signs of erosion on the northeast side. Therefore, crescent-shaped pockmarks on the Xisha Uplift are influenced by bottom currents.

In the northern part of the study area, bottom current activity is complex. When the northeastward middle-layer circulation encounters positive topography, such as a submarine platform, its flow direction is altered and accelerates around the platform (Figure 12a). The bottom currents pass over previously formed pockmark groups around the edges of the platform. These earlier-formed pockmark groups no longer release gas (Figure 12b). Due to the increased flow velocity, there is enhanced sediment transport and erosion. Intense erosion and transport cause the interconnected pockmarks traversed by the bottom currents to merge into elongated mega-pockmarks. Prolonged bottom current erosion may eventually transform these pockmark groups into narrow channels.

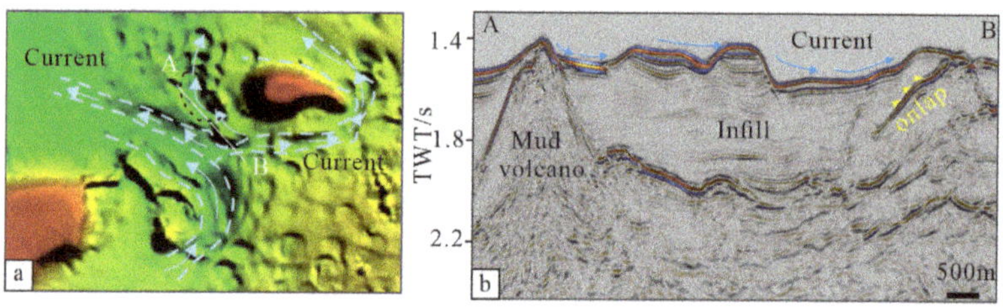

Figure 12. Bottom currents modifying pockmark morphology. Blue arrows indicate direction of bottom currents, white lines represent positions of seismic profiles, and dashed black lines encircle elongated pockmarks. The dotted line AB indicates the position of seismic section AB in the base map. See Figure 3 for specific geographical locations in diagram. (**a**) shows the topographic and geomorphic map of the dominant bottom current region, while (**b**) presents the seismic imaging and interpretation of profile AB.

6. Conclusions

(1) Based on multibeam bathymetric data and 2D seismic profiles, 261 seabed pockmarks have been identified near the Ganquan Platform. Among them, there are 69 circular pockmarks, 19 elliptical pockmarks, 40 irregular-shaped pockmarks, 28 crescent-shaped pockmarks, 27 elongated pockmarks, and 18 ring-shaped pockmarks. The diameters of these pockmarks range from 0.21 to 4.96 km, with maximum depths reaching up to 30.882 m. A large number of small-scale pockmarks arranged in strings have been discovered in the southern part of the study area.

(2) The genesis of pockmarks on the Xisha Uplift is primarily associated with the escape of gas. There are various types of escape pathways on the Xisha Uplift, such as faults, fractures, and gas chimneys. Based on the formation mechanism, pockmarks on the Xisha Uplift can be classified into four types: pockmarks formed by gas escape leading to strata collapse, pockmarks formed by gas escape causing strata deformation, pockmarks related to gas chimneys, and pockmarks related to diapirs.

(3) Bottom current activity has a significant impact on already formed pockmarks. Crescent-shaped and elongated pockmarks on the Xisha Uplift are largely the result of bottom current modifications of pre-existing pockmarks. Crescent-shaped pockmarks form due to uneven sedimentation rates within the pockmark caused by bottom currents,

while elongated pockmarks result from strong erosion by bottom currents connecting multiple pockmarks together.

Author Contributions: Conceptualization, Investigation, Data Curation, and Methodology, X.L. and X.F.; Writing—Original Draft, X.L. and X.G.; Writing—Review And Editing, F.T. All authors have read and agreed to the published version of the manuscript.

Funding: This research was funded by the National Science Foundation of China (grant number 42376078), the Project of Sanya Yazhou Bay Science and Technology City (grant number SCKJ-JYRC-2022-36), the Guangzhou Basic and Applied Basic Research Program (grant number 202201011367), and the China Geological Survey Project (grant number DD20221725).

Institutional Review Board Statement: Not applicable.

Informed Consent Statement: Not applicable.

Data Availability Statement: All data are provided in this paper. Further inquiries can be directed to the corresponding authors.

Conflicts of Interest: The authors declare no conflicts of interest.

References

1. Judd, A.; Hovland, M. *Seabed Fluid Flow: The Impact on Geology, Biology and the Marine Environment*; Cambridge University Press: Cambridge, UK, 2009.
2. Pilcher, R.; Argent, J. Mega-Pockmarks and Linear Pockmark Trains on the West African Continental Margin. *Mar. Geol.* **2007**, *244*, 15–32. [CrossRef]
3. Riboulot, V.; Cattaneo, A.; Sultan, N.; Garziglia, S.; Ker, S.; Imbert, P.; Voisset, M. Sea-Level Change and Free Gas Occurrence Influencing a Submarine Landslide and Pockmark Formation and Distribution in Deepwater Nigeria. *Earth Planet. Sci. Lett.* **2013**, *375*, 78–91. [CrossRef]
4. Xiong, P.; Cheng, C.; Kuang, Z.; Ren, J.; Liang, J.; Lai, H.; Chen, Z.; Lu, J.; Fang, X.; Jiang, T. Sedimentary Characteristics and Genetic Mechanism of the Giant Ancient Pockmarks in the Qiongdongnan Basin, Northern South China Sea. *Acta Oceanol. Sin.* **2023**, *42*, 120–133. [CrossRef]
5. King, L.H.; MacLean, B. Pockmarks on the Scotian Shelf. *Geol. Soc. Am. Bull.* **1970**, *81*, 3141–3148. [CrossRef]
6. Nelson, H.; Thor, D.; Sandstrom, M.W.; Kvenvolden, K.A. Modern Biogenic Gas-Generated Craters (Sea-Floor "Pockmarks") on the Bering Shelf, Alaska. *Geol. Soc. Am. Bull.* **1979**, *90*, 1144–1152. [CrossRef]
7. Hovland, M. Characteristics of Pockmarks in the Norwegian Trench. *Mar. Geol.* **1981**, *39*, 103–117. [CrossRef]
8. Hovland, M.; Talbot, M.R.; Qvale, H.; Olaussen, S.; Aasberg, L. Methane-Related Carbonate Cements in Pockmarks of the North Sea. *J. Sediment. Res.* **1987**, *57*, 881–892.
9. Bussmann, I.; Suess, E. Groundwater Seepage in Eckernförde Bay (Western Baltic Sea): Effect on Methane and Salinity Distribution of the Water Column. *Cont. Shelf Res.* **1998**, *18*, 1795–1806. [CrossRef]
10. Çifçi, G.; Dondurur, D.; Ergün, M. Deep and Shallow Structures of Large Pockmarks in the Turkish Shelf, Eastern Black Sea. *Geo-Mar. Lett.* **2003**, *23*, 311–322. [CrossRef]
11. Sumida, P.Y.G.; Yoshinaga, M.Y.; Madureira, L.A.S.-P.; Hovland, M. Seabed Pockmarks Associated with Deepwater Corals off SE Brazilian Continental Slope, Santos Basin. *Mar. Geol.* **2004**, *207*, 159–167. [CrossRef]
12. Dandapath, S.; Chakraborty, B.; Karisiddaiah, S.M.; Menezes, A.; Ranade, G.; Fernandes, W.; Naik, D.K.; Raju, K.P. Morphology of Pockmarks along the Western Continental Margin of India: Employing Multibeam Bathymetry and Backscatter Data. *Mar. Pet. Geol.* **2010**, *27*, 2107–2117. [CrossRef]
13. MacDonald, I.; Leifer, I.; Sassen, R.; Stine, P.; Mitchell, R.; Guinasso, N., Jr. Transfer of Hydrocarbons from Natural Seeps to the Water Column and Atmosphere. *Geofluids* **2002**, *2*, 95–107. [CrossRef]
14. Sun, Q.; Wu, S.; Hovland, M.; Luo, P.; Lu, Y.; Qu, T. The Morphologies and Genesis of Mega-Pockmarks near the Xisha Uplift, South China Sea. *Mar. Pet. Geol.* **2011**, *28*, 1146–1156. [CrossRef]
15. Xu, C.; Xu, G.; Xing, J.; Sun, Z.; Wu, N. Research Progress of Seafloor Pockmarks in Spatio-Temporal Distribution and Classification. *J. Ocean Univ. China* **2020**, *19*, 69–80. [CrossRef]
16. Ramos, R.B.; Dossantos, R.F.; Schattner, U.; Figueira, R.C.L.; Bícego, M.C.; Lobo, F.J.; de Mahiques, M.M. Deep Pockmarks as Natural Sediment Traps: A Case Study from Southern Santos Basin (SW Atlantic Upper Slope). *Geo-Mar. Lett.* **2020**, *40*, 989–999. [CrossRef]
17. Zhang, K.; Guan, Y.; Song, H.; Fan, W.; Li, H.; Kuang, Y.; Geng, M. A Preliminary Study on Morphology and Genesis of Giant and Mega Pockmarks near Andu Seamount, Nansha Region (South China Sea). *Mar. Geophys. Res.* **2020**, *41*, 2. [CrossRef]
18. Hammer, Ø.; Webb, K.E.; Depreiter, D. Numerical Simulation of Upwelling Currents in Pockmarks, and Data from the Inner Oslofjord, Norway. *Geo-Mar. Lett.* **2009**, *29*, 269–275. [CrossRef]

19. Chen, S.C.; Hsu, S.K.; Tsai, C.H.; Ku, C.Y.; Yeh, Y.C.; Wang, Y. Gas Seepage, Pockmarks and Mud Volcanoes in the near Shore of SW Taiwan. *Mar. Geophys. Res.* **2010**, *31*, 133–147. [CrossRef]
20. Di, P.; Huang, H.; Huang, B.; He, J.; Chen, D. Seabed Pockmark Formation Associated with Mud Diapir Development and Fluid Activities in the Yinggehai Basin of the South China Sea. *J. Trop. Oceanogr.* **2012**, *31*, 26–36.
21. Wang, L.J.; Zhu, J.T.; Zhou, H.T.; Sun, Z.; Sun, Z.P.; Zhang, H. Seismic Characteristics and Mechanism of Fluid Flow Structures in the Central Depression of Qiongdongnan Basin, Northern Margin of South China Sea. *Int. Geol. Rev.* **2020**, *62*, 1108–1130. [CrossRef]
22. Chen, J.; Song, H.; Guan, Y.; Yang, S.; Pinheiro, L.M.; Bai, Y.; Liu, B.; Geng, M. Morphologies, Classification and Genesis of Pockmarks, Mud Volcanoes and Associated Fluid Escape Features in the Northern Zhongjiannan Basin, South China Sea. *Deep Sea Res. Part II Top. Stud. Oceanogr.* **2015**, *122*, 106–117. [CrossRef]
23. Guo, X.R.; Zhang, M.H.; Huang, H.B.; Qiu, X.L.; Zhang, J.Z. Crustal Structure of Xisha Block and Its Tectonic Attributes. *Chin. J. Geophys.* **2016**, *59*, 288–300.
24. Taylor, B.; Hayes, D.E. Origin and History of the South China Sea Basin. *Geophys. Monogr. Ser.* **1983**, *27*, 23–56.
25. Li, C.; Zhi, P.; Ding, R.; Zhao, L.; Gong, W.; Li, Z.; Ge, J. Differences in Thermo-Rheological Structure between Qiongdongnan Basin and Pearl River Mouth Basin: Implications for the Extension Model in the Northwestern Margin of the South China Sea. *J. Mar. Sci. Eng.* **2023**, *11*, 443. [CrossRef]
26. Bai, Y.; Wu, S.; Liu, Z.; Müller, R.D.; Williams, S.E.; Zahirovic, S.; Dong, D. Full-Fit Reconstruction of the South China Sea Conjugate Margins. *Tectonophysics* **2015**, *661*, 121–135. [CrossRef]
27. Lei, C.; Alves, T.M.; Ren, J.; Tong, C. Rift Structure and Sediment Infill of Hyperextended Continental Crust: Insights from 3D Seismic and Well Data (Xisha Trough, South China Sea). *J. Geophys. Res. Solid Earth* **2020**, *125*, e2019JB018610. [CrossRef]
28. Qin, G.Q. A Preliminary Study on Foraminiferal Assemblages of Well 1, Xiyong, Xisha Islands and Their Coral Reef Formation. *Trop. Oceanol./Redai Haiyang* **1987**, *6*, 10–20.
29. Xia, S.; Fan, C.; Sun, J.; Cao, J.; Zhao, F.; Wan, K. Characteristics of Late Cenozoic Magmatic Activities on the Northern Margin of South China Sea and Their Tectonic Implications. *Mar. Geol. Quat. Geol. Chin.* **2017**, *37*, 25–33.
30. Zhu, W.; Xie, X.; Wang, Z.; Zhang, D.; Zhang, C.; Cao, L.; Shao, L. New Insights on the Origin of the Basement of the Xisha Uplift, South China Sea. *Sci. China (Earth Sci.)* **2017**, *60*, 2214–2222. [CrossRef]
31. Yang, Z.; Li, X.; Huang, L.; Wang, L.; Wu, S.; Zhang, X. Development of the Miocene Guangle Carbonate Platform in the South China Sea: Architecture and Controlling Factors. *Acta Geol. Sin.-Engl. Ed.* **2021**, *95*, 177–191. [CrossRef]
32. Li, Y.; Liu, X.; Chen, W.; Yi, L. Magnetic Properties and Initiation of Biogenic Reefs in Xisha Islands, South China Sea, at the Oligo–Miocene Boundary. *J. Mar. Sci. Eng.* **2021**, *9*, 1031. [CrossRef]
33. Wu, S.; Yang, Z.; Wang, D.; Lu, F.; Lüdmann, T.; Fulthorpe, C.; Wang, B. Architecture, Development and Geological Control of the Xisha Carbonate Platforms, Northwestern South China Sea. *Mar. Geol.* **2014**, *350*, 71–83. [CrossRef]
34. Wu, X.; Pu, R.; Chen, Y.; Qu, H.; Shen, H. Seismic Analysis of Early-Mid Miocene Carbonate Platform in the Southern Qiongdongnan Basin, South China Sea. *Acta Oceanol. Sin.* **2018**, *37*, 54–65. [CrossRef]
35. Wang, D.; Wu, S.; Qin, Z.; Spence, G.; Lü, F. Seismic Characteristics of the Huaguang Mass Transport Deposits in the Qiongdongnan Basin, South China Sea: Implications for Regional Tectonic Activity. *Mar. Geol.* **2013**, *346*, 165–182. [CrossRef]
36. Ma, Y.B.; Wu, S.G.; Lv, F.L.; Dong, D.D.; Gu, M.F. Seismic Characteristics and Development of the Xisha Carbonate Platforms, Northern Margin of the South China Sea. *J. Asian Earth Sci.* **2011**, *40*, 770–783.
37. Hovland, M.; Gardner, J.V.; Judd, A. The Significance of Pockmarks to Understanding Fluid Flow Processes and Geohazards. *Geofluids* **2002**, *2*, 127–136. [CrossRef]
38. White, R.S. Seismic Bright Spots in the Gulf of Oman. *Earth Planet. Sci. Lett.* **1977**, *37*, 29–37. [CrossRef]
39. Schroot, B.M.; Schüttenhelm, R.T. Shallow Gas and Gas Seepage: Expressions on Seismic and Otheracoustic Data from the Netherlands North Sea. *J. Geochem. Explor.* **2003**, *78*, 305–309. [CrossRef]
40. Luo, W.; Zhang, D.J.; Liu, X.Y.; Wang, Z.F.; Hu, W.Y.; Wang, Y.H. A Comprehensive Stratigraphic Study of Well XK-1 in the Xisha Area. *J. Stratigr.* **2018**, *42*, 485–498.
41. Sultan, N.; Marsset, B.; Ker, S.; Marsset, T.; Voisset, M.; Vernant, A.-M.; Bayon, G.; Cauquil, E.; Adamy, J.; Colliat, J. Hydrate Dissolution as a Potential Mechanism for Pockmark Formation in the Niger Delta. *J. Geophys. Res. Solid Earth* **2010**, *115*, B08101. [CrossRef]
42. Riboulot, V.; Thomas, Y.; Berné, S.; Jouet, G.; Cattaneo, A. Control of Quaternary Sea-level Changes on Gas Seeps. *Geophys. Res. Lett.* **2014**, *41*, 4970–4977. [CrossRef]
43. Dando, P.R.; Hara, S.C.M.; Schuster, U. Gas seepage from a carbonate-cemented sandstone reef on the Kattegat coast of Denmark. *Mar. Pet. Geol.* **1994**, *11*, 182–189. [CrossRef]
44. Serié, C.; Huuse, M.; Schødt, N.H. Gas Hydrate Pingoes: Deep Seafloor Evidence of Focused Fluid Flow on Continental Margins. *Geology* **2012**, *40*, 207–210. [CrossRef]
45. Hustoft, S.; Mienert, J.; Bünz, S.; Nouzé, H. High-Resolution 3D-Seismic Data Indicate Focussed Fluid Migration Pathways above Polygonal Fault Systems of the Mid-Norwegian Margin. *Mar. Geol.* **2007**, *245*, 89–106. [CrossRef]
46. Faulkner, D.; Jackson, C.; Lunn, R.; Schlische, R.; Shipton, Z.; Wibberley, C.; Withjack, M. A Review of Recent Developments Concerning the Structure, Mechanics and Fluid Flow Properties of Fault Zones. *J. Struct. Geol.* **2010**, *32*, 1557–1575. [CrossRef]

47. Zhong, G.; Gao, H. Sequence Characteristics of Cenozoic Stratigraphy in Zhongjiannan Basin, South China Sea. *Geotecton. Metallog.* **2005**, *29*, 403–409.
48. Liang, Q.S.; Liu, Z.; Wang, D.J.; Li, W.L.; Chang, M. Gas Chimney and Hydrocarbon Exploration. *Xinjiang Pet. Geol.* **2006**, *27*, 288.
49. Yang, T.; Lu, F.; Wang, B.; Fan, G.; Lu, Y.; Yang, Z. Geophysical Characteristics and Genetic Mechanism of the Gas Chimney in Deepwater Area of Southern Qiongdongnan Basin. *Prog. Geophys.* **2013**, *28*, 2634–2641.
50. Dillon, W.P.; Nealon, J.W.; Taylor, M.H.; Lee, M.W.; Drury, R.M.; Anton, C.H. Seafloor Collapse and Methane Venting Associated with Gas Hydrate on the Blake Ridge: Causes and Implications to Seafloor Stability and Methanerelease. *Geophys. Monogr. Ser.* **2001**, *124*, 211–233.
51. Bulat, J.; Long, D. Images of the Seabed in the Faroe-Shetland Channel from Commercial 3D Seismic Data. *Mar. Geophys. Res.* **2001**, *22*, 345–367. [CrossRef]
52. Betzler, C.; Lindhorst, S.; Hübscher, C.; Lüdmann, T.; Fürstenau, J.; Reijmer, J. Giant Pockmarks in a Carbonate Platform (Maldives, Indian Ocean). *Mar. Geol.* **2011**, *289*, 1–16. [CrossRef]
53. Paull, C.; Ussler Iii, W.; Borowski, W. Freshwater Ice Rafting: An Additional Mechanism for the Formation of Some High-Latitude Submarine Pockmarks. *Geo-Mar. Lett.* **1999**, *19*, 164–168. [CrossRef]
54. Hovland, M.; Jensen, S.; Indreiten, T. Unit Pockmarks Associated with Lophelia Coral Reefs off Mid-Norway: More Evidence of Control by 'Fertilizing' Bottom Currents. *Geo-Mar. Lett.* **2012**, *32*, 545–554. [CrossRef]
55. Michel, G.; Dupré, S.; Baltzer, A.; Ehrhold, A.; Imbert, P.; Pitel, M.; Loubrieu, B.; Scalabrin, C.; Lazure, P.; Marié, L. Pockmarks on the South Aquitaine Margin Continental Slope: The Seabed Expression of Past Fluid Circulation and Former Bottom Currents. *Comptes Rendus. Géosci.* **2017**, *349*, 391–401. [CrossRef]
56. Gan, J.; Liu, Z.; Hui, C.R. A Three-Layer Alternating Spinning Circulation in the South China Sea. *J. Phys. Oceanogr.* **2016**, *46*, 2309–2315. [CrossRef]
57. Sun, Q. Focused Fluid-Flow Escape System and Sediments Deformation in Deep-Water Basins of Northern South China Sea. Ph.D. Thesis, Chinese Academy of Sciences (Institute of Oceanology), Qingdao, China, 2011.
58. Sun, Z.; Bian, M.; Ding, J.; Liu, J.; Zhang, H.; Xu, D. Study on the Wind and Wave Environmental Conditions of the Xisha Islands in the South China Sea. *J. Mar. Sci. Eng.* **2022**, *10*, 1459. [CrossRef]

Journal of
Marine Science and Engineering

MDPI

Article

Identification of Mass Transport Deposits and Insights into Gas Hydrate Accumulation in the Qiongdongnan Sea Area, Northern South China Sea

Yuehua Gong [1,2,3], Shengxiong Yang [2], Jinqiang Liang [1,2,3], Dongmei Tian [2,*], Jing'an Lu [1,2,3], Wei Deng [1,2,3] and Miaomiao Meng [1,2,3]

1 Guangzhou Marine Geological Survey, China Geological Survey, Guangzhou 511458, China
2 Southern Marine Science and Engineering Guangdong Laboratory (Guangzhou), Guangzhou 511458, China
3 Gas Hydrate Engineering Technology Center, China Geological Survey, Guangzhou 511458, China
* Correspondence: dongmeitian@gmlab.ac.cn

Abstract: Accurately identifying the Bottom Simulating Reflector (BSR) is a crucial and fundamental task in seismic exploration and the interpretation of gas hydrates in marine areas. During our seismic interpretation and inference work on a gas hydrate survey in the Qiongdongnan Sea area, we encountered a phenomenon that closely resembled the seismic reflection characteristics of BSR. By comparing and identifying various geological phenomena, we have determined that this unique seismic reflection phenomenon is, in fact, the reflection of the depositional bottom interface known as "mass transport deposits (MTDs)" as described by previous researchers. The physical properties of the MTDs developed on the shallow surface of the seafloor are similar to those of gas hydrate reservoirs in the seismic exploration of marine areas, particularly in the northern South China Sea's Qiongdongnan Sea area. Due to the lack of active neotectonic movement in the area, most identified BSR reflection occurrences are parallel to the seafloor. Consequently, during the process of seismic interpretation of BSR in the Qiongdongnan Sea area, it may be confused with the bottom boundary reflection interface of MTDs. Accurately identifying MTDs' sedimentary bodies in gas hydrate exploration activities in this area would greatly enhance the accurate identification of BSR and support the refined evaluation of gas hydrate resources. In this paper, the structural characteristics of MTDs are compared with the reflection characteristics of seismic profiles, the reflectors are identified as MTDs rather than BSR through analysis and interpretation, and the possible mechanism of hydrate accumulation in this region is discussed.

Keywords: gas hydrate; mass transport deposits; accumulation mechanism

1. Introduction

The accurate identification of BSR in the seismic exploration of gas hydrates in marine areas is critical for the precise evaluation of gas hydrate resources. Generally, the reflection interface can only be interpreted as BSR if the following fundamental conditions are satisfied in the seismic profile. First, the water depth must meet the conditions for gas hydrate occurrence. Second, there should be strong abnormal reflection waves observed in the seismic profile. Third, a blanking zone above the abnormal reflection wave should be present, with a relatively high interval velocity. Fourth, the interval velocity below the abnormal reflection wave should be low. Fifth, the polarity of the abnormal reflection wave should be opposite to that of the seabed reflected wave [1]. However, during gas hydrate exploration in the Qiongdongnan area of the South China Sea, a geological phenomenon resembling BSR characteristics has been discovered. Through comparison and identification of the characteristics, it has been determined that this special sedimentary phenomenon highly resembles the reflection of the depositional bottom interface known as mass transport deposits (MTDs). The recognition and formation mechanism of MTDs have been discussed in previous studies [2]. Weimer (1990) initially introduced the term "mass transport complexes" to describe sedimentary stratigraphic units that erode at the base of sedimentary sequences and are overlain by channels and natural levees [3]. This sediment transport mechanism predominates on the outer shelf/upper continental slope and involves gravity flow processes such as sliding, slumping, and debris flows. Its seismic reflection features typically exhibit a hummocky shape, wavy reflection structure, weak amplitude (with local medium-strength amplitude), and poor continuity, with a chaotic interior. As it involves clastic flow plasticization, flow, dehydration, and remodeling, it exhibits characteristics such as low permeability, high resistivity, and high density [4–7]. The physical characteristics of gas hydrate reservoirs above the BSR are similar to the characteristics that are of interest in gas hydrate seismic exploration. However, the practice of gas hydrate exploration in the Qiongdongnan Sea area has shown that the neotectonic movement in this region is not active [8]. Most BSRs identified in seismic interpretation are parallel to the seafloor, and the key evidence of BSR identification, which is oblique to the strata, is largely absent. Consequently, there is a high risk of confusion with the reflection bottom interface of MTDs developed in the shallow surface of the seabed during BSR interpretation and identification research. The accurate identification of MTDs holds significant potential for accurately identifying BSR in the Qiongdongnan Sea area, thereby providing robust support for the precise assessment of gas hydrate resources in this area. Previous studies have addressed MTDs in the Qiongdongnan Basin to varying degrees, shedding light on their formation mechanisms from diverse perspectives. Wang et al. (2009) identified and investigated the MTDs occurring in the deepwater area of the Qiongdongnan Basin [9]. They highlighted that the genetic mechanism of MTDs is influenced by multiple factors, with seismic activity being a primary trigger and the gradual decomposition of gas hydrates representing a long-term influencing factor. However, the research conducted thus far has been limited to the Huaguang Sag, located in the southwest of the Qiongdongnan Basin. The relationship between MTD development and gas hydrate accumulation remains unclear. Li et al. (2013) investigated the formation of MTDs in the deepwater area of the Qiongdongnan Basin and pointed out that the Central Canyon area of the Qiongdongnan Basin developed multiple stages of MTDs [5], which constituted an important sedimentary unit in the strata since the Neogene period, and proposed that the high sediment supply rate and the activity of the Red River strike-slip fault are likely key factors contributing to the development of MTDs in the Central Canyon area. However, the study did not discuss the relationship between the MTDs observed in the Central Canyon area of the Qiongdongnan Basin and the accumulation of gas hydrates in the area.

In short, the identification of MTDs in the Qiongdongnan Basin and their relationship with gas hydrate accumulation have not been thoroughly investigated, possibly due to data limitations or divergent research focuses across disciplines. Therefore, based on previous studies, this study used the latest high-resolution 2D and 3D seismic data collected in the gas hydrate exploration in the Qiongdongnan Sea area to focus on the gas hydrate accumulation source conditions, focusing on the Lingnan low uplift and its adjacent areas in the Qiongdongnan Basin and identifying the MTDs that are prone to BSR confusion in the gas hydrate stability zone in the area. The origin mechanism of MTDs in the study area and its significance in gas hydrate accumulation are discussed.

2. Geological Background and Data

2.1. Geological Background

The Qiongdongnan Basin in the study area is a passive continental margin stretching basin. The basin is connected to the Yinggehai Basin in the northwest, the Hainan Island Uplift in the north, the Yongle Uplift in the south, and the Shenhu Ansha Uplift in the northeast, the Pearl River Mouth Basin. It is a Cenozoic sedimentary basin spreading in a NE direction, with a water depth of 150–2500 m and an area of about 4.5×10^4 km^2. The basin is distributed in the NE direction, showing a NS zoning and EW partitioning structure. From north to south, the basin can be divided into four parts, the northern depression belt, the northern uplift, the central depression belt, and the Yongle Uplift. Previous studies have shown that the stability region of gas hydrate spans several tertiary structural units from northeast to southwest, such as the Baodao Depression, Songnan Depression, Beijiao low bulge, Beijiao Depression, Lingshui Depression, and Ledong Depression.

2.2. Data and Collection Methods

The three-dimensional (3D) seismic acquisition system utilized an air gun with a capacity of 620 cu.in. as the source and covered an area of 411 km^2. The source pressure was 2000 PSI, and both the air gun and cable were at depths of 4 m and 5 m, respectively. The number of receiving channels was 768, the coverage times were 48, the spacing was 3.125 m, and the sampling rate was 1.0 ms [10,11]. Moreover, the total length of the two-dimensional (2D) seismic data survey line was 7000 km. An air gun boasting a capacity of 570 cu.in. serves as the source of the seismic acquisition system. The source pressure was 2000 PSI, the depth of the air gun was 5 m, and the depth of the cable was 6 m. The number of receiving channels was 240, the coverage times were 60, the spacing was 12.5 m, and the sampling rate was 1.0 ms. The collected 3D and 2D seismic data were processed and loaded into Geofram 4.5 software for geological interpretation. The seismic reflection phenomena indicating fluid migration and gas hydrate enrichment, such as BSR, gully source fault, gas chimney, and MTDs, were mainly identified.

3. BSR Characteristics in the Qiongdongnan Sea Area

To date, over 200 gas hydrate target areas have been identified globally, with the majority of discoveries attributed to the seismic anomaly marker for gas hydrates known as "BSR". BSR is one of the earliest, most widely used, most reliable, and most intuitive seismic markers for predicting the occurrence of gas hydrate. So far, the identification of marine gas hydrate in the world is mostly achieved through the identification of BSR on seismic profiles [12]. In addition, when the gas supply is sufficient and the reservoir is good, the distribution of the gas hydrate stability zone is solely determined by the temperature and pressure of the formation. BSR represents the lower boundary of the gas hydrate stability zone, which is an isothermal surface that resembles the seafloor in parallel and is unrelated to the occurrence of the formation. When the formation occurrence is not aligned with the seafloor, the BSR is often oblique to the formation. This is one of the main features to identify BSR [13–15]. In the field of gas hydrate exploration, previous studies have extensively investigated the BSRs that exist in the Qiongdongnan Sea region, confirming the significant distribution of BSRs within the Qiongdongnan Basin [16]. The sedimentary and structural characteristics of the Qiongdongnan Basin pose significant challenges to identifying BSR markers. Specifically, most identified BSR markers lack a fundamental criterion for BSR judgment, namely, the obliqueness of the BSR relative to the strata, which greatly increases the uncertainty of BSR identified in the Qiongdongnan Basin, and whether the horizon indicated by the blue and green arrows is really BSR needs further study (Figure 1). From a physical point of view of the formation, BSR is essentially the wave impedance difference interface on the seismic profile. Theoretically, any depositional interface with high-to-low impedance contrast has the potential to generate a reflection interface through which seismic waves can propagate and create seismic reflection characteristics similar to those of BSR. Because of this, identifying BSR on seismic profiles has multiple solutions. The difficulty of BSR identification lies in the extensive development of MTDs in the shallow surface of the Qiongdongnan Sea area.

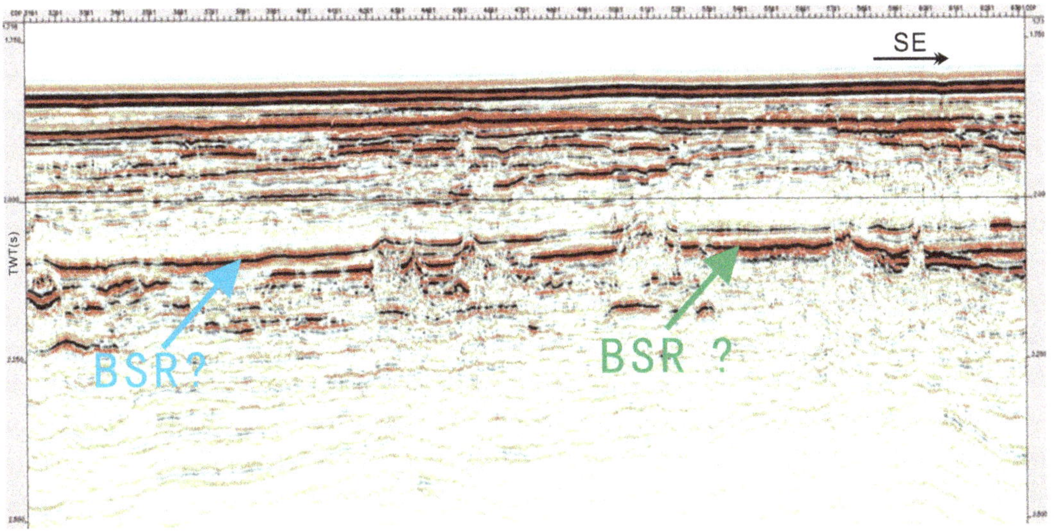

Figure 1. The seismic profile in this area exhibits ambiguous BSRs (there are two different reflectors, and either of these two reflectors could be the BSR. The location of the 2D seismic profile is shown in Figure 2).

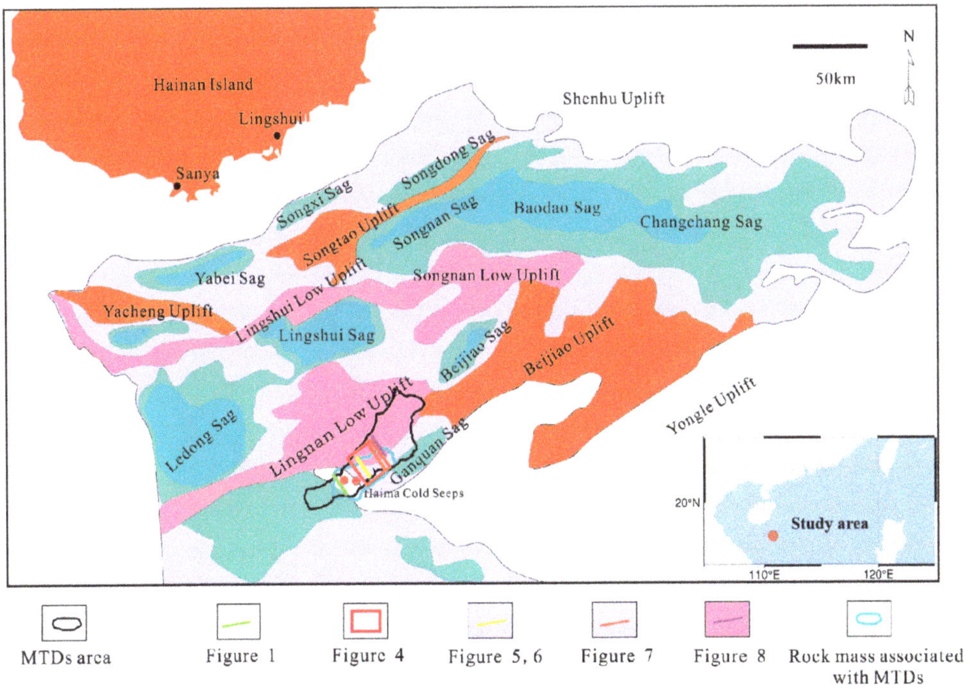

Figure 2. Location of the study area (modified from [17,18]).

4. Lingnan Low Uplift and Adjacent MTDs in Qiongdongnan Basin

4.1. Classification of Mass Transport Deposits

Moscardelli and Wood (2008) classified deep-water MTDs into two distinctive categories based on their external morphology, triggering mechanism, sedimentary–provenance relationship, and provenance location (Figure 3) [19]. The first category is the adjunct type MTDs, primarily induced by regional geological events and typically producing large-scale MTDs. Adjunct-type MTDs can be further divided into two subcategories: adjunct continental shelf-type MTDs (fed by the marginal continental delta systems and triggered by sea level changes and high sedimentation rates) and adjunct continental slope-type MTDs (caused by upper continental slope collapse in response to earthquakes, coastal currents, or storms). The second category is non-adjunct MTDs, inducing relatively small MTDs due to local geological events. They can be divided into mud volcano or unilateral salt rock uplift, micro basin margin formed by salt rock, and channel-natural dike margin [20].

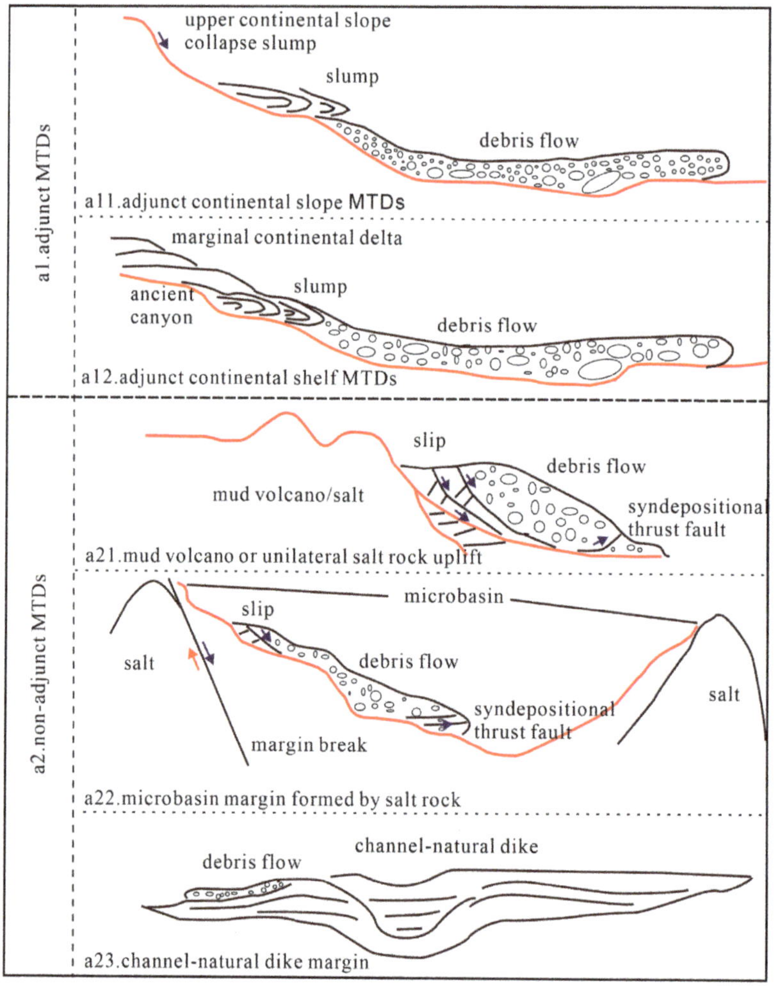

Figure 3. Classification of deep-water MTDs into two distinctive categories based on their external morphology, triggering mechanism, sedimentary–provenance relationship and provenance location (modified from [21]).

Currently, it is the most crucial approach for identifying deep-water MTDs based on seismic data [22,23]. It is based on the seismic wave response to MTD boundary and internal kinematic direction or anisotropy characteristics to identify the overall morphology of deep-water MTDs or the formation distribution characteristics in the process of transport [20]. Previous research has identified and studied the MTD sedimentary bodies developed in the Qiongdongnan Basin. Wang et al. (2009) pointed out that deep-water MTDs have been widely developed in the continental slope of the Qiongdongnan Basin since the Quaternary period [9]. According to Wang et al. (2011), MTDs are widely distributed in the gas hydrate exploration areas of the Qiongdongnan Basin, and the high deposition rate and over-pressure environment unique to this basin contribute to their formation [24]. Furthermore, MTDs observed in Quaternary sediments are located within the hydrate stability zone. Numerous previous studies have confirmed the existence of a significant number of MTDs within the Qiongdongnan Basin. Based on recently acquired high-resolution 2D and 3D seismic data in the Qiongdongnan Basin, this study focuses on the Lingnan low uplift area and adjacent regions. We present an identification study of BSR in this area, delimiting MTDs that developed during the Late Quaternary across the study area (see Figure 2). Finally, we provide a detailed explanation of the inferred formation mechanism of these MTDs.

4.2. MTDs Identification of Lingnan Low Uplift and Adjacent Areas in Qiongdongnan Basin

Previous studies have shown that MTDs can generally be divided into three structural units, namely head extension area, slip area of the main sediment body, and toe extrusion area [24]. Each segment exhibits unique seismic reflection characteristics that assist in distinguishing between them. Figure 4 displays a slice of the 3D seismic reflection amplitude at a depth of 2100 ms in our study area (as indicated by the red box in Figure 2). Since the head extension area occurs within MTDs and lies in a stretching environment, its seismic reflection is characterized by inconsistent amplitudes and sudden changes in intensity (as depicted in the lower part of Figure 4). The slip area of the main sediment body represents the long-distance transport of sliding blocks, where sediments experience mixtures, compaction, and dehydration by the overlying sedimentary layer during later-stage transport. These factors result in a relatively single lithology, with seismic reflection frequently appearing as a blank zone on the plane (refer to the middle section of Figure 4). The toe compression zone belongs to a compression stress environment, and the seismic reflection also presents as a fuzzy zone. However, its physical properties should be stratified relative to the body slip zone deposits because its force is weaker than that of the body slip zone, the degree of sediment mixing is weak, and the seismic plane properties are enhanced compared to the amplitude of the body slip zone deposits (see the upper part of Figure 4).

4.2.1. Head Extension Area

The head extension area of MTDs is readily identifiable during its current stage of development. This region generally develops a back wall characterized by one or more cliffs, which serves as a clear indicator of the extension range for the MTDs' head. In the area of the development of the back wall, the seabed topography will generally change, the up-dip strata of the cliff usually will not be deformed, and the thickness of the strata in the downdip direction of the cliff will change, or even miss. Due to the tensile stress environment, a series of plow normal faults and small tensile faults commonly develop within the head tension area, which has been described in detail in prior research literature [25]. However, MTDs developed during certain geological periods are often buried beneath newer strata with a certain depth, leading to compaction or erosion of plow-type normal faults and small tensile faults within the head stretching area. The identification of these features can be challenging without high-resolution 3D seismic data. Nevertheless, the tension stress environment leaves distinctive characteristics that remain visible (Figure 5, with section location depicted in Figure 2), and the existence of vertical tensioning small faults is evident from Figure 5.

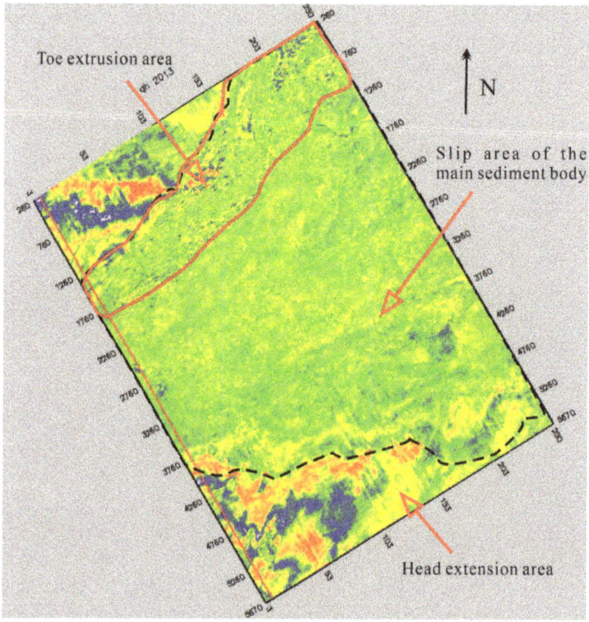

Figure 4. The slice of 3D seismic reflection amplitude at 2100 ms in the study area, which divides the head extension area, slip area of the main sediment body, and toe extrusion area (the location of the 3D seismic zone is shown in Figure 2).

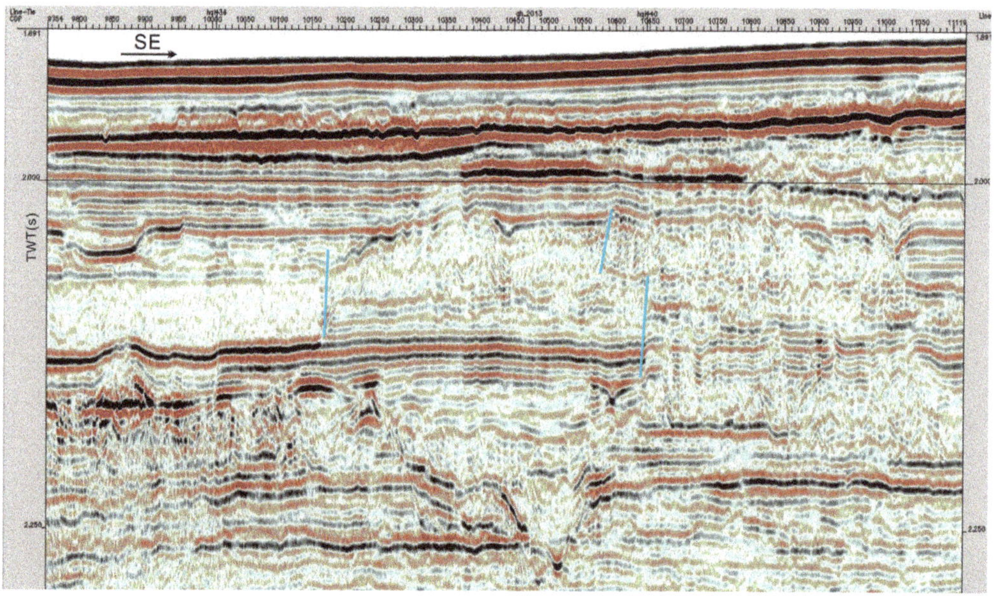

Figure 5. The seismic profile of the head extension area. Small tensile faults in the head extension area of MTDs due to the tension stress environment (marked in blue near the vertical line) (the location of the 2D seismic profile is shown in Figure 2).

4.2.2. Slip Area of the Main Sediment Body

MTDs are multi-phase sedimentary bodies that originate from unstable geological features situated near continental slopes, canyons, channels, flanks of uplift structures, and channel sidewalls. These features move down the slope via overall "freezing" under gravitational influence for varied reasons [26]. It can be seen that the transport process of MTDs has a certain gravitational potential energy and can form a certain scale effect, which is mainly reflected in the slip area of the main sediment body. Due to the strong transport capacity of MTDs, they are often able to form a certain scale. The plane development area of MTDs identified in the study area in Figure 2 (the range of thick black solid lines) is preliminarily estimated to cover an area of 1240 km^2 and the slip area of the main sediment body accounts for more than 75% of the total area (Figure 4). During the long-distance transport of MTDs, due to the uneven strength of sediments overturning and mixing, there are often some unmodified undisturbed strata and residual blocks in MTDs, which become one of the important features to identify the slip area of the main sediment body on seismic profiles (Figure 6, seismic profile location given in Figure 2). In seismic profiles, the physical properties between the residual block and the stirred sediment are different. The strata inside the residual block are continuous, with good stratification and strong amplitude, which is very easy to identify. Moreover, MTDs are severely deformed during transport and may also be subjected to modification by channels and underflows, resulting in an often irregular upper interface that is then filled and altered by various types of sediment fed into the basin, such as channel systems and underflows. The overlying sediments generally appear as channels, floodplains, and possibly matted sand. As shown in Figure 6, the upper interface of the block transport body has been identified in the study area as eroded into irregular grooves with non-uniform erosion degrees across different parts. The closer to the toe area, the stronger the erosion effect is, which may be related to the relatively fine toe area sediment and which is more susceptible to external forces.

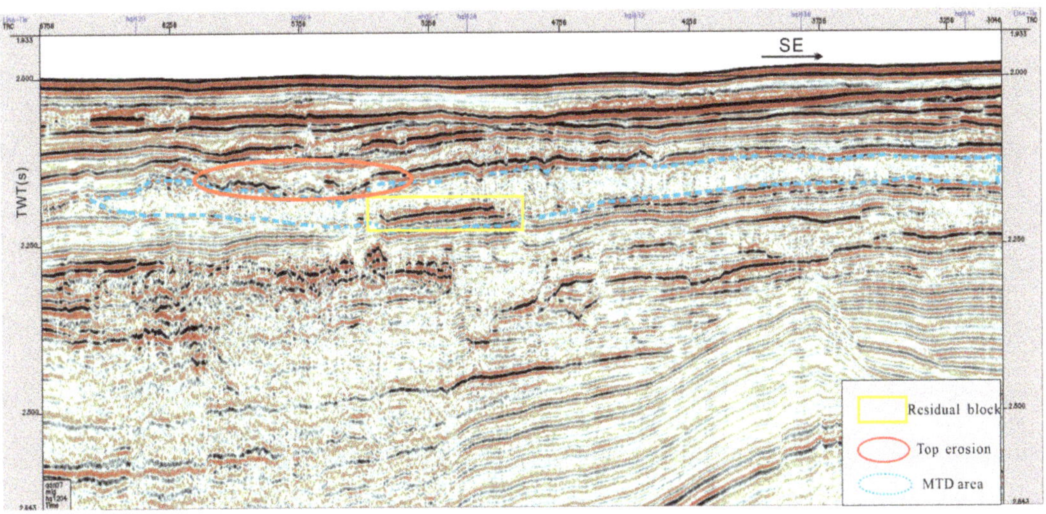

Figure 6. The seismic profile of slip area of the main sediment body. Residual block and top erosion channel of MTD (the location of the 2D seismic profile is shown in Figure 2).

In addition, during the transportation of MTDs, rapid migration, uneven sediment particle sizes, and energetic migration processes often cause varying degrees of erosion on the bottom shear plane, leading to obviously stepped outlines in local areas with severe erosion. The bottom shear plane is characterized by multiple types of erosion grooves (Figure 7, the seismic profile location is shown in Figure 2).

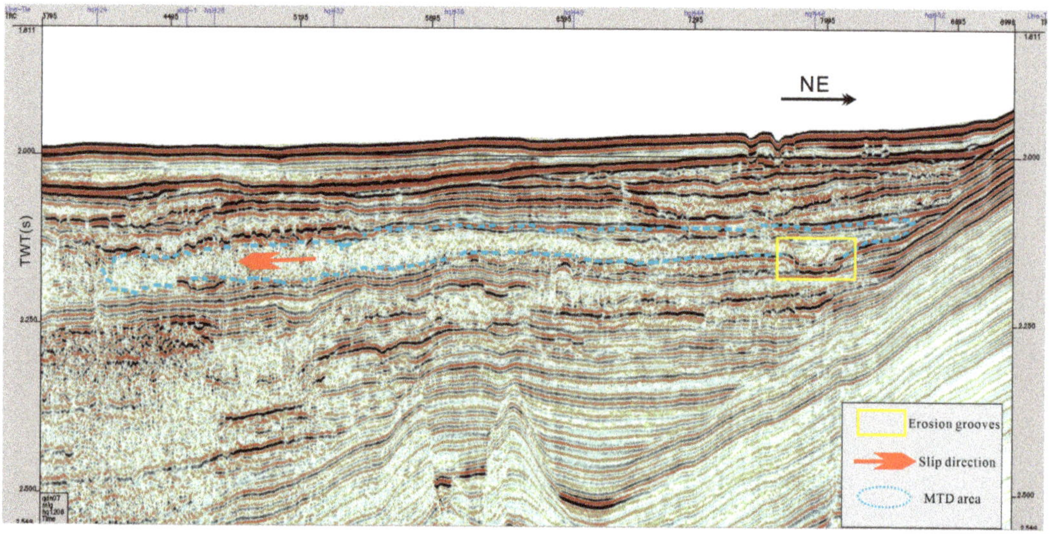

Figure 7. The seismic profile of slip area of the main sediment body. Characteristics of erosion grooves in the lower part of MTD (the location of the 2D seismic profile is shown in Figure 2).

4.2.3. Toe Extrusion Area

Following the occurrence of MTDs, the sediment therein undergoes long-distance transport resulting in weakened gravitational potential energy. If the transport path encounters obstacles such as topographic protrusions, or if the movement's energy is depleted, MTDs will cease migration and form a strong extrusion effect on the topographic protrusions. A variety of extrusion structures are developed along the MTD's toe, characterized by parallel and continuous reflections. As shown in Figure 8 (the seismic profile location is shown in Figure 2), at the end of the marked MTDs (part of which is located inside the lateral wall of the MTDs), imbricated reflection features appear on the seismic profile, showing a typical thrust fault structure.

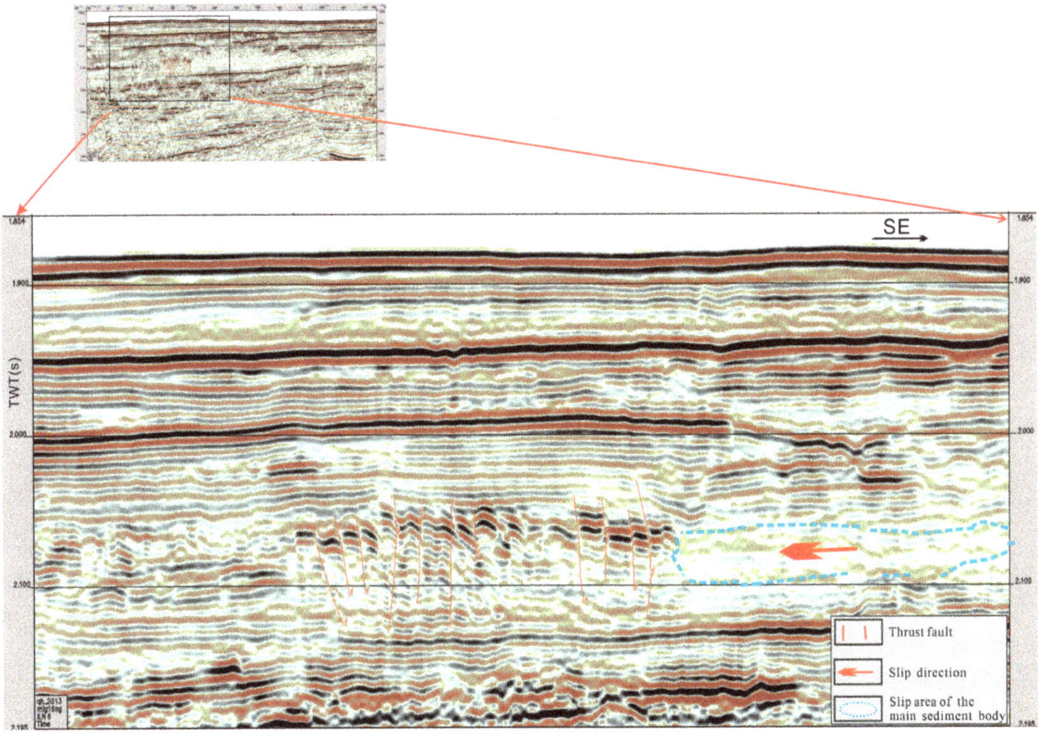

Figure 8. The seismic profile of the toe extrusion area. The thrust fault structure developed in the toe of MTD (the location of the 2D seismic profile is shown in Figure 2).

Therefore, from the above analysis of the three structural regions, it can be confirmed that this region is definitely developed as MTDs. Although the characteristics are similar to BSR, from the multi-angle analysis on the seismic profile it can be found that the development of MTDs is caused by local structures. Figure 9 points out the velocity response with the sediments. Singular velocity information is not enough to identify MTDs, and seismic profiles are needed to assist in the identification.

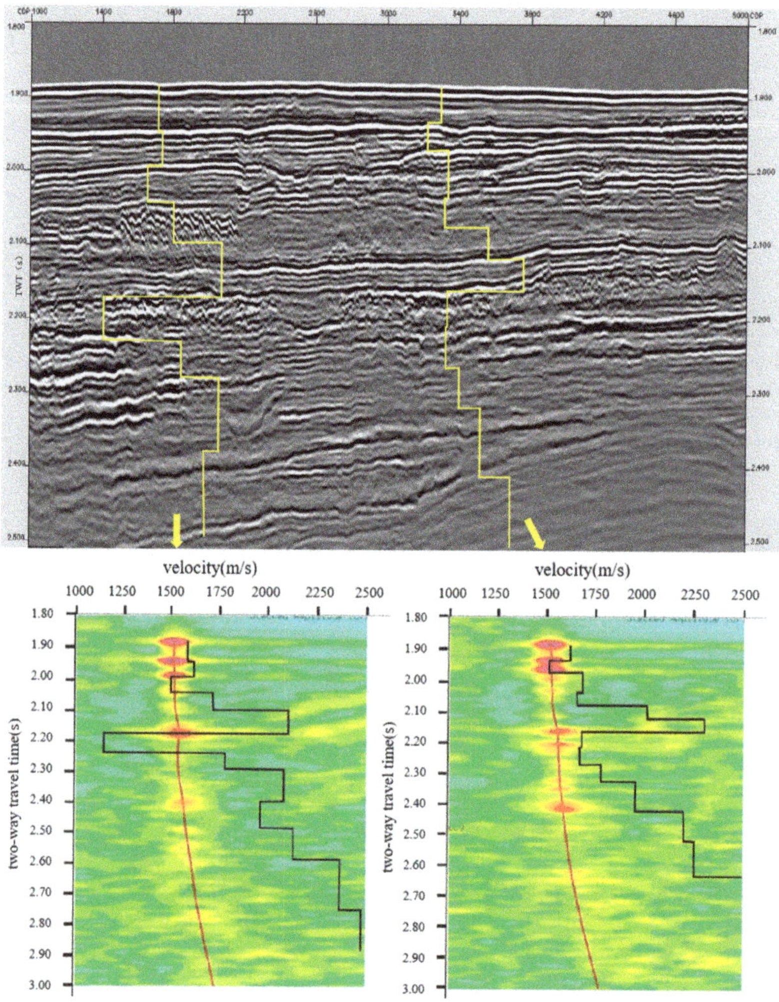

Figure 9. The seismic profile of toe extrusion area beside the location in Figure 8 and seismic velocity (including the interval velocity).

5. Discussion

5.1. MTDs Trigger Mechanism

At present, the trigger mechanisms of deep-water MTDs are mainly qualitatively characterized [17,26,27]. According to the genetic classification method of deep-water MTDs mentioned above, adjunct MTDs can be further divided into shelf adjuncts and slope adjuncts, which are based on different trigger mechanisms. Shelf adjunct MTDs are mainly affected by three factors, namely deltaic sediment supply shelf margin, fluctuations in sea level, and high sedimentation rates. The study area located in the deep-water continental slope region of the Qiongdongnan Basin does not align with this type. The upper slope collapse type (continental slope dependent type) MTDs are mainly triggered by earthquakes, coastal currents, storms, and other mechanisms. Due to its location in the lower slope environment, it is unlikely for the study area to be associated with this MTD type. Non-adjunct MTDs are triggered by local geological events and can be divided into subcategories such as mud volcanoes, unilateral salt uplifts, and micro-basin margins formed by salt rocks and channel-natural dike margins [26]. According to Xia et al. (2008), the Cenozoic tectonic activities in the Qiongdongnan Basin migrated from north to south as a whole [28], with intense subsidence occurring during the Oligocene (E_3) and Pliocene–Quaternary (N_2-Q) periods. The latest research also shows that the neotectonic activity in the Qiongdongnan Basin also conforms to this rule, that is, the tectonic activity is early in the north and late in the south [29]. Seismic interpretations of numerous cross-cutting basins indicate that from the Pliocene to the Quaternary period, tectonic activities were particularly pronounced in the southern region of the Qiongdongnan Basin. This caused the uplift and formation of the Yongle uplift block while also leading to the absence of Pliocene and Quaternary strata in several magmatic active zones. Some magma even penetrated the seabed directly, resulting in the creation of volcanoes [30]. Comparisons between Figures 2 and 7 reveal that the Qiongdongnan Basin's Yongle uplift tectonic belt lies at the southeastern end of the study area. The area has active magmatic activity and was still in the tectonically active stage until the late Quaternary period (Figures 2 and 7). Therefore, we infer that the study area was affected by the late basement uplift (Pliocene–Early Quaternary) in the southeastern Qiongdongnan Basin, which caused the pre-deposited strata to slip and extrude to the northwest direction, finally forming large-scale MTDs in the study area.

5.2. Contribution of MTDs to Gas Hydrate Accumulation Mechanism in the Study Area

Studies on the gas hydrate accumulation mechanism have shown that low temperature and high-pressure environments, adequate gas supply, favorable migration pathways, and good spatiotemporal matching of suitable reservoirs are sufficient and necessary conditions for marine gas hydrate accumulation mechanism. Given the distinctive geological backgrounds of different regions, each of the four reservoir-forming elements has its specific key requisites. With respect to the Qiongdongnan Sea area, recent studies on the temperature and pressure conditions of the gas hydrate accumulation mechanism in the Qiongdongnan Sea area have shown that most areas in the deep-water area of the southeast Qiongdongnan Sea area satisfy the temperature and pressure conditions of gas hydrate formation [30–33]. In addition, the deep-water oil and gas survey in the Qiongdongnan Basin has shown that the Qiongdongnan Sea area is located in the northwest atmospheric area of the South China Sea. In recent years, oil and gas exploration companies such as China National Offshore Oil Corporation (CNOOC) have successively discovered the LS17-2, LS25-1, and LS18-1 gas fields, as well as the BD21 high-production gas well in the deep-water and ultra-deep-water areas of the Qiongdongnan Basin, which have confirmed the ample gas sources in this area [34–36]. The potential gas source foundation of the gas hydrate accumulation mechanism is sufficient. However, the research results of gas hydrate reservoirs, especially a large number of gas hydrate drill-coring studies in the northern South China Sea in recent years, show that gas hydrates may occur in both muddy silty sand and sandy deposits, and the quality of gas hydrate reservoirs is not the key control factor for the gas hydrate accumulation mechanism in this region. Therefore, a good gas migration channel is the key factor for the gas hydrate accumulation mechanism in the Qiongdongnan Sea area.

The fluid transport system of the gas hydrate accumulation mechanism has been the subject of extensive discussion among experts from various fields in the Qiongdongnan Sea area. Wang et al. (2011) first pointed out that polygonal faults, tubular structures, and diapir structures destroyed the sealing property of the huge thick mudstone strata in the thermal settlement stage of the post-fracture in the Qiongdongnan Basin [24], causing a large number of fluids to migrate vertically or nearly vertically upward, providing a good gas migration channel for gas hydrate formation in this area. However, polygonal faults are not universally developed in the Qiongdongnan Basin. Moreover, it is only a small-scale fluid migration channel. Li et al. (2005) showed that the Paleogene fault activity in the Qiongdongnan Basin was strong and controlled sedimentary development [37]. In the early Neogene period, the basin underwent a peak in hydrocarbon generation and expulsion with clear evidence of fault-related fluid migration and hydrocarbon accumulation. However, during the late Miocene and Quaternary rapid subsidence, fault activities exhibited relatively weak vertical fluid transport capacity. The latest high-resolution seismic analysis of gas hydrate also shows that there have been few faults developed in the Qiongdongnan Basin, especially in large depressions, since the late Miocene and most unconformities were developed during the Paleogene era. Therefore, the gas channels of faults and unconformities should not be the dominant fluid migration modes of gas hydrate sources in this area. However, in recent years, there have been recent reports on the discovery of gas hydrates in the Qiongdongnan Sea area, among which a large number of leaking gas hydrates have been found [38]. Moreover, the existence of diffused gas hydrate has also been found [39]. Different scholars have discussed the fluid transport system in the gas hydrate discovery area to different degrees. On the whole, it is widely recognized that gas chimney is the dominant fluid transport mode in the area. However, studies have shown that the gas chimney development area is not completely consistent with the gas hydrate discovery areas. Furthermore, the gas migrating to the gas hydrate stability area relying on the gas chimney also needs a certain mechanism to aggregate to form the possibility of gas hydrate deposits.

Based on previous studies and MTD analysis in this paper, it is proven that MTDs are a widely developed sedimentary mode in the Qiongdongnan Sea area. Moreover, studies have shown that MTDs are a kind of plasticization and clastic flow, during which dehydration and remodeling occur, and differential compaction results in a higher impedance value relative to the surrounding rock, giving it characteristics of low permeability, high resistivity, and high density. The formation with "low permeability and high density characteristics" is a good cap layer for shallow deep-sea fluids. It has a good blocking effect on the upward migration of free gas in the lower stratum, which is conducive to the gas accumulation of hydrate accumulation in the lower stratum. When the fluid pressure blocked under MTDs is high enough, the MTDs' sedimentary layer will be impaled and broken by the high-pressure hot fluid. At this time, the gas hydrate may be formed in the fracture zone of MTDs, forming a near-vertical fracture along the gas escape channel, enlarging the gas hydrate accumulation space and forming high-quality leaking gas hydrate. Even the fluid continues to move upward to the seafloor surface, forming a "cold seeps" phenomenon (Figure 10).

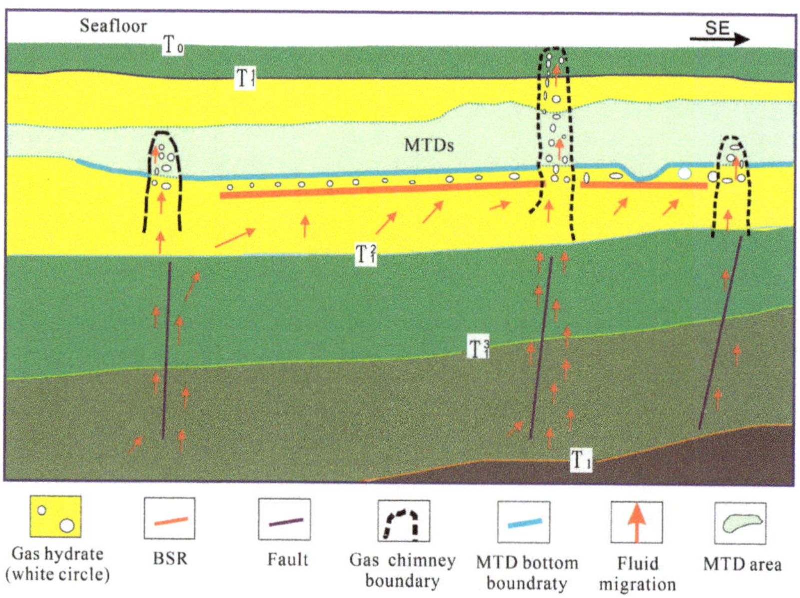

Figure 10. The relationship between MTDs and gas hydrate accumulation.

6. Conclusions

(1) The MTDs of the study area are identified and delineated, and it is believed that the MTDs of the study area were caused by the late uplift of the southern Qiongdongnan Basin and the early platform uplift, which caused the large-scale collapse and slip of sediments in the central depression of the basin, plasticized and flowed in the long-distance transport process, and finally formed MTDs.

(2) The physical properties of MTDs generally depend on the original sediment type before their formation, which is determined by the deep-water environment in which they are formed, and most of them are mainly muddy. As they are formed by plasticization and clastic flow, dehydration and remodeling will occur in the flow, which determines its characteristics of low permeability and high density. It can form a good blocking effect on the upward migration of free gas in the lower layer and is favorable for gas accumulation of gas hydrate under it.

(3) When the fluid pressure under the MTDs is high enough, the MTDs' sedimentary layer will be pierced by the high-pressure hot fluid. In this case, the gas hydrate may form in the MTDs' layer or the entire fluid escape channel, forming high-saturation leakage gas hydrate, or the fluid will even continue to migrate upward to the seabed surface, forming "cold seeps".

(4) In the process of BSR interpretation and inferential analysis, if the sedimentary background study suggests that MTDs may be formed in the gas hydrate stable zone in the study area, the MTDs' sedimentary bodies must be accurately identified. Otherwise, they are easy to be confused with the bottom interface of MTDs. The accurate identification of MTD deposits is of great significance to the accurate evaluation of gas hydrate resources in the Qiongdongnan Sea area.

Author Contributions: Conceptualization, J.L. (Jinqiang Liang); Methodology, S.Y.; Formal analysis, W.D.; Investigation, J.L. (Jing'an Lu); Resources, M.M.; Writing—original draft, Y.G.; Writing—review & editing, D.T. All authors have read and agreed to the published version of the manuscript.

Funding: This research was funded by the National Natural Science Foundation of China (No. 42327901); Guangdong Major Project of Basic and Applied Basic Research (2020B0301030003); the China Geological Survey Project (No. DD20221700); the National Development Program of China (No. 2021YFC2800901); the National Natural Science Foundation of China (No. U2244224); the Guangdong Basic and Applied Basic Research Foundation (No. 2022A1515110764); the Basic Foundation of Guangzhou (No. 2023A04J0916); the project funded by China Postdoctoral Science Foundation (No. 2022M720887); National Natural Science Foundation of China (No. 42102144); Guangzhou Science and Technology Planning Project (No. 202201011409).

Institutional Review Board Statement: Not applicable.

Informed Consent Statement: Not applicable.

Data Availability Statement: Data are contained within the article.

Conflicts of Interest: The authors declare that they have no known competing financial interests or personal relationships that could have appeared to influence the work reported in this paper.

References

1. Bangs, N.; Sawyer, D.S.; Golovchenko, X. Free gas at the base of the gas hydrate zone in the vicinity of the Chile triple junction. *Geology* **1993**, *21*, 905–908. [CrossRef]
2. Nwoko, J.; Kane, I.; Huuse, M. Mass transport deposit (MTD) relief as a control on post-MTD sedimentation: Insights from the Taranaki Basin, offshore New Zealand. *Mar. Pet. Geol.* **2020**, *120*, 104489. [CrossRef]
3. Weimer, P. Sequence stratigraphy, facies geometries, and depositional history of the Mis sissippi Fan, Gulf of Mexico. *AAPG Bull.* **1990**, *74*, 425–453. [CrossRef]
4. Piper, D.J.W.; Pirmez, C.; Manley, P.L.; Long, D.; Flood, R.D.; Normark, W.R.; Showers, W. Mass-transport deposits of the Amazon Fan. In *Proceedings ODP, Scientific Results*; Ocean Drilling Program: College Station, TX, USA, 1997. [CrossRef]
5. Li, W.; Wu, S.G.; Wang, X.J.; Wang, D.; Zhao, F. Seimic characteristics and distribution of pliocene mass transport deposits in central canyon of Qiongdongnan basion. *Mar. Geol. Quat. Geol.* **2013**, *33*, 9–15. [CrossRef]

6. Azadbakht, F.; Monfared, M.S.; Radfar, A. New insights into the geometry of gas chimneys in the Gorgan plain through seismic attribute integration. *Acta Geophys.* **2023**, 1–16. [CrossRef]
7. Soltani, P.; Soleimani, M.; Aghajani, H. Faults and fractures detection in 2D seismic data based on principal component analysis. *Int. J. Min. Geo-Eng.* **2017**, *51*, 199–207. [CrossRef]
8. Gong, Y.H.; Yang, S.X.; Wang, H.B.; Liang, J.Q.; Liang, J. Prospect of Gas Hydrate Resources in Qiong Dong-Nan Basin. *J. Jilin Univ. (Earth Sci. Ed.)* **2018**, *48*, 1030–1043. [CrossRef]
9. Wang, D.W.; Wu, S.G.; Dong, D.D.; Yao, G.; Cao, Q. Seimic characteristics of quaternary mass transport deposits in Qiongdongnan basin. *Mar. Geol. Quat. Geol.* **2009**, *29*, 69–74. [CrossRef]
10. Ren, J.; Cheng, C.; Xiong, P.; Kuang, Z.; Liang, J.; Lai, H.; Chen, Z.; Chen, Y.; Li, T.; Jiang, T. Sand-rich gas hydrate and shallow gas systems in the Qiongdongnan Basin, northern South China Sea. *J. Pet. Sci. Eng.* **2022**, *215*, 110630. [CrossRef]
11. Geng, M.; Zhang, R.; Yang, S.; Guo, J.; Chen, Z. Focused Fluid Flow, Shallow Gas Hydrate, and Cold Seep in the Qiongdongnan Basin, Northwestern South China Sea. *Geofluids* **2021**, *2021*, 5594980. [CrossRef]
12. Liang, J.; Zhang, Z.; Su, P.; Sha, Z.; Yang, S. Evaluation of gas hydrate-bearing sediments below the conventional bottom-simulating reflection on the northern slope of the South China Sea. *Interpretation* **2017**, *5*, SM61–SM74. [CrossRef]
13. Monteleone, V.; Marin-Moreno, H.; Bayrakci, G. Seismic characterization and modelling of the gas hydrate system in the northern Bay of Bengal, offshore Bangladesh. *Mar. Pet. Geol.* **2022**, *141*, 105690. [CrossRef]
14. Mosher, D.C. A margin-wide BSR gas hydrate assessment: Canada's Atlantic margin. *Mar. Pet. Geol.* **2011**, *28*, 1540–1553. [CrossRef]
15. Shipley, T.H.; Houston, M.H.; Buffler, R.T.; Shaub, F.J.; Mcmillen, K.J.; Laod, J.W.; Worzel, J.L. Seismic evidence for widespread possible gas hydrate horizons on continental slopes and rises. *Am. Assoc. Pet. Geol. Bull.* **1979**, *63*, 2204–2213. [CrossRef]
16. Liang, J.; Zhang, W.; Lu, J.; Wei, J.; Kuang, Z.; He, Y. Geological occurrence and accumulation mechanism of natural gas hydrates in the eastern Qiongdongnan Basin of the South China Sea: Insights from site GMGS5-W9-2018. *Mar. Geol.* **2019**, *418*, 106042. [CrossRef]
17. Du, H.; Shi, W.Z.; Liang, J.Q.; Wang, R.; He, Y.L.; Xu, L.T. Genesis of mass transport deposits and their effect on gas hydrate accumulation in the Qiongdondnan Basin. *Oil Geophys. Prospect.* **2021**, *56*, 869–881. (In Chinese)
18. Zhang, W.; Liang, J.Q.; Lu, J.A.; Meng, M.M.; He, Y.L.; Deng, W.; Feng, J. Characteristics and controlling mechanism of typical leakage gas hydrate reservoirforming system in the Qiongdongnan Basin, northern South China Sea. *Nat. Gas Ind.* **2020**, *40*, 90–99.
19. Moscardelli, L.; Wood, L. New classification system for mass transport complexes in offshore Trinidad. *Basin Res.* **2008**, *20*, 73–98. [CrossRef]
20. Posamentier, H.W.; Walker, R.G. Deep-Water Turbidites and Submarine Fans. In *Facies Models Revisited*; SEPM (Society for Sedimentary Geology): Claremore, OK, USA, 2011; pp. 399–520. [CrossRef]
21. Qin, Y.Q.; Wan, L.K.; Ji, Z.F.; Li, F.H.; Xu, H.L.; Ba, D. Progress of research on deep-water mass-transport deposits. *Oil Gas Geol.* **2018**, *39*, 140–152. [CrossRef]
22. Ratzov, G.; Collot, J.-Y.; Sosson, M.; Migeon, S. Mass-transport deposits in the northern Ecuador subduction trench: Result of frontal erosion over multiple seismic cycles. *Earth Planet. Sci. Lett.* **2010**, *296*, 89–102. [CrossRef]
23. Bull, S.; Cartwright, J.; Huuse, M. A review of kinematic indicatiors from mass-transport complexes using 3D seismic data. *Mar. Pet. Geol.* **2009**, *26*, 1132–1151. [CrossRef]
24. Wang, X.; Wu, S.; Dong, D.; Guo, Y.; Deborah, H. Control of mass transport deposits over the occurrenceof gas hydrate in Qiong Dongnan Basin. *Geol. Quat. Geol.* **2011**, *31*, 109–118. [CrossRef]
25. Wu, S.G.; Qin, Z.L.; Wang, D.W.; Peng, X.C.; Wang, Z.J.; Yao, G.S. Seismic charaetersties and triggering mechanism analysis of mass transport deposits in the northern continental slope of the South China Sea. *Chin. J. Geophsics* **2011**, *54*, 3184–3195. (In Chinese)
26. Qin, L.; Mao, J.X.; Ni, F.L.; Xu, S.; Li, S.; Cai, C.; Shang, W.; Liu, J. A Brief Introduction to Deep-Water Mass-Transport Complexes: Structures, Genetic Classifications and Identification Methods. *Adv. Earth Sci.* **2020**, *35*, 632–642. [CrossRef]
27. Jenner, K.A.; Piper, D.J.W.; Campbell, D.C.; Mosher, D.C. Lithofacies and origin of late Quaternary mass transport deposits in submarine canyons, central Scotian Slope, Canada. *Sedimentology* **2007**, *54*, 19–38. [CrossRef]
28. Xia, B.; Lv, B.F.; Wu, G.G.; Chen, G.W.; Li, W.Q.; Wang, R. The cenozoc tectonic transport and its control on the source rock in the northen south china sea. *Nat. Gas Geosci.* **2007**, *18*, 629–634. [CrossRef]
29. Wang, Z.Z.; Zhu, J.T.; Li, A.Q.; Hu, Q.W.; Mao, X.L.; Yin, H.W. Differential control of structures over reservoirs and its significance in Qiongdongnan Basin. *Geol. Quat. Geol.* **2021**, *41*, 157–169. [CrossRef]
30. Fan, Q.; Li, Q.; Li, L.; Lyu, X.; Zhu, Z.; Sun, T. Gas hydrate accimilation model in the Central Canyon of Southeast Basin of Hainan. *J. Northeast. Pet. Univ.* **2021**, *45*, 13–20. [CrossRef]
31. Chen, D.F.; Li, X.X.; Xia, B. Distribution of gas hydrate stable zones and resource prediction in the Qiongdongnan Basin of the South China Sea. *Chin. J. Geophsics* **2004**, *47*, 548–555.
32. Zhu, J.T.; Deng, Y.; Guo, M.G.; Song, P.; Xiong, X.F.; Mao, X.F.; Li, F.; Huang, S.Z.; He, Y.; Ren, T. Mineralization conditions and accumulation pattern of the gas hydrate in Qiongdongnan basin floor plain. *China Offshore Oil Gas* **2020**, *32*, 11–19.
33. Ren, J.; Xu, L.; Shi, W.; Yang, W.; Wang, R.; He, Y.; Du, H. Shallow Overpressure Formation in the Deep Water Area of the Qiongdongnan Basin. China. *Front. Earth Sci.* **2022**, *10*, 922802. [CrossRef]

34. Huang, B.; Huang, H.; Wang, Z.; Huang, Y.; Sun, Z. Kinetics and model of gas generation of source rocks in the deepwater area, Qiongdongnan Basin. *Acta Oceanol. Sin.* **2015**, *34*, 11–18. [CrossRef]
35. Zhu, W.; Huang, B.; Mi, L.; Wilkins, R.W.T.; Fu, N.; Xiao, X. Geochemistry, origin, and deep-water exploration potential of natural gases in the Pearl River Mouth and Qiongdongnan basins, South China Sea. *AAPG Bull.* **2009**, *93*, 741–761. [CrossRef]
36. Zhu, Y.S.; Wang, Y.Q.; Feng, X.Z.; Wang, D.; Wang, S. Surface sediments and their geotechnical characteristics in the development area of deepwater gas field LS17-2. *Mar. Geol. Quat. Geol.* **2022**, *42*, 45–56. [CrossRef]
37. Li, X.X.; Zhu, G.H. The fault system and its hydrocarbon carrier sigificance in Qiongdongnan basin. *China Offshore Oil Gas* **2005**, *17*, 1–7. [CrossRef]
38. Wei, J.; Liang, J.; Lu, J.; Zhang, W.; He, Y. Characteristics and dynamics of gas hydrate systems in the northwestern South China Sea—Results of the fifth gas hydrate drilling expedition. *Mar. Pet. Geol.* **2019**, *110*, 287–298. [CrossRef]
39. Hu, Y.; Luo, M.; Liang, Q.; Chen, L.; Feng, D.; Yang, S.; Liang, J.; Chen, D. Pore fluid compositions and inferred fluid flow patterns at the Haima cold seeps of the South China Sea. *Mar. Pet. Geol.* **2019**, *103*, 29–40. [CrossRef]

Journal of
Marine Science and Engineering

MDPI

Article

Acid-Extracted Hydrocarbon Anomalies and Significance in the Chaoshan Depression of the Northern South China Sea

Guangjian Zhong [1], Jing Zhao [1,*], Zhongquan Zhao [1,*], Kangshou Zhang [1,*], Junhui Yu [2], Chunjiang Shang [1], Guanghong Tu [1] and Changmao Feng [1]

1 Guangzhou Marine Geological Survey, Guangzhou 510301, China; zguangjian@mail.cgs.gov.cn (G.Z.); shangcj@mail.cgs.gov.cn (C.S.); tguanghong@mail.cgs.gov.cn (G.T.); fengchangm@163.com (C.F.)

2 Key Laboratory of Ocean and Marginal Sea Geology, South China Sea Institute of Oceanology, Chinese Academy of Sciences, Guangzhou 510301, China

* Correspondence: zhaojing_0315@163.com (J.Z.); zzqhello@163.com (Z.Z.); zkangshou@mail.cgs.gov.cn (K.Z.)

Abstract: To predict the favorable zones and the types of reservoirs, acid extraction has been used in the Chaoshan depression to detect trace amounts of light hydrocarbons, heavy hydrocarbons, and the $\delta\ ^{13}C$ (‰) of methane. As a result, two integration anomalous zones for exploration activity were blocked out in the northeastern and southwestern parts of the Chaoshan Depression, respectively. By analyzing the differentiation law and structural characteristics of hydrocarbon gases, as well as the stable carbon isotope ratio of methane, the underlying reservoirs were predicted to be gas reservoirs, and the seismically interpreted Dongsha-A (DS-A) structure was predicted to be a gas-rich structure. By correlating the seismic profile and geochemical anomalies, it was determined that fault planes and micro-fractures are the main controlling factors for the occurrence of the seabed's geochemical anomalies. A composite formation mechanism of "lower generation, upper accumulation and micro fractures leaking" is proposed for the control of the underlying petroleum reservoirs, as well as for the micro-fracture control of permeability and surface adsorption control. Acid-extracted hydrocarbon anomalies have favorable indicating significance for exploration activity.

Keywords: South China Sea; Chaoshan Depression; surface geochemistry; acid-extracted hydrocarbon; methane carbon isotope

1. Introduction

The Chaoshan Depression is a Mesozoic residual depocenter in the northeast of the South China Sea. The deposit has favorable exploration activity potential but complex oil and gas accumulation. Accurate identification of favorable exploration areas and reservoirs is the key to successful exploration. Drilling data in the South China Sea reveal a marine Mesozoic development [1], making it an ideal area for studying fossils in the South China Sea. The tectonic framework of the South China Sea exerts strong control over the expansion and evolution of the Cenozoic South China Sea, and also determines the generation, migration, accumulation, and preservation processes of oil and gas. Understanding the geological structure of fossils in the South China Sea is of great significance for a comprehensive understanding of the Southeast Asian marginal seas, especially regarding the Mesozoic and Neozoic tectonic evolution of the South China Sea. The Chaoshan Depression is a Mesozoic and Cenozoic superimposed depression in the northern South China Sea (Figure 1), characterized by polyphase basin formation and structural transformation. It developed thick marine Mesozoic strata and has strong potential for exploration activity [2–4]; however, oil and gas have not yet been discovered in this region through drilling. The physical properties of the South China Sea's Mesozoic reservoir are complex and the predrilling predictions do not match the results of LF35-1-1 [5–9]. In recent years, extensive investigations and studies have been conducted in this area, and a lot of trapped structures rich in organic matter have been discovered [10–13]. To ensure further successful exploration, the effective identification and prediction of favorable exploration areas and petroleum reservoirs is key.

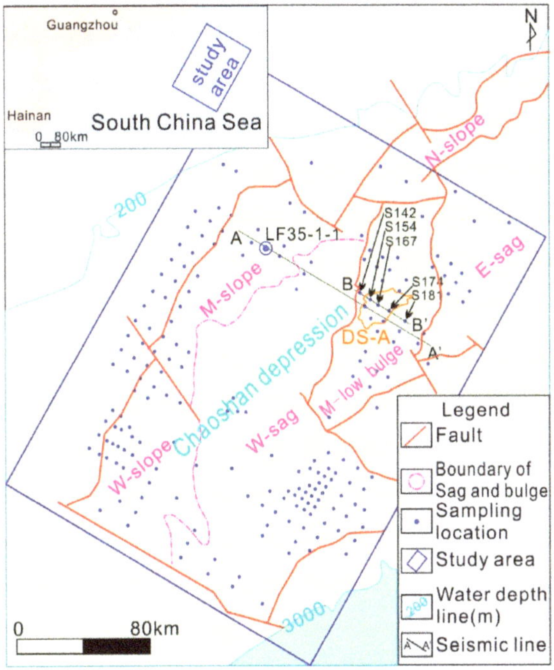

Figure 1. Location map of the study area and structural zoning map.

Micro-leakage in oil and gas reservoirs are a common natural phenomenon and form geochemical anomalies on the land surface or the seafloor. Geochemical anomalies in seafloor samples can be used to quickly delineate favorable exploration zones in underexplored basins.

Surface geochemistry (SG) is a rapid method for predicting favorable zones of exploration activity and the types of petroleum reservoirs. Driven by the pressure of oil and gas, petroleum migrates upward along complex micro-fractures, fault planes, and other migration channels in the form of micro-bubbles or continuous gas phases. These hydrocarbons are adsorbed by clay minerals and encapsulated by secondary carbonates. As a result, the adsorbed hydrocarbon anomalies corresponding to the underlying reservoir accumulate in the seafloor sediments above the reservoirs. By detecting such anomalies, potential reservoirs can be predicted. Based on the direct correlation between latent petroleum reservoirs and the hydrocarbon concentrations of the gas adsorbed in sea bed sediments, the hydrocarbons' internal composition can be used to reliably forecast the properties of the migrated hydrocarbons and forecast the development characteristics of the underlying petroleum reservoir and distinguish its overall properties [14–28].

Marine SG mainly employs the analysis of gaseous alkanes (C_1–C_5) present in seabed sediments to predict the characteristics of the underlying petroleum reservoir [29]. These bound gases are believed to be attached to organic and/or mineral surfaces, which are entrapped in water structures or authigenic carbonate inclusions. Horvitz (1981, 1985) pioneered the concept of bound sediment gas analysis with his method of acid-extraction adsorbed gas analysis [30,31]. Migrated thermogenic hydrocarbon gases readily bind to near-surface fine-grained sediments (e.g., clays) due to the highly adsorptive nature of these sediments towards hydrocarbons. He also believed there was a preferential adsorption of the migrating thermogenic hydrocarbons relative to the in situ-derived microbial gases. Whiticar (2002) describes how structured water creates a relatively impermeable membrane of organized water molecules that entrap the migrated thermogenic gases, and the in situ-generated interstitial microbial gases will have little or no exchange with the entrapped/sorbed phase [32]. He describes a contiguous coating or network of structured water whereby thermogenic hydrocarbons will migrate vertically within the sedimentary column in a contiguous structured water network, a process known as "handshake migration" [29].

However, gaseous alkanes can also be affected by factors such as disturbances to seabed microorganisms and the interference of biogenic methane, which increases the difficulty of predicting the location of underlying petroleum reservoirs and exploration risks. Therefore, analyzing and identifying whether hydrocarbon anomalies in seafloor sediments originate from the hydrocarbons of the underlying petroleum reservoir is an important factor in predicting the success or failure of oil and gas traps, keeping in mind that the near-surface petroleum seepage signature can be generated from microbial as well as thermogenic sources. This study qualitatively and quantitatively analyzes the acid-extracted hydrocarbon anomalies in the Chaoshan Depression in the northeast of the South China Sea. Using anomalies in the abundance, structure, and distribution characteristics of light and heavy hydrocarbons, this study predicts the attributes of the Mesozoic Petroleum system. The findings delineate the northeastern and southwestern parts of the Chaoshan Depression as comprehensive anomalous zones for oil and gas exploration, and these are also the most favorable prospective oil and gas areas. The DS-A structure is determined as being a predominantly gas-bearing reservoir.

2. Geological Setting

The Dongsha Sea is an area in the northeast of the South China Sea comprising of thick Mesozoic strata, and the Chaoshan Depression is the largest residual depression in the area (Figure 1). The depression has undergone six stages of tectonic evolution in the Mesozoic and Cenozoic eras: (1) In the late Triassic rifting period, semi enclosed bay began to develop in the Mesozoic basin in the northeast of South China Sea; (2) During the Jurassic depression period, the largest scale transgression occurred, and the basin expanded to its maximum extent; (3) In the Late Jurassic tectonic uplift period, compaction fold and structural unconformity occurred under compression; (4) In the Early Cretaceous re-subsidence period, volcanic debris, lava flows, and continental margin debris began to rapidly fill and deposit in the basin; (5) In the Late Cretaceous tectonic inversion period, under the influence of four episodes of the Yanshan movement, the Chaoshan Depression was uplifted and denuded again, forming the second regional unconformity; and (6) In the Neogene regional thermal subsidence period, the central basin of the South China Sea reached its peak of expansion during the Miocene, and the Cenozoic basins in the northern continental margin of the South China Sea successively entered a stage of post-rift depression evolution; large-scale transgression also occurred, mainly accepting coastal and shallow marine deposits [33–38]. The DS-A structure is located in the M-low bulge and is a fault anticline structure (Figure 1), controlled by NE-, NNE-, and NW-trending faults.

The 1005–1369 m interval of LF35-1-1 is a Cretaceous sandstone mudstone interval and the 1369–1698 m interval contains Cretaceous tuff. The 1698–2412 m interval is Jurassic sandstone and mudstone, in which the 1698–1940 m interval is grayish-black laminated mudstone and argillaceous siltstone mixed with siliceous rock, containing a small amount of micrite limestone and pyroclastic materials. The 1940–2412 m interval is grayish-black laminated mudstone and argillaceous siltstone mixed with sandstone and limestone. The mudstone is rich in organic debris. Radiolarian fossils were found in the 1716 m to 1839 m interval, indicating a deep-water environment, dating back to the late Jurassic. Fossils of benthic foraminifera were found in the 2049–2112 m interval, indicating a shallow tropical marine environment. Spore and pollen fossils can be found in the 2187 m to 2268 m interval, mainly consisting of a combination of Carassous pollen and Alsophila spinulosa spores.

This set of spore and pollen fossils appeared in the Middle and Upper Jurassic period of southern China, indicating a coastal swamp environment [1]. A granite granodiorite intrusion with a small amount of diorite appears in the well interval below 2412 m, and its invasion period is mainly in the Late Cretaceous.

Well LF35-1-1 reveals two source rock strata in the 1940–2022 m interval and the 2100–2412 m interval, wherein the 1940–2022 m rock is poor-to-medium source, and the 2100–2400 m rock is medium-to-good source. The organic matter of these two source rocks is mainly of type III and a small amount of type II2, which are stable in a considerable range. The minimum thickness of a single layer of source rock is 6 m, the maximum thickness is >40 m, and the average thickness is about 20 m. The average thickness of the single layer of the lower source rock is greater than the upper layer. The results of well seismic tracking and further research and comparisons of regional geology show that there are two sets of source rock developed in the Chaoshan Depression, including the upper Triassic–lower Jurassic shallow sea bathyal mudstone and lower Jurassic shallow sea shelf facies mudstone. According to the drilling strata of well LF35-1-1 and seismic data, there are two sets of reservoirs developed in the DS-A structure in the Upper and Middle Jurassic series, respectively. Among them, the top of the Middle Jurassic series is a set of stable shallow sea shelf facies reef limestone with thickness, while the Upper Jurassic is a set of basin floor fan sandstones (Figure 2).

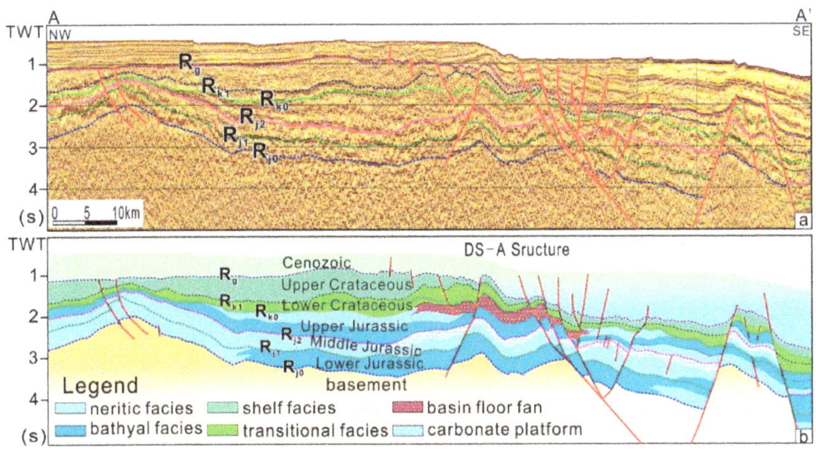

Figure 2. Seismic interpretation profile (See Figure 1 for location). ((**a**). preliminary interpretation of seismic profile; (**b**). geological interpretation section).

3. Sampling and Methods

The samples used are sourced from the petroleum surveys of 200 stations collected by the Guangzhou Marine Geological Survey in the Dongsha Sea Area in recent years (Figure 1). Sampling grid points with a size of 8 km × 16 km are mainly distributed in the Chaoshan Depression. Samples at some grid points could not be collected due to the presence of submarine cables. Moreover, due to the large number of submarine optical cables in this area, sampling is limited by cable protection requirements, making it impossible to conduct encrypted investigations and detailed target investigations. The samples were collected using an open barrel gravity corer, with a length of 1.3 m. Sediment samples were taken at a depth of 85 cm–100 cm from the top of the columnar sample, acid extraction was performed on the hydrocarbon content, and δ13C (‰) testing and analysis was performed as well.

Acid-extraction hydrocarbon testing was performed as follows: under vacuum conditions and constant temperature, the hydrocarbons (mainly adsorbed hydrocarbons) present in the carbonate mineral lattice and its cement in the sample were separated by means of pressure reduction, heating, and chemical treatment (acid addition). The content of acid-extracted hydrocarbons was detected using a gas chromatography instrument.

Before formally measuring the test material, the standard gas was measured 3–5 times. Only when the relative error of the standard gas measurement was $\leq 3\%$ could the test material be measured. During the measurement process, it was necessary to regularly check the controlled status of the instrument with standard gas.

The procedure for GC testing is as follows: First, 500 μL $\pm$ 0.5 μL of gas was accurately extracted using a 1 mL syringe. The gas was quickly injected into the gas chromatograph. The measuring instrument was then started, chromatograph charts were drawn, and the data were collected. After the chromatographic peak was reached and the recombination components were "driven" at high temperature according to the set temperature program, qualitative and quantitative analysis was conducted.

Result Calculation: Response factor calculation: $F_i = C_{si} \cdot V_s / A_{si}$

Fi—Response factor of component i, μL/(mV•s);

Csi—concentration of standard gas component i, φ (i)/10^{-2};

Vs—Standard gas injection volume, μL;

Asi—Peak area of component i in standard gas, mV•s.

Calculation of component content in the sample: $X_i = \frac{F_i \cdot V_t \cdot A_i}{m \cdot V_j}$

Xi—The content of acid-hydrolyzed hydrocarbon component i in the sample (if the integer number is less than 3, retain 1 decimal place; if the integer number is ≥ 3, do not retain decimals), μL/kg;

Vt—degassing volume, mL;

Ai—peak area of acid-hydrolyzed hydrocarbon component i in the test sample, mV•s;

Vj—Injection volume of test sample, mL; and

M—Weighing mass of the sample, kg.

The analysis and data-processing methods include probability distribution statistics, data structure analysis (where the mean represents the average level of the data, the standard deviation represents the size of data dispersion, and the coefficient of variation represents the size of data relative fluctuations), and background value and anomaly determination (via the mean method and Kriging method). The results of comprehensive analysis of the acid-extracted hydrocarbon and methane carbon isotope test data are then used to determine the underlying oil and gas attributes [39–43].

4. Results

4.1. Acid-Extracted Hydrocarbon Composition

Acid-extracted C_1–C_5 hydrocarbons were detected in the samples, with methane (CH_4), ethane (C_2H_6), and propane (C_3H_8) being the main components. The content of butane- and pentane-series hydrocarbons was extremely low, especially for pentane, with only a few samples being detected. The acid-extracted hydrocarbon index of the samples at each station had the characteristic of $CH_4 > C_2H_6 > C_3H_8$. The variation in methane content in the acid-extracted hydrocarbons ranged from 13.34~1799.31 μL/kg, with an average value of 138.25 μL/kg and a median value of 94.01 μL/kg. The variation in ethane content in the acid-extracted hydrocarbons ranged from 0.66~195.26 μL/kg, with an average value of 21.61 μL/kg and a median value of 9.54 μL/kg, which was significantly lower than the content of methane. The variation in propane content in the acid-extracted hydrocarbons ranged from 0.26~80.60 μL/kg, with an average value of 8.17 μL/kg and a median value of 3.57 μL/kg. Detailed data are presented in Table 1.

Table 1. The abundance anomaly of acid-extracted hydrocarbons.

	Minimum (μL/kg)	Maximum (μL/kg)	Median (μL/kg)	Mean (μL/kg)	SD	CV	K
SC_1	13.34	1799.31	94.01	138.25	163.23	1.18	1.47
SC_2	0.66	195.26	9.54	21.61	29.99	1.39	2.27
SC_3	0.26	80.64	3.57	8.17	11.79	1.44	2.29
SiC_4	0.10	67.37	2.14	5.97	9.61	1.61	2.80
SnC_4	0.06	32.53	1.25	2.89	4.37	1.51	2.32
SiC_5	0.02	87.15	2.47	7.38	12.29	1.66	2.99
SnC_5	0.01	9.74	0.22	0.53	0.99	1.87	2.47
SC_{2g}	1.23	465.18	19.74	46.56	68.47	1.47	2.36

4.2. Acid-Extracted Hydrocarbon Distribution

The high values of methane in the acid-extracted hydrocarbons are mainly distributed in the northern part of the study area; other high values are distributed in the central–southern part of the study area. The high values of heavy hydrocarbons in the acid-extracted hydrocarbons are also mainly distributed in the northern part of the study area, and secondary high values are mainly distributed in the central–southern part of the study area. The anomalies in the acid-extracted methane and heavy hydrocarbons have high positive correlation (Figures 3 and 4).

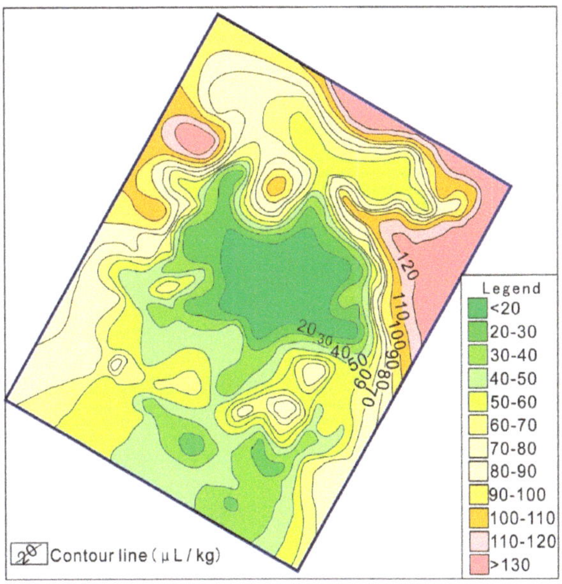

Figure 3. Isogram of acid-extracted methane.

4.3. δ ^{13}C (‰) of Acid-Extracted Hydrocarbon Methane

The variation in δ^{13}C (‰) content in the acid-extracted hydrocarbons ranged from −34‰~−48‰, with a maximum value of −34.45‰, a minimum value of −47.11‰, a median value of −37.5‰, and a standard deviation of 2.48. The distribution is characterized by a high north–south distribution and low middle distribution (Figure 5).

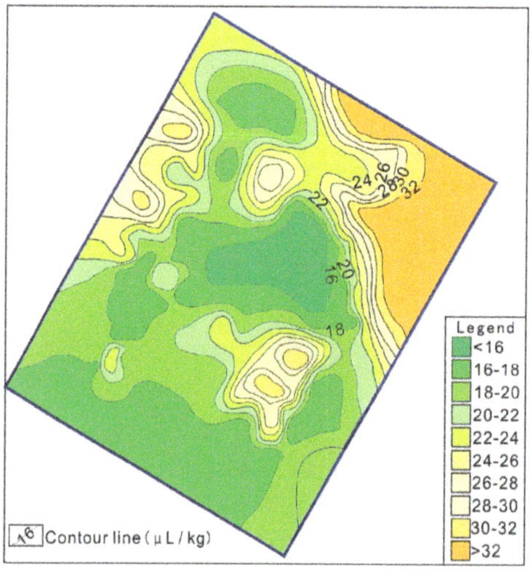

Figure 4. Isogram of acid-extracted heavy hydrocarbons.

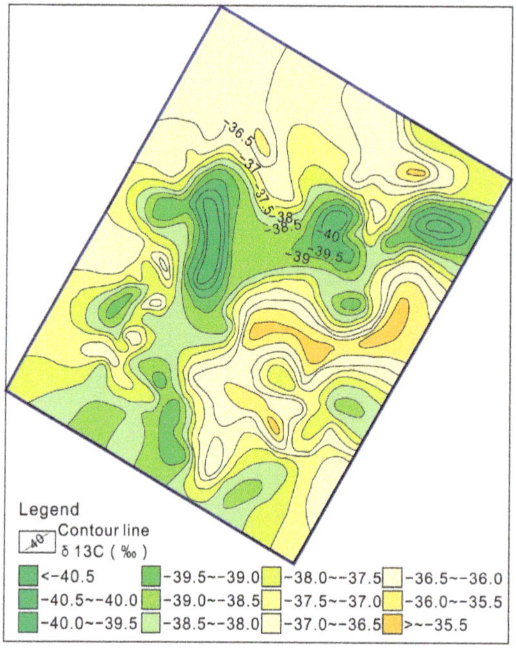

Figure 5. Isogram of δ ^{13}C (‰).

5. Discussion

5.1. Variations in Acid-Extracted Hydrocarbon Abundance

From the perspective of data statistics, the mean reflects the average level of the data, the standard deviation (SD) reflects the size of the data dispersion, and the coefficient of variation (CV) reflects the size of the relative fluctuations in the data. From a geological perspective, the coefficient of variation and intensity (K) reflects the impact of later geological processes. The coefficient of variation is calculated as the standard deviation/mean and the intensity coefficient as the mean/median. The coefficients of variation of the indicators of geochemical exploration in the Chaoshan Depression are all greater than 1, and the intensity factors of later superimposed actions are all greater than 2, indicating that the geochemical exploration indicators in the region are greatly affected by later geological processes. More details are presented in Table 1.

The acid-extracted hydrocarbons of the Chaoshan Depression belong to the medium-abundance geochemical background and are characterized by heterogeneous fields and homogeneous fields [44,45]. Altered carbonates belong to a medium abundance geochemical background and possess homogeneous field characteristics. The interference of seabed microbial disturbances and biogenic methane on indicators of oil and gas geochemical exploration is relatively small [45], and all geochemical exploration indicators are affected by later oil and gas interactions.

5.2. Acid-Extracted Hydrocarbon Structural Indicators

Under normal circumstances, the sample data of a single parent obey a normal distribution; however, when multiple parent characteristics are present, the data structure characteristics, for the most part, do not obey a normal distribution. The ideal separation mode between the background and an anomaly is shown in Figure 6a, which features a trailing phenomenon.

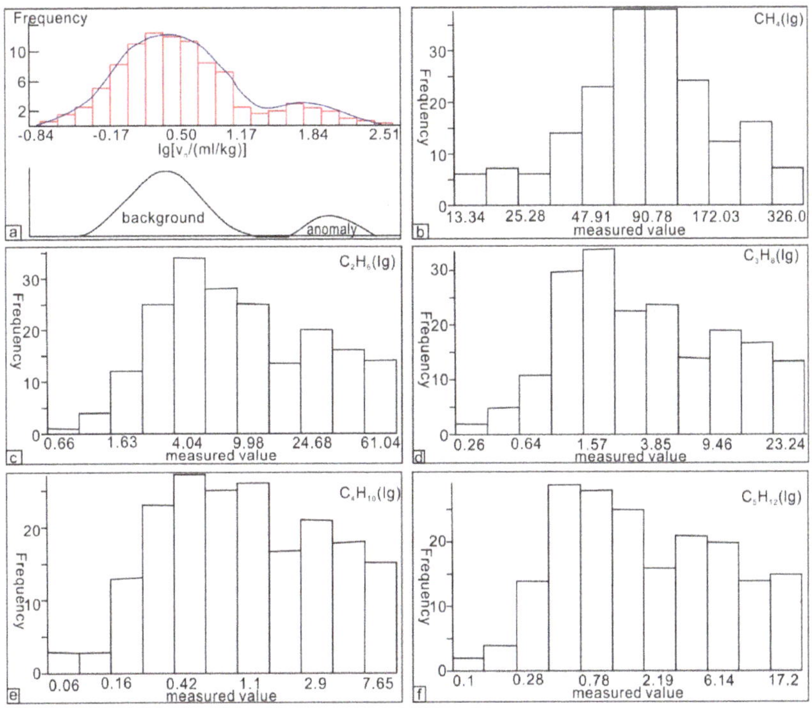

Figure 6. Background and anomaly separation mode diagram, and characteristic diagrams of the acid-extraction hydrocarbon indicators' stacking analysis. ((**a**). Background and anomaly separation mode diagram; (**b**). characteristic diagram of methane; (**c**). characteristic diagram of ethane; (**d**). characteristic diagram of propane; (**e**). characteristic diagram of butane; (**f**). characteristic diagram of pentane).

Figure 6b–f show the results of the iteration analysis of the main indicators of oil and gas geochemical exploration in the Dongsha area. The iteration results of each indicator have similar non-normal distribution characteristics, and all indicators have tailing phenomenon or multiple parent backgrounds, suggesting the effect of late superposition. Moreover, the intensity factors of the later superimposed actions are all greater than 1, indicating that all the indicators are affected by the geological factors of the later stage (oil–gas, sediment sources, environment, etc.).

5.3. The Fabric Characteristics of Hydrocarbons

The composition of adsorbed gas near the surface has a favorable tracing effect for distinguishing oil and gas properties and indicating the presence of gas reservoirs. The ratio of gas components in the hydrocarbons, such as $C_1/\sum C$, C_1/C_2, C_3/C_1, reflect this differentiation effect. The range of variation in the composition and characteristics of the hydrocarbon series indicators in the sediments above petroleum reservoirs are shown in Table 2.

Table 2. The range of gas composition changes in the overlying sediments of petroleum reservoirs.

Type	Ratio of Hydrocarbon		
	$C_1/\sum C \times 100$	C_1/C_2	$(C_3/C_1) \times 1000$
Oil	75~50	10~4	60~500
Gas	>75	>10	<60

The $C1/\sum C \times 100$ ratio ranges from 51.55 to 97.06, with an average of 78.43; C_1/C_2 ranges from 2.66 to 52.71, with a mean of 10.58; and $(C_3/C_1) \times 1000$ ranges from 5.65 to 163.88, with an average of 55.36. Based on the characteristics of the hydrocarbon ratio, the petroleum reservoirs in the Chaoshan Depression are determined to be mainly composed of gas (Figures 7–9).

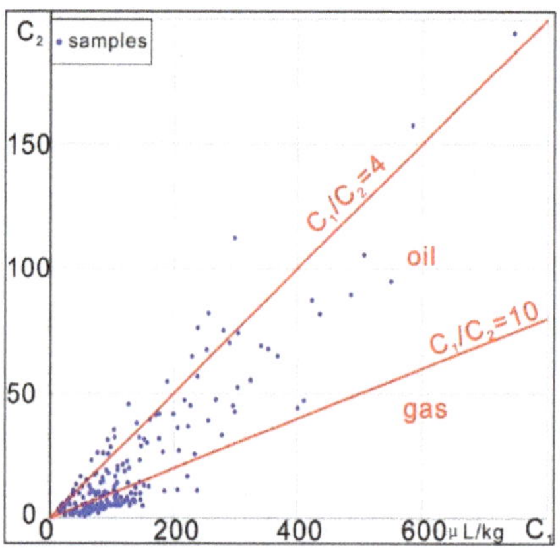

Figure 7. Cross plot of C_1 vs. C_2.

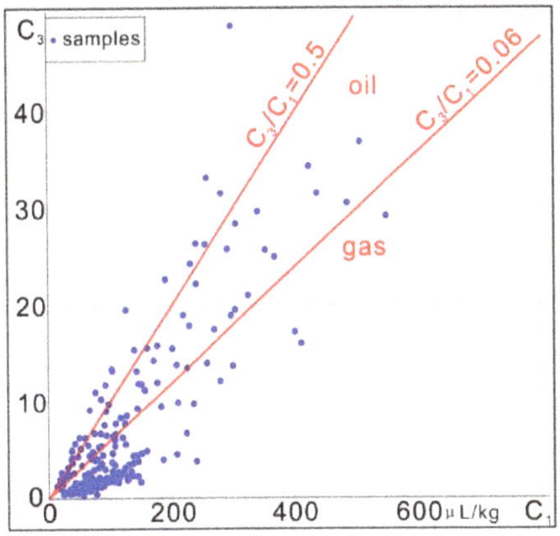

Figure 8. Cross plot of C_1 vs. C_3.

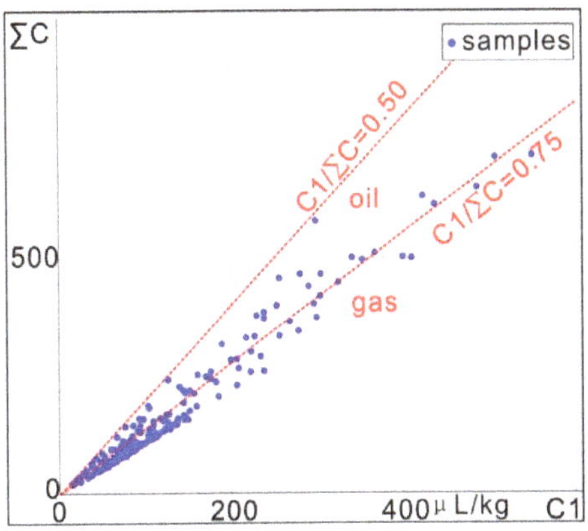

Figure 9. Cross plot of C1 vs. ∑C.

5.4. Differentiation Characteristics of Hydrocarbon

Due to the differences in the chemical and physical properties of oil and natural gas, the differentiation of hydrocarbons occurs during the micro-leakage of gas into seafloor sediments, changing the proportion and dynamic balance of the original hydrocarbons and forming significant differences between oil, gas, and non-oil-and-gas zones in seafloor sediments. As shown in the differentiation diagram of heavy and light hydrocarbons (Figure 10), 198 sites out of 200 have $C_2 + C_5/C_1$ ratios greater than 5%, thus belonging to wet gas.

The C_1–C_4 ratio represents the content of methane to butane ($nC_4 + iC_4$) in the acid-extracted hydrocarbons adsorbed from the underlying oil- and gas-bearing structures that leak into the seafloor sediments. The $C_1/(C_2 + C_3)$ ratio represents the degree of evolution (or maturity) of the hydrocarbon-generating organic matter. As the maturity increases, this value increases exponentially, indicating a gradual transition from source rock to crude oil, to condensate oil and gas, and to dry gas in the underlying strata. The $C_2/(C_3 + C_4)$ ratio reflects the interrelationships between heavy components in the acid-extracted light hydrocarbons, and to some extent, it can also reflect the degree of evolution of the hydrocarbon-generating parent materials. The ratio of $C_1/(C_2 + C_3)$ to $C_2/(C_3 + C_4)$ can be used to identify the types of oil and gas reservoir from which of the acid-extracted adsorbed hydrocarbons originate.

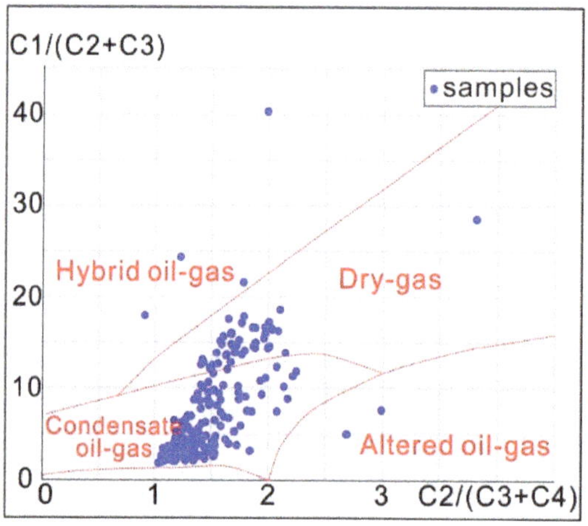

Figure 10. Cross plot of $C_1/(C_2 + C_3)$ vs. $C_2/(C_3 + C_4)$.

5.5. Carbon Isotope of Methane ($\delta^{13}C$) Features

The stable carbon isotope of methane in the acid-extracted hydrocarbons is an important indicator for distinguishing oil and gas properties. The average $\delta^{13}C$ value of biogenic gas is less than −54‰, the average $\delta^{13}C$ value of oil field gas ranges from −54‰ to −40‰, and the average $\delta^{13}C$ value of mature gas or coal-type gas is greater than −40‰ (Figure 11) [32]. The stable carbon isotope values of acid-extracted hydrocarbon methane in the Chaoshan Depression range from −34‰ to −48‰, with a maximum value of −34.45‰, a minimum value of −47.11‰, a median value of −37.50‰, and a standard deviation of 2.48. Its oil and gas characteristics are of an organic origin and are in the over-mature stage (Figure 11).

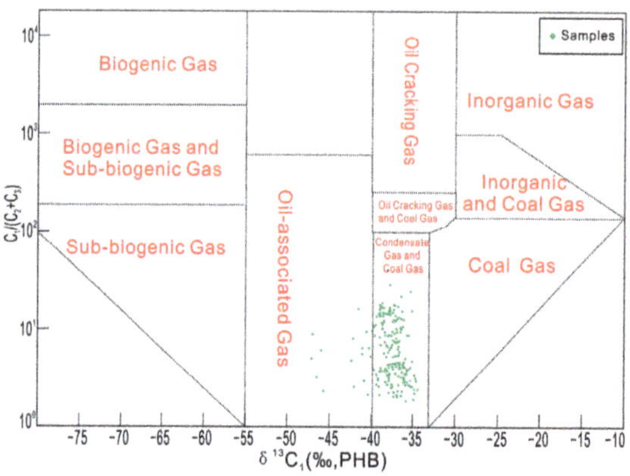

Figure 11. Cross plot of $\delta^{13}C_1$ vs. $C_1/(C_2 + C_3)$.

The planar anomalies of carbon isotopes in methane are high in the south and the north of the region, and low in the middle area (Figure 12).

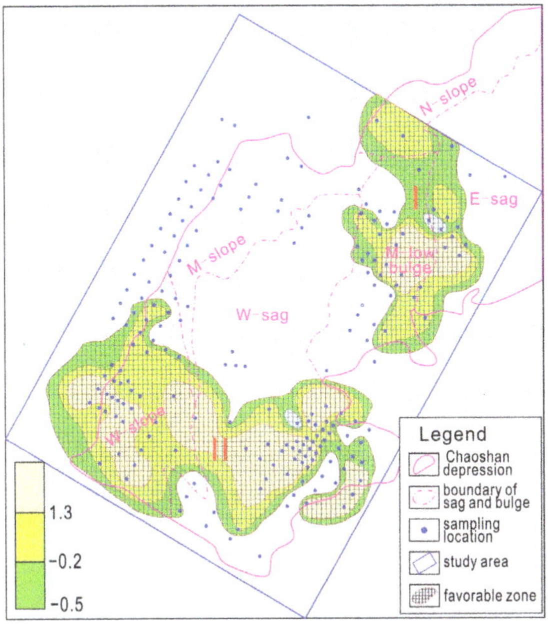

Figure 12. Favorable prospective oil and gas areas of the Chaoshan Depression.

5.6. Evaluation of Favorable Oil and Gas Areas

Through acid-extracted hydrocarbons anomaly analysis of the composition, abundance, structure, planar distribution, and planar distribution of methane stable isotope values of acid-extracted hydrocarbons in the Chaoshan Depression, it is believed that anomalous zones are distributed in the central and southwestern parts of the depression. After normalizing the oil and gas exploration indicators, the moving average method and Kriging method were used to delineate the northeast (I) and southwest (II) parts of the Chaoshan Depression as the most favorable prospective oil and gas areas (Figure 12).

From the seismic profile crossing the northern abnormal area, it can be observed that the methane present in the acid-extracted hydrocarbons is abnormally developed in the seafloor sediments above the reservoir. However, this anomaly gradually decreases as it approaches the edge of the reservoir, tending towards the background value. There is a favorable correspondence between the oil reservoir and the anomaly (Figure 13), and the micro-leakage of oil and gas from the underlying reservoir has been determined as the main reason for the formation of these geochemical anomalies. The formation mechanism of geochemical anomalies in the Chaoshan Depression is as follows: gas from the underlying reservoir is driven upwards by pressure and leaks into the seabed through fault planes and micro-fractures. The underlying petroleum reservoir is controlled by the source, the micro-fractures control the permeability, and the surface sediments adsorb and trap the leaked hydrocarbons, forming a "lower generation, upper accumulation and micro fractures leaking "model. This model would benefit from understanding the mechanism of hydrate accumulation in the Dongsha area also.

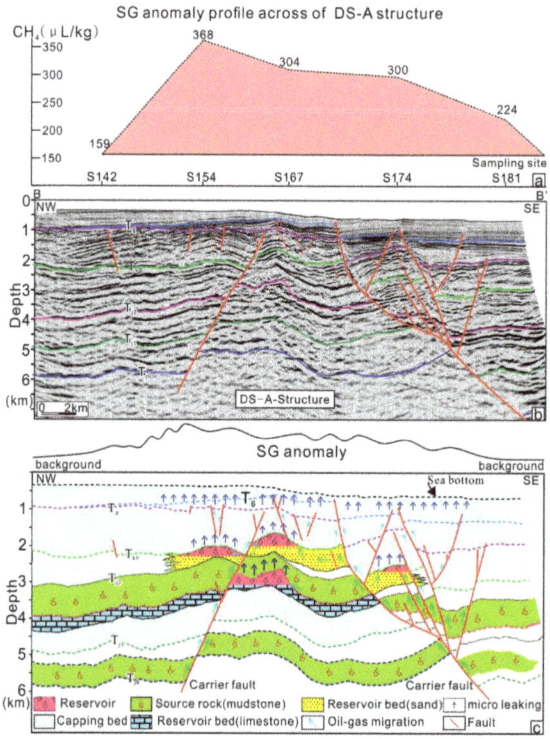

Figure 13. The model of downward migration and upward accumulation: (**a**) shows SG anomaly profile across DS-A structure, (**b**) shows the seismic profile across the DS-A structure, (**c**) is the formation model of SG anomaly.

6. Conclusions

1. The oil and gas reservoir in the Chaoshan Depression is mainly composed of condensate gas (oil). The reservoir is of an organic thermal origin. The source rock is dominated by marine sedimentation, with highly mature organic matter.
2. The favorable oil and gas prospects in the Chaoshan Depression are divided into two regions, respectively favorable zone I and favorable zone II, mainly located at the intersection of the central low bulge, the western depression, and the western slope of the depression; the DS-A structure indicates an oil and gas reservoir dominated by gas.
3. The "lower generation, upper accumulation and micro fractures leaking" model is the mechanism explaining the formation of oil and gas anomalies in the Chaoshan Depression, in which hydrocarbons are controlled by the underlying petroleum reservoir, leak through micro-fractures, and adsorb on the seabed's surface.

Author Contributions: Data processing: J.Z.; writing—original draft preparation: G.Z., Z.Z., K.Z., G.T. and J.Z.; data analysis: J.Y., C.S. and C.F. All authors have read and agreed to the published version of the manuscript.

Funding: This research was funded by the National Natural Science Foundation of China (No. 42076218, U1901217, and 91855101).

Institutional Review Board Statement: Not applicable.

Informed Consent Statement: Not applicable.

Data Availability Statement: Data are available on request from the author.

Acknowledgments: We are grateful for the efforts of all who participated in data acquisition and processing. We thank the reviewers for helpful suggestions that improved the clarity of this paper.

Conflicts of Interest: The authors declare no conflicts of interest.

References

1. Shao, L.; You, H.; Hao, H.; Wu, G.; Qiao, P.; Lei, Y. Petrology and depositional environments of Mesozoic strata in the Northeastern South China Sea. *Geol. Rev.* **2007**, *53*, 164–169. Available online: https://en.cnki.com.cn/Article_en/CJFDTOTAL-DZLP200702004.htm (accessed on 5 March 2022).
2. Hao, H.; Shi, H.; Zhang, X.; Wang, T.; Tang, S. Mesozoic sediments and their petroleum geology conditions in Chaoshan sag: A discussion based on drilling results from the exploratory well LF35-1-1. *China Offshore Oil Gas* **2009**, *21*, 151–156. Available online: https://en.cnki.com.cn/Article_en/CJFDTOTAL-ZHSD200903001.htm (accessed on 5 May 2020).
3. Zhang, L.; Geng, A.; Wang, L.; Liao, Y.; Xu, G.; Wei, Z. Assessment of Mesozoic source rocks at the Margin of South China Continent. *Mar. Geol. Quat. Geol.* **2012**, *32*, 99–108. [CrossRef]
4. Yang, S.; Tong, Z.; He, Q.; Hao, J. Mesozoic hydrocarbon generation history and igneous intrusion impacts in Chaoshan depression. South China Sea: A case of LF35-1-1 well. *China Offshore Oil Gas* **2008**, *20*, 152–156. (In Chinese with English abstract)
5. Hu, T.; Wu, G.; Xu, Z.; Pang, X.; Liu, Y.; Yu, S. Potential resources of conventional, tight, and shale oil and gas from Paleogene Wenchang Formation source rocks in the Huizhou Depression. *Adv. Geo-Energy Res.* **2022**, *6*, 402–414. [CrossRef]
6. Guo, H.; Wang, Z.; Wang, B.; Zhang, Y.; Meng, H.; Sui, H. Molecular dynamics simulations of oil recovery from dolomite slit nanopores enhanced by CO_2 and N_2 injection. *Adv. Geo-Energy Res.* **2022**, *6*, 306–313. [CrossRef]
7. Yang, Z.; Zou, C.; Gu, Z.; Yang, F.; Li, J.; Wang, X. Geological characteristics and main challenges of onshore deep oil and gas development in China. *Adv. Geo-Energy Res.* **2022**, *6*, 264–266. [CrossRef]
8. Hao, H.; Wang, R.; Zhang, X.; Xie, H.; Chen, Z. Mesozoic marine sediment identification and distribution in the eastern Pearl River Mouth Basin. *China Offshore Oil Gas* **2004**, *16*, 84–88.
9. Duan, J.; Mi, H. Seismic facies and sedimentary facies study of Mesozoic in Chaoshan sag. *Resour. Ind.* **2012**, *14*, 100–105. (In Chinese with English abstract)
10. Yan, P.; Wang, Y.; Jin, Y.; Zhao, M.; Zhong, G. Deep-water coral of multiple benthal strategies discovered from mounds in the Dongsha Waters in the South China Sea. *Earth Sci. Front.* **2021**, *29*, 202–210. [CrossRef]
11. Zhang, Q.; Zhang, H.; Zhang, X. The upper Cretaceous prototype basin of the Chaoshan depression in the northern South China Sea and its tectonic setting. *Chin. J. Geophys.* **2018**, *61*, 4308–4321. [CrossRef]
12. Wang, P.; Xia, K.; Huang, C. The Mesozoic Marine sediment distribution and Geology-Geophysics characteristic at the North-Eastern of South China Continent. *J. Trop. Oceanogr.* **2000**, *19*, 28–35.
13. Wu, N.; Zhang, G.; Liang, J.; Su, Z.; Wu, D.; Lu, H. Progress of gas hydrates research in northern South China Sea. *Adv. New Renew. Energy* **2013**, *1*, 80–94.
14. Brown, A. Evaluation of possible gas micro seepage mechanisms. *AAPG Bull.* **2000**, *84*, 1775–1789. [CrossRef]
15. Ramkumar, M.; Nagarajan, R.; Santosh, M. Advances in sediment geochemistry and chemostratigraphy for reservoir characterization. *Energy Geosci.* **2021**, *2*, 308–326. [CrossRef]
16. Decker, J.; Teas, P.; Orange, D.; Bernard, B.B. Sea bottom characteristics and geochemistry of oil and gas seeps in the Gulf of Mexico. *Interpretation* **2022**, *10*, SB49–SB62. [CrossRef]
17. Zavatsky, M.D.; Veduta, O.V.; Naumenko, V.O. Prospecting terrain for surface geochemical exploration of oil and gas in West Siberia. *Geol. Ecol. Landsc.* **2022**. [CrossRef]
18. Rice, G.K. Vertical migration in theory and in practice. *Interpretation* **2021**, *10*, SB17–SB26. [CrossRef]
19. Jatiault, R.; Loncke, L.; Dhont, D.; Imbert, P.; Dubucq, D. Geophysical characterization of active thermogenic oil seeps in the salt province of the lower Congo basin. Part II: A regional validation. *Mar. Pet. Geol.* **2019**, *103*, 773–791. [CrossRef]
20. Hosseinpour, M. Relationship between hydrocarbon micro-seepages and structures by detection of altered minerals using ASTER remote sensing data in the West of Coastal Fars, Zagros, Iran. *Arab. J. Geosci.* **2020**, *13*, 511. [CrossRef]
21. Cameron, N.; White, K. Surface geochemical exploration continues to progress global Deepwater Frontiers. In Proceedings of the IBC "Worldwide Deepwater Technologies" Forum, London, UK, 11–12 April 2000; pp. 11–12. Available online: https://www.tdi-bi.com/wp-content/uploads/2017/01/SGE-Progresses-DW-Frontiers.pdf (accessed on 23 May 2024).
22. Belt, J.Q., Jr.; Rice, G.K. Application of statistical quality control measures for near—Surface geochemical petroleum exploration. *Comput. Geosci.* **2002**, *28*, 243–260. [CrossRef]
23. Kumar, B.; Patil, D.J.; Kalpana, G.; Vishnu Vardhan, C. Geochemical prospecting of hydrocarbons in Frontier basins of India. In Proceedings of the American Associa-tion of Petroleum Geologists Annual Convention, AAPG, Dallas, TX, USA, 18–21 April 2004; pp. 18–21. Available online: https://www.semanticscholar.org/paper/Geochemical-Prospecting-of-Hydrocarbons-in-Frontier-Kumar-Kalpana/884ab57ad958193c280683b2d9485f5e4747cd9d (accessed on 23 May 2024).
24. Abrams, M.A. Significance of hydrocarbon seepage relative to petroleum generation and entrapment. *Mar. Pet. Geol.* **2005**, *22*, 457–477. [CrossRef]

25. Hovland, M.; Jensen, S.; Fichler, C. Methane and minor oil macro-seep systems—Their complexity and environmental significance. *Mar. Geol.* **2012**, *332–334*, 163–173. [CrossRef]
26. Al-Areeq, N.M.; Maky, A. Organic geochemical characteristics of crude oils and oil-source rock correlation in the Sunah oilfield, Masila Region, Eastern Yemen. *Mar. Pet. Geol.* **2015**, *63*, 17–27. [CrossRef]
27. Bougi, M.S.M.; Baklouti, S.; Rasheed, M.A.; Pls, R.; Hasan, S.Z.; Ksibi, M.; Gharbi, J. Integrated geo-microbial and trace metal anomalies for recognition of hydrocarbon microseepage in petroliferous regions: Case of El Hajeb oilfield in the South East of Tunisia. *Pet. Sci. Technol.* **2018**, *36*, 2029–2036. [CrossRef]
28. Anderson, J.; Romanak, K.; Alfi, M.; Hovorka, S. Light hydrocarbon and noble gas migration as an analogue for potential CO_2 leakage: Numerical simulations and field data from three hydrocarbon systems. *Greenh. Gases Sci. Technol.* **2019**, *9*, 226–244. [CrossRef]
29. Abrams, M.A. Evaluation of near-surface gases in marine sediments to assess subsurface petroleum gas generation and entrapment. *Geoscience* **2017**, *7*, 35. [CrossRef]
30. Horvitz, L. Hydrocarbon prospecting after forty years. In *Unconventional Exploration for Petroleum and Natural Gas II*; Gottlieb, B.M., Ed.; Southern Methodist University Press: Dallas, TX, USA, 1981; pp. 83–95.
31. Horvitz, L. Geochemical exploration for petroleum. *Science* **1985**, *229*, 812–827. [CrossRef]
32. Whiticar, M.J. Characterization and Application of Sorbed Gas by Microdesorption CF-IRMS. In Proceedings of the Near-Surface Hydrocarbon Migration: Mechanisms and Seepage Rates, American Association Petroleum Geology Hedberg Conference, Vancouver, BC, Canada, 7–10 April 2002.
33. Su, P.; Liang, J.; Zhang, W.; Liu, F.; Wang, F.; Li, T. Natural gas hydrate accumulation system in the Shenhu sea area of the northern South China Sea. *Nat. Gas Ind.* **2020**, *40*, 77–88. [CrossRef]
34. Zhang, J.; Sun, Z.; Zhang, S. Analysis of Mesozoic tectonic deformation in the Chaoshan Depression of Pearl River Mouth Basin. *J. Trop. Oceanogr.* **2014**, *33*, 41–49. Available online: http://www.jto.ac.cn/EN/lexeme/showArticleByLexeme.do?articleID=578 (accessed on 5 March 2021).
35. Ye, Q.; Mei, L.; Shi, H. The Late Cretaceous tectonic evolution of the South China Sea area: An overview, and new perspectives from 3D seismic reflection data. *Earth-Sci. Rev.* **2018**, *187*, 186–204. [CrossRef]
36. Hao, H.; Zhao, J.; Liu, H. Prediction of oil and gas reservoir traps by aromatic hydrocarbons from seabed sediments in Chaoshan depression, South China Sea. *Acta Pet. Sin.* **2018**, *39*, 528–540. [CrossRef]
37. Zhou, D.; Wang, W.; Wang, J. Mesozoic subduction-accretion zone in northeastern South China Sea inferred from geophysical interpretations. Sciences China Earth, zone in northeastern South China Sea inferred. *Sci. China Earth* **2006**, *49*, 471–482. [CrossRef]
38. Yan, Y.; Liao, J.; Yu, J.; Chang, L.; Guangjian, Z.; Yanlin, W. Velocity Structure Revealing a Likely Mud Volcano off the Dongsha Island, the Northern South China Sea. *Energies* **2022**, *15*, 195. [CrossRef]
39. Blumenberg, M.; Lutz, R.; Schlömer, S.; Krüger, M.; Scheeder, G.; Berglar, K.; Heyde, I.; Weniger, P. Hydrocarbons from near-surface sediments of the Barents Sea north of Svalbard—Indication of subsurface hydrocarbon generation? *Mar. Pet. Geol. Bull.* **2016**, *76*, 432–444. [CrossRef]
40. Abrams, M.A. Geophysical and geochemical evidence for subsurface hydrocarbon leakage in the Bering Sea, Alaska. *Mar. Pet. Geol. Bull.* **1992**, *9*, 208–221. [CrossRef]
41. Abrams, M.A. Distribution of subsurface hydrocarbon seepage in near-surface marine sediments. In *Hydrocarbon Migration and Its Near Surface Effects*; American Association Petroleum Geology Memoir No. 66; Schumacher, D., Abrams, M.A., Eds.; American Association of Petroleum Geologists: Tulsa, OK, USA, 1996; Chapter 1, pp. 1–14.
42. Abrams, M.A. Marine seepage variability and its impact on evaluating the surface migrated hydrocarbon seep signal. *Mar. Pet. Geol.* **2020**, *121*, 104600. [CrossRef]
43. Klusman, R.W.; Saeed, M.A. A comparison of light hydrocarbon micoseepage mechanisms. In *Hydrocarbon Migration and Its Near Surface Effects*; Schumacher, D., Arams, M.A., Eds.; American Association of Petroleum Geologist Memoil: Tulsa, OK, USA, 1996; Volume 66, pp. 157–168.
44. Dai, J.; Fei, X.; Qi, H. *China Gas-Geology*; Petroleum Industry Press: Tulsa, OK, USA, 1992.
45. Zhao, J.; Liang, Q.; Zhang, L.; Zhong, G. Oil and gas reservoir attribute discrimination based on surface sediment acid-extraction hydrocarbon in the western depression of Taiwan Strait Basin. *Geophys. Geochem. Explor.* **2018**, *42*, 436–441. [CrossRef]

Journal of
Marine Science and Engineering

MDPI

Article

The Mesozoic Subduction Zone over the Dongsha Waters of the South China Sea and Its Significance in Gas Hydrate Accumulation

Pibo Su [1,2,3], Zhongquan Zhao [1,*] and Kangshou Zhang [1,*]

1 Guangzhou Marine Geological Survey, Guangzhou 510301, China; spb_525@sina.com
2 Sanya Institute of South China Sea Geology, Guangzhou Marine Geological Survey, Sanya 572025, China
3 National Engineering Research Center for Gas Hydrate Exploration and Development, Guangzhou 511458, China
* Correspondence: zzqhello@163.com (Z.Z.); zkangshou@mail.cgs.gov.cn (K.Z.)

Abstract: The Mesozoic subduction zone over the Dongsha Waters (DSWs) of the South China Sea (SCS) is a part of the westward subduction of the ancient Pacific plate. Based on the comprehensive interpretation of deep reflection seismic profile data and polar magnetic anomaly data, and the zircon dating results of igneous rocks drilled from well LF35-1-1, the Mesozoic subduction zone in the northeast SCS is accurately identified, and a Mesozoic subduction model is proposed. The accretion wedges, trenches, and igneous rock zones together form the Mesozoic subduction zone. The evolution of the Mesozoic subduction zone can be divided into two stages: continental subduction during the Late Jurassic and continental collision during the late Cretaceous. The Mesozoic subduction zone controlled the structural pattern and evolution of the Chaoshan depression (CSD) during the Mesozoic and Neogene eras. The gas source of the hydrate comes from thermogenic gas, which is accompanied by mud diapir activity and migrates along the fault. The gas accumulates to form gas hydrates at the bottom of the stable domain; BSR can be seen above the mud diapir structure; that is, hydrate deposits are formed under the influence of mud diapir structures, belonging to a typical leakage type genesis model.

Keywords: South China Sea; Chaoshan depression; Mesozoic; subduction zone; gas hydrate

1. Introduction

The westward subduction of the ancient Pacific plate was an important tectonic event of the East Asian continental margin during the Mesozoic era. Research on the manifestation of this zone and its impact on hydrate accumulation is relatively weak, so strengthening the relevant research will help enrich evidence on the westward subduction of the ancient Pacific plate, as well as expand on our knowledge of the types of hydrate accumulation in the northern SCS. Studying the subduction zone of the East Asian continental margin is thus crucial for understanding the Mesozoic tectonic evolution of East Asia and helps in understanding the accumulation of gas hydrate in the South China continental margin. The Dongsha Waters (DSWs) is one of three favorable exploration areas for gas hydrates in the northern South China Sea (SCS). However, unlike the other two favorable exploration areas for gas hydrates in Shenhu and Qiongdongnan, which are distributed in large Cenozoic basins, the DSWs are distributed on a Mesozoic residual depression (the Chaoshan depression (CSD), Figure 1), the gas hydrate accumulation of which is closely related to the tectonic evolution of the Mesozoic and Cenozoic eras. The westward subduction of the ancient Pacific plate on the edge of the East Asian continent began at least in the Early Triassic (approximately 251 Ma) and continued until the Late Cretaceous, exhibiting a multi-stage, Andean-type accretion pattern [1–21]. The ancient Pacific plate subducted along the edges of the South China and Sunda continents, forming a late Mesozoic magmatic rock belt that extends for thousands of kilometers, from Japan in the north and to Sumatra in the south [22–41]. In the Late Cretaceous, due to the subduction and retreat of the paleo-Pacific plate, the edges of the South China and Sunda continents experienced significant tension, resulting in the development of a series of rift basins.

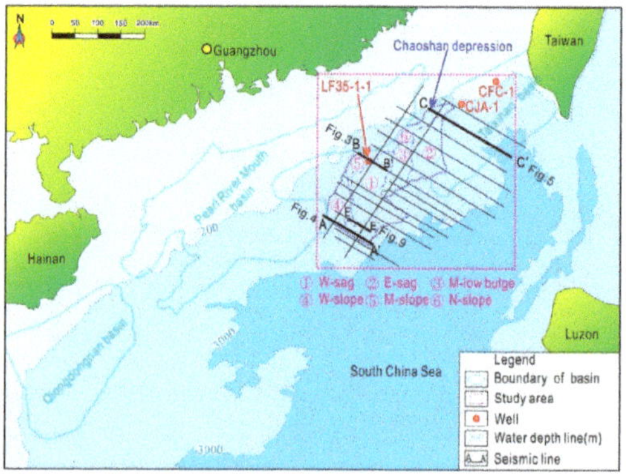

Figure 1. Map of the study area (Location of CSC, and tectonic division. Seismic line used for Isobaths of Moho compiled, and Seismic profiles in the following figures. Location of Well LF35-1-1).

The existence of a Mesozoic subduction zone in the northern SCS has been discussed by scholars [42–52]. There are different opinions on whether there is a Mesozoic subduction zone in the northern of the CSC. Some scholars believe that a paleo-subduction occurred from the Late Indosinian to the Early Yanshanian and that an accretion zone existed from Hainan Island to the northern shelf of the SCS. The Indosinian Movement was a tectonic movement between the Late Permian and Triassic periods, which further overlapped three different land blocks in eastern Asia (Yangtze, China Korea, and Siberia). The Yanshan Movement was a widespread crustal movement that occurred in China during the Late Triassic to Cretaceous periods, resulting in many northeast-facing or -trending folded fault mountains and numerous small fault basins, accompanied by magmatic activity, especially in the southeastern coastal areas, where granite intrusion and volcanic eruption are particularly severe. The DSW Mesozoic subduction zone is part of the westward subduction of the ancient Pacific plate. Based on the thickening of the crust beneath the Dongsha uplift, the large amount of magnetic anomalies, and the existence of a fault zone in the lower slope, Yao et al. believed that the subduction accretion zone extends along the lower slope in a NEE to SWW direction and reaches the Xisha Trough [43]. Based on gravity, magnetic, and wide-angle seismic data, Zhou et al. Inferred in 2006 that a Mesozoic subduction and accretion zone exists roughly 45° NE from the southwestern Taiwan Basin to the northern edge of the deep-sea basin in the northern SCS [44]. Yao et al. classified the Dongsha uplift and other areas as volcanic arcs, thus classifying the CSD as a pre-arc basin and the Hanjiang depression as a post-arc basin [45]. However, some scholars believe that the northern SCS is the result of the eastward extension of Tethys, and no Mesozoic subduction zone exists in the DSWs. Cai believed that a "residual Tethys" was present in the northern continental margin of the SCS, based on the marine Mesozoic strata on the south side of the Dongsha Islands and the marine shale-containing nannofossils in the CFC-1 well; he further pointed out that a "residual Tethys" could have existed in the Late Indosinian to the Early Yanshanian and believes that this "residual Tethys" started from the northern side of the Himalaya, passed through the southern Honghe River in Vietnam, continued along the northwest to Yinggehai, and entered the northern continental slope trough of the SCS [46]. Cai et al. believed that the CSD is part of the superposition of the Neocraton marine sedimentary depression basin on the Paleozoic folded basement and the fore-arc basin [47].

The reason for the above controversy is the lack of direct evidence of subduction. Past elucidations of the subduction of the paleo-Pacific were mainly based on studies of Mesozoic igneous rocks in the eastern coastal areas of China, so further exploration on the specific location and manifestation of the subduction zone on the northern margin of the SCS is urgently needed [36]. In recent years, the Guangzhou Marine Geological Survey (GMGS) has acquired rich survey data on the DSWs, especially deep seismic reflection profile and magnetic data, providing new evidence for in-depth research on the Mesozoic subduction zone in the DSWs.

What is the significance of subduction zone for hydrate accumulation? The gas source of hydrate in the DSWs are derived Mesozoic fissure gas, and the relationship between its accumulation and the subduction zone is contingent upon the diversion of gas leakage by faults and mud diapirs. The Mesozoic subduction zone is accompanied by a large number of mud diapir structures, and mud diapir structures and mud volcanoes are important causes of gas hydrate accumulation. Milkov et al. proposed four models of hydrate accumulation based on several gas migration control factors: fault structure type, mud volcano type, stratigraphic control type, and tectonic stratigraphic type [53]. Mud diapir structures are well developed around the world [54] and are formed by rapidly deposited, undercompacted mudstone arching or even piercing the overlying strata due to gravity, or by shale softening and arching under high-temperature and -pressure conditions, piercing the overlying strata. Mud diapir structures are always accompanied by deep fluid migration, therefore allowing for it to serve as not only a good representation of deep structural and fluid activity but also an important indicator of the existence and distribution of oil and gas reservoirs. They are thus of great significance in understanding the formation of oil and gas reservoirs [55–58]. In Cenozoic sedimentary basins such as the Qiongdongnan Basin and the Pearl River Mouth Basin in the north SCS, many mud diapirs or mud volcanoes are active. The massive undercompacted mud shale rapidly deposited and filled in the depression center is the material basis for the formation of mud diapirs. The high-temperature overpressure potential formed by the hydrocarbon generation and pressurization of organic matter in the source rock is the power source for the formation of mud diapirs. The regional tectonic dynamic background and ductile bed are the external conditions for the formation of mud diapirs. The area where mud diapirs and gas chimneys are concentrated is a favorable target for exploration activity in a deep-water continental slope area [59–62]. The DSWs are an area where the Mesozoic and Cenozoic superimposed basins are developed, with relatively thick Mesozoic and thin Cenozoic strata that have undergone changes in marine continental marine sedimentation during the Mesozoic and Cenozoic eras. Specifically, the Mesozoic strata underwent significant uplift and deformation during the Mesozoic and Cenozoic eras, forming a large number of seamounts, many of which have the characteristics of mud volcanoes, reflecting the fact that the area has sufficient gas sources to supply mud volcanic activities. However, relatively little research has been conducted on the mud diapir structures associated with subduction zones, due to the lack of high-resolution and deep reflection seismic profile data. Therefore, the genesis mechanism of mud volcanoes in the DSWs and their impact on gas hydrates need to be further studied.

Based on a comprehensive analysis of magnetic anomalies, high-resolution deep reflection seismic data, and drilling data, in this article, the distribution patterns of the Mesozoic igneous rock zone, the Mohorovičić discontinuity (or Moho) buried depth turning zone, and the subduction zone accretion wedge in the DSWs are proposed. Additionally, the characteristics of the Mesozoic subduction zone are systematically analyzed, and the specific location and subduction time of the suture line in the Mesozoic subduction zone are determined. Furthermore, the Mesozoic subduction model is proposed, and the influence of the Mesozoic subduction zone on the formation of gas hydrates is further analyzed. This helps to enrich the evidence on the westward subduction of the ancient Pacific plate and to expand upon our knowledge on the accumulation of gas hydrates in the northern SCS.

2. Geological Setting

The northeast SCS is located in the central part of the East Asian continental margin, structurally situated on the southern edge of the South China landmass, with the central basin of the Cenozoic SCS to the south; the northern South China Sea has complex rifting, magmatic styles [63]. A large number of geophysical surveys and drillings have confirmed the presence of the marine Mesozoic in the northeast SCS, such as the CFC-1 and CJA-1 wells drilled into the Lower Cretaceous strata [64,65] and the LF35-1-1 well drilled into the Middle Upper Jurassic strata [66,67]; the Mesozoic is widely distributed in the northeast SCS (Figures 1 and 2), with an area of 100,000 km^2. The northeast SCS has gone through six stages of tectonic evolution in the Mesozoic and Cenozoic eras, namely, a rifting period (the late Triassic), a depression period (the Jurassic), a tectonic uplift period (the Late Jurassic), a re-subsidence period (the Early Cretaceous), a tectonic inversion period (the Late Cretaceous), and a regional thermal subsidence period (the Neogene). Mesozoic basins have multiple stages of basin formation and transformation and are large, superimposed basins formed by stacking prototype basins with different evolutionary characteristics [68–71]. The CSD is the largest superimposed depression in the DSWs, distributed in a northeast direction. Structurally, the CSD consists of six secondary units: the northern slope, the central slope, the western slope, the western depression, the eastern depression, and the middle low bulge (Figure 1) [69]. The maximum thickness of its sedimentary layer exceeds 7000 m (Figure 2) [56]. The CSD mainly develops a set of Mesozoic strata, which are mostly in contact with the overlying Cenozoic at an angle. The LF35-1-1 well (completed at a depth of 2412 m) revealed that the strata below 977 m belong to the Mesozoic, with 977 m to 1698 m being the Cretaceous, 1698 m to 2400 m being the Jurassic, and 2400–2412 m being granite. The Cretaceous strata are composed of gray-green, purple-red, gray-purple, and gray-white clastic rocks and comprise a set of fluvial to lacustrine sedimentary layers. The Jurassic strata are composed of gray and gray black sandstone and mudstone, interbedded with oolitic limestone, and are a set of shallow to deep sea sedimentary layers. Fossils of radiolarian combination indicate that the strata (1860–1725 m) were formed in a deep-sea island arc environment, dating back to the Late Jurassic, and fossils of benthic foraminifera indicate that the strata (2112–2049 m) were formed in a shallow tropical marine environment. The fossil assemblage of Kelasuo pollen and Alsophila spores indicates that the strata (2268–2187 m) were formed in a coastal swamp environment, dating from the Middle to Late Jurassic. The granite section is mainly composed of granite and granodiorite intrusions, with the intrusion period mainly occurring in the Late Cretaceous. The CSD has developed two sets of source rocks, namely, the Upper Triassic–Lower Jurassic bathyal facies mudstone and the Upper Jurassic shallow sea shelf mudstone. The source rock conditions are good, and the LF35-1-1 well confirms the existence of marine source rocks [66]. The DSWs are rich in gas hydrates and are a favorable area for the accumulation of gas hydrates [72–91].

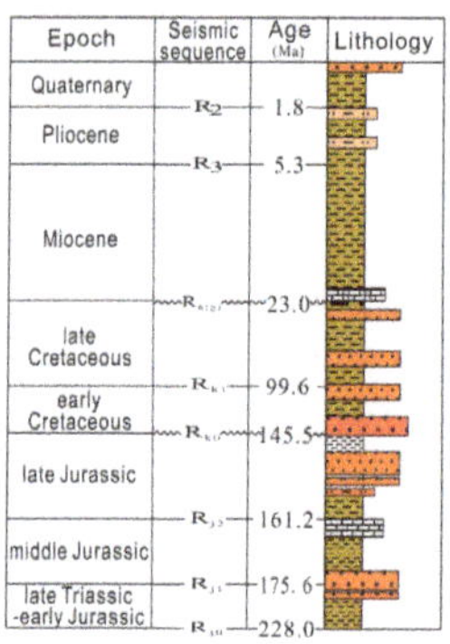

Figure 2. Stratigraphy of the CSD compiled from Geological data of northern South China Sea and well LF35-1-1, indicating seismic sequence, geological age, and lithology.

3. Data and Methods

3.1. Data Acquisition and Processing

The GMGS has conducted a number of new, long-array seismic surveys in recent years. The 2D seismic streamer has 480 channels spaced 12.5 m apart, and shooting occurs at 37.5 m intervals, using a source capacity of 5080 cubic inches. The coverage is up to 80-fold. Seismic data are processed with routine industry procedures and pre-stacked time migration. The focus is on the suppression of turbulence noise, various random noises, linear interferences, and different types of multiples. With processing, the signals from the shallow, medium, and deep layers are readily discernable, and the wave train characteristics and amplitude characteristics are much clearer than those from previous 2D seismic surveys. This improved seismic imaging poses as an excellent basis for further analysis of complex geological structures. All of the magnetic data used in this study were acquired by the GMGS in recent years.

3.2. Interpretation

Two-dimensional seismic data from an area with a length of about 2000 km were used for this study. The Schlumberger Geoframe Software Platform was used for structural and stratigraphic interpretation of the seismic data. The LF35-1-1 well was selected to provide the stratigraphic and lithological constraints for seismic interpretation. Seven reflecting boundaries were identified based on the recognition of unconformities, reflection terminations, prominent seismic characteristics, and lateral continuity (Figure 3); the interpretation was also controlled by closing loops within the seismic grid to eliminate misties.

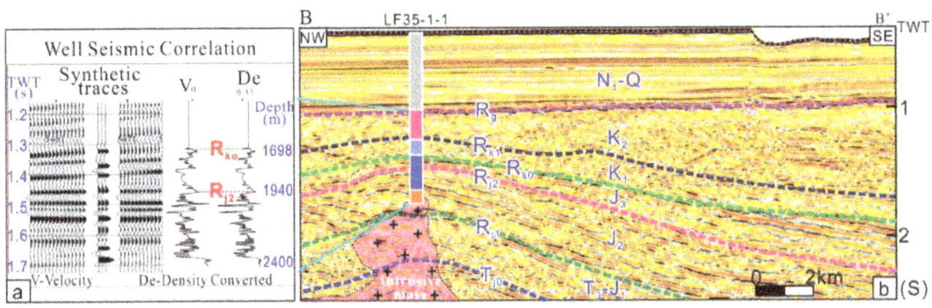

Figure 3. Seismic profile (BB′) across the northern SCS, running from northwest to southeast: (**a**) well seismic correlation of LF35-1-1; (**b**) reflecting the boundaries calibrated by synthetic traces (see Figure 1 for the location).

Horizon R_g corresponds to the top of the Mesozoic, expressed by erosion of the Mesozoic. Horizon R_g is an unconformable reflection of high amplitude and a laterally continuous reflection, calibrated by the LF35-1-1 well at a depth of 977 m or TWT 1060 ms.

Horizon R_{k1} corresponds to the top of the Upper Cretaceous, expressed by erosion of the Lower Cretaceous. Horizon R_{k1} is an unconformable reflection of low amplitude and a laterally discontinuous reflection, calibrated by the LF35-1-1 well at a depth of 1369 m.

Horizon R_{k0} corresponds to the top of the Jurassic, expressed by erosion of the Jurassic. Horizon R_{k0} is an unconformable reflection of medium amplitude and a laterally semi-continuous reflection, calibrated by the LF35-1-1 well at a depth of 1698 m.

The interface R_{j2} between J_3 and J_2 occurs at a depth of 1940 m, calibrated by the LF35-1-1 well, corresponding to a time of 1436 ms. Below the interface is a set of carbonate deposits, and above it is a set of sandstone and mudstone deposits. The upper and lower strata of this interface exhibit an integrated contact relationship, and due to differences in lithology, a strong reflection interface is formed on the seismic profile.

The interface R_{j1} between J_2 and J_1 has not been encountered by this well. However, based on an analysis of the regional sedimentary environment, the upper and lower strata of the interface should be continuous sedimentation and should have undergone a transition from the Late Jurassic semi-deep sea facies to the mid Jurassic coastal shallow sea facies. Therefore, a stable set of semi-deep sea facies mudstone should exist below this interface, with mudstone as the main component above this interface. However, the sandstone content shows a significantly increase, and the upper and lower strata show an integrated contact relationship.

The bottom interface R_{j0} of the Upper Triassic shows an unconformity contact relationship between the upper and lower parts of the interface, with a significant difference in seismic velocity compared with the pre-Mesozoic basement.

The upper part of the Mesozoic is the Cretaceous, and the lower part is the Jurassic Upper–Triassic. The Mesozoic exhibits a typical double-layered structure. The Cretaceous is characterized by wedge-shaped, medium-frequency, medium–low-amplitude, and medium continuous seismic facies, with obvious erosion at the top and upward superimposition dominating the bottom. The lower part of the Cretaceous is filled with a set of wedge-shaped alluvial fan sediments, which are lacustrine upward sediments. The Cretaceous has typical fault depression-filling characteristics. The Jurassic is characterized by parallel or sub-parallel, low-frequency, and medium–low continuous seismic facies. This sedimentary body has a relatively stable sedimentary environment and a relatively stable regional distribution, has a non-significant spatial variation in stratigraphic thickness, and is a shallow sea–semi-deep sea–deep sea sediment. An angular unconformity contact occurs between the Cretaceous and Jurassic, and erosion occurs in high parts of the structure. The plate subduction in the Late Jurassic is the reason for the unconformity between the Cretaceous and Jurassic (Figure 3).

4. Results

4.1. Mesozoic Subduction Accretion Wedges and Ancient Suture Lines

From the deep seismic reflection profile AA′ (Figure 4), typical subduction accretion wedges can be observed. The accretion wedges are composed of stacked thrust faults and structural wedges (①~⑤), which fault the Upper Triassic Jurassic. The detachment layer extends along the bottom of the Upper Triassic, and due to the compression of the structural wedges, strong folding and faulting occur in the Upper Triassic Jurassic (Figure 4). R_{k0} and R_g are two important unconformity surfaces, with unconformity contact between the upper and lower strata, and erosion occurring at the top of the Jurassic and Cretaceous. Following the cessation of subduction activity during the Mesozoic era, it underwent a transformation as a result of subsequent tectonic movements. The northwest side of the Dongsha detachment fault had strong folding and compression in the Mesozoic era, while the southeast side had relatively weak deformation in the Mesozoic era. A certain inheritance relationship occurred between the tectonic activity in the Cenozoic era and the structure in the Mesozoic era: before the stretching and thinning in the Cenozoic era, the zone was weak, and the Cenozoic detachment fault developed along the weak zone.

From profile CC′ (Figure 5), we can observe that between 8 s and 11 s, a moderately continuous and strong amplitude seismic reflection interface gradually deepens from southeast to northwest, and the seismic velocity above and below the interface undergoes a sudden change, which is interpreted as a Moho. A large detachment fault, the Dongsha Detachment Fault, developed in the middle of the profile. This fault is an important structural boundary that extends downwards and merges with the Moho. The Moho merges with the detachment fault to form a steep deepening slope break. The slope break zone is also the boundary between the oceanic crust and the continental crust, with the Dongsha detachment fault as the boundary. The northwest section develops a thicker Mesozoic and a thin Cenozoic, which are invaded by rock masses. The maximum crustal thickness exceeds 30 km, and strong folding and compression occur in the Mesozoic near the Dongsha fault; the crustal thickness in the southeast section is relatively thin, with a minimum thickness of less than 10 km. The Dongsha Fault is a large detachment fault deep in the Moho, which is a manifestation of the reactivation of ancient suture lines in the Mesozoic subduction zone during the Cenozoic era (Figure 5).

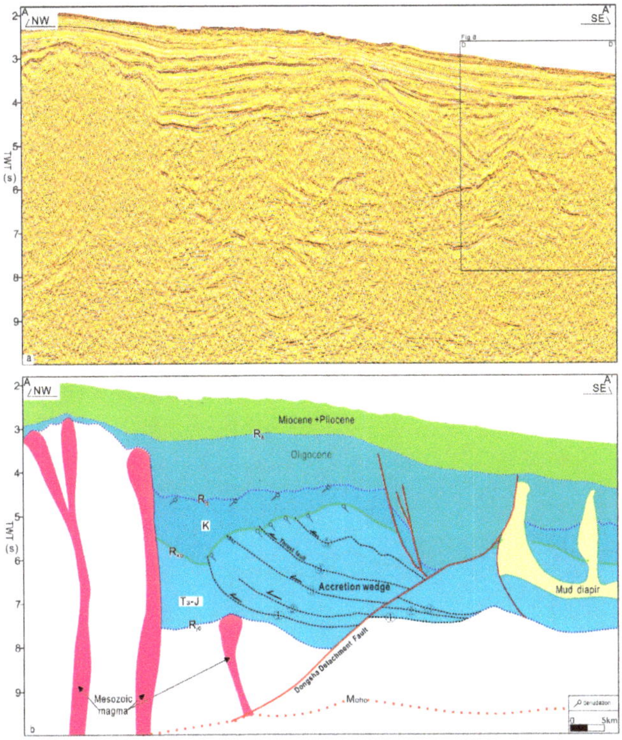

Figure 4. Seismic profile AA′ across the northern SCS, running from northwest to southeast, showing accretion wedges, mud diapers, and detachment fault ((**a**) seismic profile; (**b**) interpretation profile showing erosion at the top of the Jurassic and Cretaceous, accretion wedges, and mud diaper).

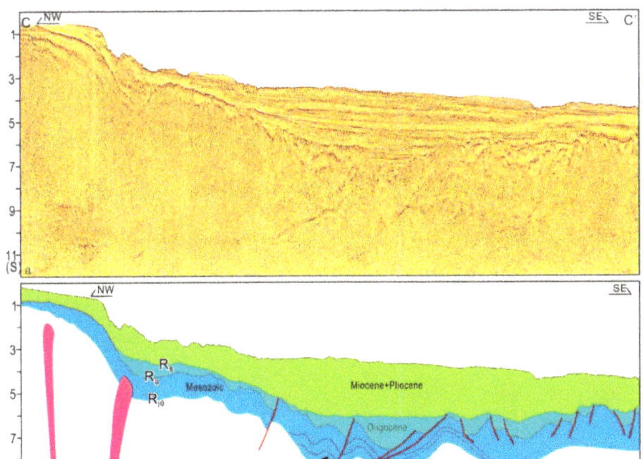

Figure 5. Seismic profile CC′ across the northern SCS, running from northwest to southeast; Moho can be traced distinctly, presenting a laterally semi-continuous reflection ((**a**) seismic profile; (**b**) Moho, with a compressive fold and Dongsha fault able to be interpreted).

4.2. Mesozoic Igneous Rocks

The amplitude of polar magnetic anomalies in the study area ranges from −80 to 250 nT, exhibiting a large number of magnetic anomalies from the Dongsha uplift to the Penghu uplift, forming a northeast-oriented zone with a large number of magnetic anomalies (Figure 6). The CSD is mainly characterized by negative magnetic anomalies, and the southeast CSD also exhibits many magnetic anomalies, but their amplitudes are relatively low. Overall, a clear zoning feature of "high in the north and south, and low in the middle" can be seen (Figure 6). Based on magnetic anomaly wavelet decomposition and seismic interpretation, the distribution of igneous rocks is delineated. The igneous rocks on the northwest side of the CSD have source depths that are mostly greater than 2 km, intrude into the Mesozoic, and are mainly medium acidic, while the igneous rocks on the southeast side of the CSD are mostly exposed to the seabed in the form of eruption and are mainly medium basic; the distribution of igneous rocks in the depression is relatively rare. The Dongsha high magnetic anomaly field is a reflection of the Mesozoic subduction and accretion of intermediate acidic volcanic rocks with high residual magnetic strength (Figure 7), which is the result of the Mesozoic subduction process.

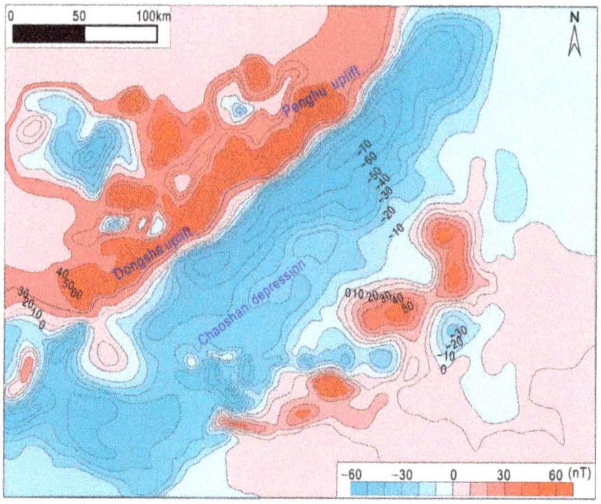

Figure 6. A magnetic anomaly map reduced to the magnetic pole.

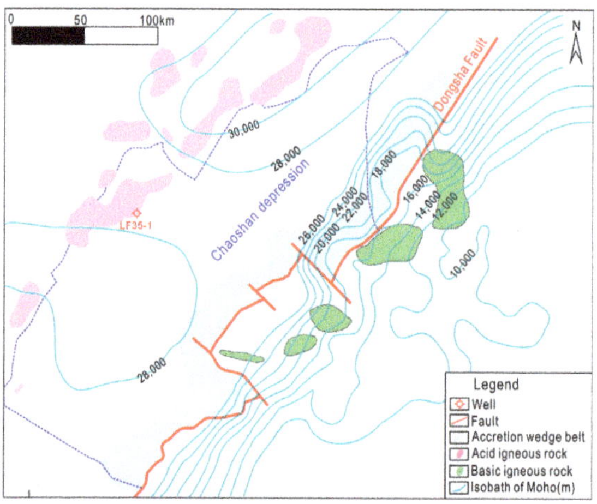

Figure 7. Isobaths of Moho and igneous rock distribution.

The zircon dating results of LF35-1-1 igneous rocks reveals that the age of the granite at the bottom of the well ranges from 90.7 ± 1.3 to 103.3 ± 1.4 Ma, and the igneous rocks are from the Late Jurassic to the Early Cretaceous. Therefore, we can infer that the initial subduction occurred in the Late Jurassic.

4.3. Mesozoic Mud Diapir Structure

In the seismic profile, the plastic body in the core of the diapir structure exhibits chaotic reflections, which can be completely separated from the reflection of the surrounding rock. The reflection of the surrounding rock strata has good continuity in the same phase axis and stops when it extends to the core. The chaotic reflections in the core significantly thicken and can be traced to both sides, but the thickness rapidly decreases. Due to the intrusion of the diapir structure into the overlying strata, the reflection phase axis of the overlying strata thins towards the top of the mud structure on both wings, reducing the number of phases and accompanying complex fault systems at the top of the structure. Due to the constant velocity of mudstone layers at various depths, a significant difference exists in the velocity between the plastic bodies of the mud diapir structures and the underlying strata, resulting in a large impedance interface difference and strong reflection. In addition, in the seismic profile, the lower mud layer exhibits a medium- to strong-amplitude reflection with good continuity, showing a low-amplitude uplift state (Figures 8 and 9).

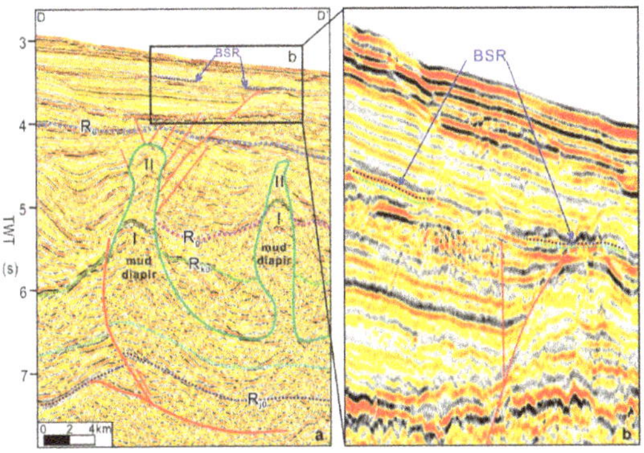

Figure 8. Seismic profile DD′ showing a mud diaper ((**a**) biphasic development of a mud diaper; (**b**) is the enlargement of box b in the left figure, showing seismic track polarity inversion of BSR (blue line) and normal seismic tracks (green line)).

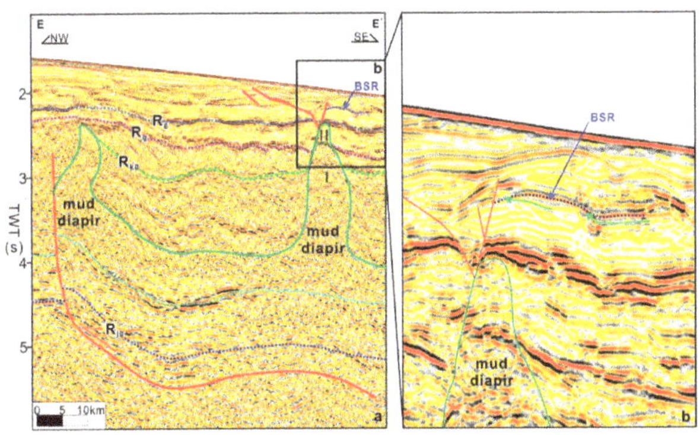

Figure 9. Seismic profile EE′ showing a mud diapir ((**a**) biphasic development of a mud diaper; (**b**) is the enlargement of box b in the left figure, showing seismic track of a BSR (blue line) and truncated normal seismic tracks (green line)).

5. Discussion

5.1. Properties of Dongsha High-Magnetic-Anomaly Zones

Much debate surrounds the geological origin of the Dongsha high-magnetic-anomaly zone. Dai believed that this high-magnetic-anomaly zone is likely to be the same magmatic tectonic zone as the Zhejiang Fujian coastal volcanic rock zone, which separated from the Zhejiang Fujian coastal volcanic rock zone during the later compression and extension processes [90]. Xia et al. believed that this high-magnetic-anomaly zone is consistent with the thinning of the upper crust, thickening of the lower crust, and uplift of the lower crust [91]. It is not a part of the South China continent or a volcanic arc, but a composite tectonic zone formed by the combined action of Cenozoic tension and pre-Cenozoic compression. The main magnetic source is the ultramafic and magnetic material deep in the lower crust. Wu et al. linked the Penghu uplift and the Diaoyu Islands uplift, suggesting that both may be the Late Cretaceous ancient Ryukyu Island Arc, which was terminated by a NW-trending strike slip fault at 118° to the west [92]. Zhou et al. comprehensively speculated that the high-magnetic-anomaly zone in the northern SCS has the same origin as the volcanic rock belt in the eastern Zhejiang Fujian region, representing a volcanic arc associated with the Mesozoic subduction and accretion, and was left laterally displaced by NW-trending faults [44]. This large amount of magnetic anomaly was formed by a medium acidic "volcanic rock mass" with a thickness of up to 6km. This belt is a crustal remelting magmatic rock belt formed by the collision of continental crustal fragments that migrated northward in the Late Cretaceous and the southeastern edge of the Asian continent. In fact, the magnetic anomaly in the Diaoyu Islands uplift zone is also comparable in shape and scale with the high-magnetic-anomaly zone in the northern SCS. Yao Bochu et al. identified the Dongsha uplift and other volcanic arcs as a reflection of the Jurassic Cretaceous volcanic arc in eastern Asia. Although the genesis of the high-magnetic-anomaly zone in the northern SCS varies, the most common consensus is that it is related to magmatic activity during subduction or collision in the Late Mesozoic [45]. The geochemical characteristics of the intermediate acidic intrusive granite encountered during LF35-1-1 drilling indicate that this zone is an active continental margin island arc environment. Furthermore, paleontological data from the LF35-1-1 well indicate a deep-sea island arc environment during the Late Jurassic [66]. The gravity and magnetic field characteristics of the Dongsha uplift, as well as the inverted lithological characteristics, are similar to the island arc characteristics formed by modern subduction and are comparable [93]. The seismic profile of DSRP2002 highlighted possible Paleo subduction fragments and subduction sutures in the area [16]. Based on the comprehensive characteristics of structure and magnetic anomalies, the high-magnetic-anomaly zone of the Dongsha uplift in the northern SCS can be inferred to be a product of the subduction of the paleo-Pacific. Magnetic materials are mainly located in the upper crust, and the strong magnetic differences on both sides of the boundary under oblique magnetization conditions are the main reason for the formation of the large amount of magnetic anomalies.

5.2. Mesozoic Tectonic Evolution and Subduction Model in the DSWs

Mesozoic tectonic evolution is closely related to subduction. The DSWs have gone through six stages of tectonic evolution in the Mesozoic and Cenozoic eras, namely, a rifting period (T_3), a depression period (J), a tectonic uplift period (J_3), a re-subsidence period (K_1), a tectonic inversion period (K_2), and a regional thermal subsidence period (the Neogene) [93–98]. The ancient Pacific lithosphere subducted northwest beneath the lithosphere of the South China continent. In front of the volcanic arc, a forearc basin developed, while behind the volcanic arc, a small post-arc extensional basin developed. In the late Early Cretaceous, the ancient SCS oceanic lithosphere had almost subducted, and the Nansha and South China landmasses collided and continued to compress in the northwest direction (Figure 10). The stratigraphic unconformity between J and K is the result of the Dongsha subduction. The subduction began in the Late Jurassic and collided with the South China landmass during the Late Cretaceous. The northeastern SCS uplifted, and the Middle and Upper Jurassic eroded, forming an angular unconformity interface between the Jurassic and Cretaceous. The evolution of the Dongsha subduction zone can be divided into two stages: (1) in the Late Jurassic, the ancient Pacific plate began to subduct in the NWW direction, and the Dongsha subduction zone already existed as a large subduction zone boundary; (2) in the Late Cretaceous, the subduction continued, and the Reed Bank landmass collided with the South China landmass, causing the ancient SCS to contract and disappear. The corresponding part of the accretion zone has a steep, downward sloping Moho towards the north, with the buried depth of the Moho dropping sharply from 16 km to 26 km. The Dongsha high-magnetic-anomaly zone on the northwest side of the subduction accretion zone is an igneous rock zone that is associated with the Mesozoic subduction.

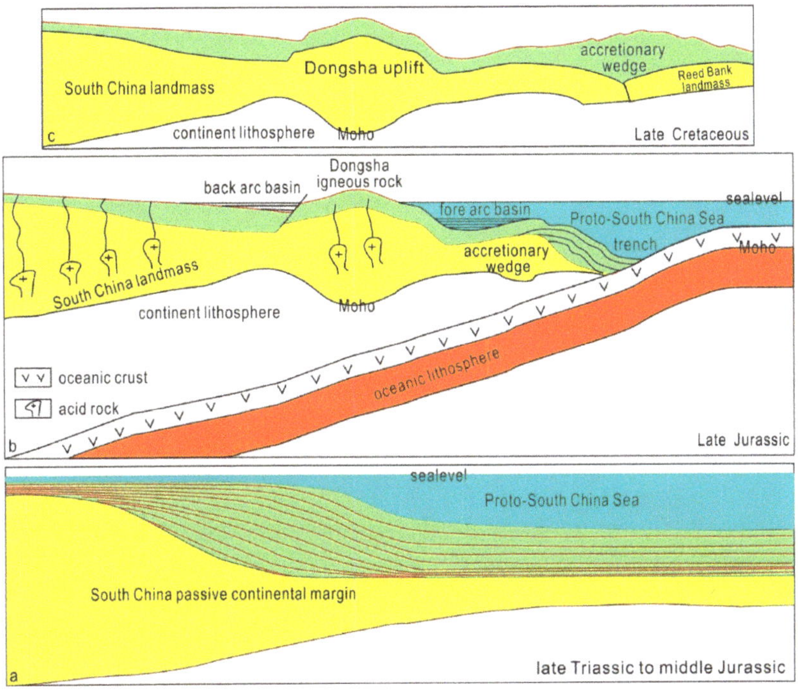

Figure 10. Subduction model of the DSWs. ((**a**) Passive continental margin period from Late Triassic to Middle Jurassic; (**b**) primary subduction period during Late Jurassic; (**c**) collision period during Late Cretaceous).

5.3. Dongsha Subduction Significance in Gas Hydrate Accumulation

5.3.1. Controlling the Gas Source of Hydrate

The Upper Triassic–Lower Jurassic bay sediments and the eastern passive continental margin sediments, as well as the Late Jurassic–Early Cretaceous forearc basins, are important strata for exploring gas hydrates today, such as in the southern part of the Tainan Basin, the CSD, and the Pear River Mouth Basin. Particularly, they are important places for exploring Mesozoic gas hydrates. Affected by subduction, the northern SCS transformed from a passive continental margin during the Late Triassic–Middle Jurassic to an active continental margin during the Late Jurassic. Starting from the Late Triassic, semi-enclosed bay-filling sedimentation began to develop in the Mesozoic basin in the DSWs. At the end of the Late Triassic, the prototype of the basin formed, mainly in an open continental shelf environment. In the Early Jurassic period, the sea invasion further expanded, expanding to the northern part of Guangdong, and the basin entered its peak period; in the late

Early Jurassic, a regression occurred, with seawater retreating to the northern SCS and developing coastal and shallow marine sediments. In the late Middle Jurassic, the water gradually deepened, and by the end of the Late Jurassic, it was affected by subduction and uplifted. In the passive continental margin stage, Jurassic semi-enclosed bay source rocks developed. These source rocks have relatively high organic matter contents and abundances compared with open sea source rocks, and their quality is also relatively good. Thus, the main source layers within CSD comprise the T_3-J_1 and J_3 mudstones, especially the thick semi-enclosed bay source rocks, which were found to have fairly high organic matter abundance. In the present era, source rocks have undergone significant maturation and become the gas source foundation for the formation of thermogenic gases. As the deep thermogenic gas interacts with microbial gas, it eventually contributes to gas hydrates [98].

5.3.2. Controlling the Structural Pattern of Hydrate Accumulation

Under the influence of subduction, the CSD formed a structural framework of two depressions and one low bulge under northwest–southeast compressive stress (Figures 1 and 10), and a series of northeast-oriented trap structures were formed in the mid-low bulge. The extensional activity of the northern margin of the SCS in the Cenozoic era was activated along the suture line of the subduction zone, forming the Dongsha detachment fault and controlling the development of the graben. The formation of natural gas hydrate reservoirs in the DSWs is closely related to the Dongsha detachment fault activity. The natural gas overflowing from gas pools within the trap structure is vertically transported to the shallow seabed through faults and micro-cracks and accumulates into hydrates under high-pressure and low-temperature conditions [69,89]. Fluid diapirs are large continuously spreading bright reflection zones in shallow surface layers and are downwards converging high-variance ribbons in middle and shallow layers, all of which are en echelons spreading in the NW–NNW-oriented direction on the plane, with a total area of up to several hundred square kilometers. The profile shows a conical or mushroom-shaped shape that converges from shallow to deep and is characterized by a compound gas chimney fuzzy zone [98].

5.3.3. Controlling the Leakage Channel of Gas Hydrates

During the Cretaceous period, under the westward subduction of the ancient Pacific plate, the plastic mudstone layer in the northern SCS arched upwards under compression, piercing through the overlying rock layers and forming a diapir structure. Affected by subduction, the Mesozoic mudstone softened under high-temperature and -pressure conditions, forming a mud diapir structure by arching upwards. Thus, the mud diapir structure went through two stages of development: (1) At the end of the Cretaceous, the structure was strengthened by the Mesozoic subduction and compression, and the Jurassic mudstone underwent plastic flow, invading the Cretaceous system, forming the stage I mud diapir structure. (2) Under differential gravity caused by the subsidence of the continental slope in the Miocene, the early developed mud diapir structures were reactivated, and the Mesozoic mud diapir invaded the Cenozoic, forming the stage II diapir structure.

Liang et al. believed that there were three genetic models: a diffusion type (pore-filling type), a leakage type (fracture-filling type), and a composite type [78]. In the DSW area, deep gas is regulated by mud diapirs, gas chimneys, and fault systems. During the upward migration of gas, low-flux methane gas forms a diffusion-type hydrate at the bottom of the stability zone, driven by a concentration differential, pressure, and capillary force. Meanwhile, high-flux methane leaks upward in the form of free leakage. It migrates along the fracture and accumulates in the upper part of the stability zone, forming a leakage-type hydrate, which also represents a typical genetic model of a composite type [80].

In fact, many more mud volcanoes have been reported over the DSWs, indicating widespread leakage of oil and gas from the substrates [27,28,35,69,99,100]. The extensional activity of the northern margin of the SCS in the Cenozoic era was activated along the suture line of the subduction zone, forming the Dongsha detachment fault and controlling the development of the graben. The fault fractures of the deep region facilitate communication between the temperature and pressure stability region of the hydrate and the Mesozoic hydrocarbon source rock. Then, the deep thermogenic gas migrates along the transport system to the overlying formation, where it is mixed with the microbial gas, eventually forming hydrates. The fault plays a pivotal role in the process of hydrate accumulation in the region. The Mesozoic fault development area represents the primary seepage area for thermogenic gas [98].

Hydrate stability is, however, a function of pressure, temperature, and concentrations of all hydrate building components in all co-existing phases. Practically, this implies that hydrate can nucleate towards mineral surfaces; hydrate can form from hydrate formers dissolved in water. What is, however, very important is hydrate dissociation towards water that is under-saturated with hydrate former. Hydrates in sediments cannot reach equilibrium [101,102] because there are too many co-existing phases that affect hydrate stability. All over the world, hydrates dissociate continuously due to inflow of seawater (almost no CH_4) through fracture systems. In offshore Norway, one example is the Nyegga region [102], in which there are fairly many leakage regions. Even inside a reservoir with several hydrate zones internal movement and changes in fluids may lead to hydrate instability.

6. Conclusions

(1) The Mesozoic subduction zone over the Dongsha Waters (DSWs) of the South China Sea (SCS) is distributed southeast of the CSD in an NE45° direction, and composed of accretion wedges, trenches, and igneous rock zones. The igneous rock formed during the period between the Late Jurassic and the Early Cretaceous.

(2) The evolution of the Dongsha subduction zone can be divided into two stages: ① In the Late Jurassic, the ancient Pacific plate began to subduct in the NWW direction, and the Dongsha subduction zone, as a large subduction zone boundary, already existed. ② In the Late Cretaceous, the subduction continued, and the Reed Bank landmass collided with the South China landmass, causing the contraction and disappearance of the ancient SCS, ultimately forming an igneous rock belt and an accretion wedge system.

(3) Mud diapir and Dongsha detachment fault activity are the main factors controlling the formation of hydrate accumulation in the DSWs. Mud diapir and Dongsha detachment fault activity are the main factors controlling the formation of natural gas hydrate reservoirs in the DSWs. The extensional activity of the northern margin of the SCS in the Cenozoic era was activated along the suture line of the subduction zone, forming the Dongsha detachment fault and controlling the development of the Cenozoic fault depression. The gas source of the hydrate in the DSWs mainly comes from thermogenic gas, which is accompanied by mud diapir activity and migrates along the fault. The gas accumulates to form gas hydrates at the bottom of the stable domain, and BSR can be seen above the mud diapir structure; that is, hydrate deposits are formed under the influence of mud diapir structures, belonging to a typical leakage type genesis model.

Author Contributions: Data processing: K.Z.; writing—original draft preparation: P.S. and Z.Z. All authors have read and agreed to the published version of the manuscript.

Funding: This research was jointly supported by Hainan Province key research and development project (ZDYF2024GXJS002), and the National Natural Science Foundation of China (42376222 and U22A20581).

Institutional Review Board Statement: Not applicable.

Informed Consent Statement: Not applicable.

Data Availability Statement: Data are available on request from the author.

Acknowledgments: We are grateful for the efforts of all who participated in data acquisition and processing. We thank the reviewers for helpful suggestions that improved the clarity of this paper.

Conflicts of Interest: The authors declare no conflicts of interest.

References

1. Almeida, J.; Kiel, N.; Rosas, F.M.; Duarte, J.C.; Schellart, W.P. Polarity-reversal subduction zone initiation triggered by buoyant plateau obstruction. *Earth Planet. Sci. Lett.* **2022**, *577*, 117195. [CrossRef]
2. Arfai, J.; Franke, D.; Gaedicke, C.; Lutz, R.; Schnabel, M.; Ladage, S.; Berglar, K.; Aurelio, M.; Montano, J.; Pellejera, N. Geological evolution of the West Luzon Basin (South China Sea, Philippines). *Mar. Geophys. Res.* **2011**, *32*, 349–362. [CrossRef]
3. Barckhausen, U.; Lngels, M.; Franke, D.; Ladage, S.; Pubellier, M. Evolution of the South China Sea: Revised ages for breakup and seafloor spreading. *Mar. Pet. Geol.* **2014**, *58*, 599–611. [CrossRef]
4. Bao, H.; Guo, Z.; Zhang, L.; Huang, Y. Tectonic dynamics of eastern China since the formation of the Pacific Plate. *Adv. Earth Sci.* **2013**, *28*, 337–346, (In Chinese with English abstract).
5. Bautista, B.; Bautista, M.; Oike, K.; Wu, F.; Punongbayan, R. A new insight on the geometry of subducting slabs in northern Luzon, Philippines. *Tectonophysics* **2001**, *339*, 279–310. [CrossRef]
6. Briai, A.; Patriat, P.; Tapponnier, P. Updated interpretation of magnetic anomalies and seafloor spreading stages in the South China Sea: Implications for the Tertiary tectonics of Southeast Asia. *J. Geophys. Res. Solid Earth* **1993**, *98*, 6299–6328. [CrossRef]
7. Camanni, G.; Ye, Q. The significance of fault reactivation on the V1′71SOn Cycle undergone by the northern South China Sea area in the last 60 Myr. *Earth Sci. Rev.* **2022**, *225*, 103893. [CrossRef]
8. Hall, H. The subduction initiation stage of the Wilson cycle. *Geol. Soc. Lond. Spec. Publ.* **2019**, *470*, 415–437. [CrossRef]
9. Kiss, D.; Candiotti, L.; Duretz, T.; Schmalholz, S. Thermal softening induced subduction initiation at a passive margin. *Geophys. J. Int.* **2020**, *220*, 2068–2073. [CrossRef]
10. Lshizuka, O.; Taylor, R.; Ohara, Y.; Yuasa, M. Upwelling, rifting, and age-progressive magmatism from the Oki-Daito mantle plume. *Geology* **2013**, *41*, 1011–1014. [CrossRef]
11. Wang, Y.; Wu, S.; Qian, X.; Junaidi, B.A.; Wang, Y.; Zhang, Y.; Feng, Q.; Wang, W.; Zhang, P. Early Triassic-Late Cretaceous Paleo-Pacific subduction zone along the East Asia continental margin: Fore-arc igneous records in Sabah of NE Borneo. *China Sci. Bull.* **2023**, *68*, 954–971. (In Chinese) [CrossRef]
12. Antonijevic, S.K.; Wagner, L.S.; Kumar, A.; Beck, S.L.; Long, M.D.; Zandt, G.; Tavera, H.; Condori, C. The role of ridges in the formation and longevity of flat slabs. *Nature* **2015**, *524*, 212–215. [CrossRef] [PubMed]
13. Agrusta, R.; Goes, S.; van Hunen, J. Subducting-slab transition-zone interaction: Stagnation, penetration and mode switches. *Earth Planet. Sci. Lett.* **2017**, *464*, 10–23. [CrossRef]
14. Chen, C.; Yan, P.; Yu, J.; Zhong, G. Seismically Imaged Crustal Breakup in the Southwest Taiwan Basin of the Northeastern South China Sea Margin. *Geochem. Geophys. Geosyst.* **2023**, *24*, e2023GC010918. [CrossRef]
15. Ding, W.; Zhu, R.; Wan, B.; Zhao, L.; Niu, X.; Zhao, P.; Sun, B.; Zhao, Y. Geodynamic processes of the southeastern Neo-Tethys Ocean and the formation mechanism of the curved subduction system in Southeast Asia. *Sci. China Earth Sci.* **2023**, *66*, 703–717. [CrossRef]
16. Huang, C.; Zhou, D.; Sun, Z. Deep crustal structure of baiyun sag, northern South China Sea revealed from deep seismic reflection profile. *Chin. Sci. Bull.* **2005**, *50*, 1131–1138. Available online: www.scichina.com (accessed on 15 May 2024). [CrossRef]
17. Hui, Z. Evolution of the Subducting Slab Morphology and its Controlling Factors. Ph.D. Thesis, University of Science and Technology of China, Hefei, China, 1 May 2023. [CrossRef]
18. Ikuhiro, K.; Yusuke, H.; Taro, S. Safe Performance of Track Dilation and Bile Aspiration with ERCP Catheter in EUS-Guided Hepaticogastrostomy with Plastic Stents: A Retrospective Multicenter Study. *J. Clin. Med.* **2022**, *11*, 4986. [CrossRef]
19. Luan, X.; Wang, J.; Liu, H.; Zhang, L.; Lu, Y. A Discussion on Tethys in Northern Margin of South China Sea. *Earth Sci.* **2021**, *46*, 866–884. [CrossRef]
20. Mar, M.; Qi, J.; Ma, J.; Peng, H.; Lei, L.; Song, Q.; Bai, M. Cenozoic Subsidence History of the Northern South China Sea: Examples from the Qiongdongnan and Yinggehai Basins. *Processes* **2023**, *3*, 956. [CrossRef]
21. Peng, G.; Liu, P.; Chen, W.; Li, Z.; Wang, X.; Zhang, Z.; Hao, G. Tectonic Transformation, Magmatic Activity and Subsidence Centre Migration of Eocene Half-Grabens: A Case Study of the Northern Pearl River Mouth Basin (PRMB) in the Northern South China Sea. *J. Mar. Sci. Eng.* **2023**, *3*, 579. [CrossRef]
22. Ye, Q.; Mei, L.; Shi, H. The Late Cretaceous tectonic evolution of the South China Sea area: An overview, and new perspectives from 3D seismic reflection data. *Earth-Sci. Rev.* **2018**, *187*, 186–204. [CrossRef]
23. Sun, T. A new map showing the distribution of granites in South China and its explanatory notes. *Geol. Bull. China* **2006**, *25*, 332–333.
24. Suo, Y.; Li, S.; Peng, G.; Du, X.; Zhou, J.; Wang, C.; Cao, X. Cenozoic basement-involved rifting of the northern South China Sea margin. *Gondwana Res.* **2023**, *120*, 20–30. [CrossRef]
25. Tian, Z. Tectonic Property of Proto-South China Sea-Constraints from Petrogeochemistry, Zircon U-Pb Geochronology and Paleomagnetic Evidence from Sabah and Sarawak. Master's Thesis, Jilin University, Changchun, China, 1 June 2021. [CrossRef]
26. Xing, T.; Zhong, G.; Zhan, W.; Zhao, Z.; Chen, X. Oil-gas reservoir in the Mesozoic strata in the Chaoshan depression, northern South China Sea: A new insight from long offset seismic data. *J. Oceanol. Limnol.* **2022**, *4*, 1377–1387. [CrossRef]
27. Yan, P.; Wang, Y.; Liu, J.; Zhong, G.; Liu, X. Discovery of the southwest Dongsha Island mud volcanoes amid the northern margin of the South China Sea. *Mar. Pet. Geol.* **2017**, *88*, 858–870. [CrossRef]
28. Yan, Y.; Liao, J.; Yu, J.; Chen, C.; Zhong, G.; Wang, Y.; Wang, L. Velocity Structure Revealing a Likely Mud Volcano off the Dongsha Island, the Northern South China Sea. *Energies* **2021**, *1*, 195. [CrossRef]

29. Zhan, H.; Suo, Y.; Zhu, J.; Li, S.; Wang, P.; Wang, G.; Zhou, J.; Wang, X. Closure mechanism of the South China Sea: Insights from subduction initiation along the Manila Trench. *Actor Petrol. Sin.* **2023**, *39*, 2569–2582. [CrossRef]
30. Zhu, R.; Yao, Y.; Liu, H.; Xu, Z.; Nie, X.; Du, W. Tectonic Contact Relationship of Continental Margins of the Southwest Sub-Basin, South China Sea in Late Mesozoic. *Earth Sci.* **2021**, *46*, 885–898. [CrossRef]
31. Li, W.; Zhou, X. Late Mesozoic subduction zone of southeastern China. *Geol. J. China Univ.* **1999**, *5*, 164–169.
32. Li, F.; Sun, Z.; Yang, H. Possible spatial distribution of the Mesozoic volcanic arc in the present-day South China Sea continental margin and its tectonic implications. *J. Geophys. Res. Solid Earth* **2018**, *123*, 6215–6235. [CrossRef]
33. Hutchison, C.S. *Geology of North-West Borneo: Sarawak, Brunei and Sabah*; Oxford University Press: Oxford, UK, 2005.
34. Hutchison, C.S. Oroclines and paleomagnetism in Borneo and south-east Asia. *Tectonophysits* **2010**, *496*, 53–67. [CrossRef]
35. Xu, C.; Shi, H.; Barnes, C. Tracing a late Mesozoic magmatic arc along the Southeast Asian margin from the granitoids drilled from the northern South China Sea. *Int. Geol. Rev.* **2016**, *X8*, 71–94. [CrossRef]
36. Tang, J.; Xu, W.; Wang, F. Subduction history of the Paleo-Pacific slab beneath Eurasian continent: Mesozoic-Paleogene magmatic records in Northeast Asia. *Sci. China Earth Sci.* **2018**, *61*, 527–559. (In Chinese) [CrossRef]
37. Breitfeld, H.T.; Davies, L.; Hall, R.; Armstrong, R.; Forster, M.; Lister, G.; Thirlwall, M.; Grassineau, N.; Hennig-Breitfeld, J.; van Hattum, M.W.A. Mesozoic Paleo-Pacific subduction beneath SW Borneo: U-Pb geochronology of the Schwaner granitoids and the pinoh metamorphic group. *Front. Earth Sci.* **2020**, *8*, 536. [CrossRef]
38. Gan, C.; Wang, Y.; Barry, T. Spatial and temporal influence of Pacific subduction on South China: Geochemical migration of Cretaceous mafic-intermediate rocks. *J. Geol. Soc.* **2020**, *177*, 1013–1024. [CrossRef]
39. Guo, F.; Wu, Y.; Zhang, B. Magmatic responses to Cretaceous subduction and tearing of the Paleo-Pacific plate in SE China: An overview. *Earth-Sci. Rev.* **2021**, *212*, 103448. [CrossRef]
40. Metcalfe, I. Multiple Tethyan Ocean basins and orogenic belts in Asia. *Gondwana Res.* **2021**, *100*, 87–130. [CrossRef]
41. Gan, C.; Qian, X.; Wang, Y. Late Cretaceous granitoids along the Northern Kuching Zone: Implications for the Paleo-Pacific subduction in Borneo. *Lithosphere* **2022**, *1*, 3310613. [CrossRef]
42. Li, Z.; Qiu, J.; Yang, X. A review of the geochronology and geochemistry of Late Yanshanian (Cretaceous) plutons along the Fujian coastal area of southeastern China: Implications for magma evolution related to slab break-off and rollback in the Cretaceous. *Earth-Sci. Rev.* **2014**, *128*, 232–248. [CrossRef]
43. Yao, B.; Zeng, W.; Cheng, Y.; Zhang, X. The crustal structure in the eastern part of the northern margin of the South China Sea. *Acta Geophys. Sin.* **1994**, *37*, 27–35.
44. Zhou, D.; Wang, W.; Pang, X. Mesozoic subduction-accretion zone in the northeastern South China Sea inferred from geophysical interpretations. *Sci. China (Ser. D)* **2006**, *36*, 209–218. [CrossRef]
45. Yao, B.; Zhang, L.; Wei, Z.; Yi, H.; Lin, Z.; Wan, L.; Wang, W.; Qiu, Z.; Li, C. The Mesozoic tectonic characteristics and sedimentary basins in the eastern margin of South China. *Mar. Geol. Quat. Geol.* **2011**, *31*, 47–60. [CrossRef]
46. Cai, Q. "The conjecture of residual Tethys", from the discovery of marine Mesozoic-Paleocene near China. *Geol. China* **1998**, *4*, 39–41.
47. Cai, D.; Fenq, X.; Gao, L.; Lianq, J.; Li, Y.; Liu, X.; Duan, J.; Lu, J. Petroleum potential and exploration direction of pre-Tertiary remnant basins in Offshore China. *China Offshore Oil Gas* **2004**, *16*, 74–83.
48. Ding, L.; Qasim, M.; Jadoon, I.; Khan, M.; Xu, Q.; Cai, F.; Wang, H.; Baral, U.; Yue, Y. The India-Asia collision in North Pakistan: Insight from the U-Pb detrital zircon provenance of Cenozoic foreland basin. *Earth Planet. Sci. Lett.* **2016**, *455*, 49–61. [CrossRef]
49. Ding, L.; Maksatbek, S.; Cai, F.; Wang, H.; Song, P.; Ji, W.; Xu, Q.; Zhang, L.; Muhammad, Q.; Upendra, B. Processes of initial collision and suturing between India and Asia. *Sci. China (Earth Sci.)* **2017**, *60*, 635–651. [CrossRef]
50. Ding, W.; Yang, S.; Chen, H.; Cheng, X.; Wu, N. Arc-continent collision orogeny in offshore Taiwan during Neogene. *Chin. J. Geol.* **2006**, *41*, 195–201, (In Chinese with English abstract).
51. Ding, W.; Sun, Z.; Dadd, K.; Fang, Y.; Li, J. Structures within the oceanic crust of the central South China Sea Basin and their implications for oceanic accretionary processes. *Earth Planet. Sci. Lett.* **2018**, *488*, 115–125. [CrossRef]
52. Eakin, D.; Mcintosh, K.; Van Avendonk, H.; Lavier, L.; Lester, R.; Liu, C.; Lee, C. Crustal-scale seismic profiles across the Manila subduction zone: The transition from intraoceanic subduction to incipient collision. *J. Geophys. Res. Solid Earth.* **2014**, *119*, 1–17. [CrossRef]
53. Milkov, A.V.; Sassen, R. Economic geology of offshore gas hydrate accumulations and provinces. *Mar. Pet. Geol.* **2002**, *19*, 1–11. [CrossRef]
54. Milkov, A.V. Worldwide distribution of submarine mud volcanoes and associated gas hydrates. *Mar. Geol.* **2000**, *167*, 29–42. [CrossRef]
55. Reed, D.L.; Silver, E.A.; Tagudin, J.E. Relations between mud volcanoes, thrust deformation, slope sedimentation, and gas hydrate, offshore north Panama. *Mar. Pet. Geol.* **1990**, *7*, 44–54. [CrossRef]
56. Dimitrov, L.I. Mud volcanoes-the most important pathway for degassing deeply buried sediments. *Earth Sci. Rev.* **2002**, *59*, 49–76. [CrossRef]
57. Kholodov, V.N. Mud volcanoes, their distribution regularities and genesis: Communication 1. Mud volcanic provinces and morphology of mud volcanoes. *Lithol. Miner. Resour.* **2002**, *37*, 197–209. [CrossRef]
58. Yang, X.; Zhong, S.; Wang, Z. The thermodynamics of mud diapir/volcano fluid and its influence on gas hydrate occurrence. *Mar. Geol. Front.* **2018**, *4*, 15–23. [CrossRef]

59. Zhang, W.; Liang, J.; He, J.; Cong, X.; Su, P. Characteristics of mud diapir and gas chimney and their relationship with reservoir forming for petroleum and natural gas hydrate on northern slope of the South China Sea. *Mar. Geol. Front.* **2017**, *33*, 11–23. [CrossRef]
60. Huang, C.; Chen, K.; Li, S. Periodicities of diapiric rise in the Yinggehai basin. *Pet. Explor. Dev.* **2002**, *29*, 44–46. Available online: https://www.semanticscholar.org/paper/Periodicities-of-diapiric-rise-in-the-Yinggehai-Chun/3b39c6e84a9f8cc27976b805f8bf9622c33900e4 (accessed on 29 December 2023). (In Chinese with English abstract).
61. He, J.; Xia, B.; Zhang, S.; Yan, P.; Liu, H. Origin and distribution of mud diapirs in the Yinggehai basin and their relation to the migration and accumulation of natural gas. *Geol. China* **2006**, *33*, 1336–1344. Available online: http://geochina.cgs.gov.cn/geochina/geochinaen/article/abstract/20060617?st=article_issue (accessed on 29 January 2024). (In Chinese with English abstract).
62. Yan, P.; Wang, Y.; Yu, J.; Luo, W.; Liu, X.; Jin, Y.; Li, P.; Liu, J.; Zhong, G.; Yi, H. Surveying mud volcanoes over the Dongsha Waters in the South China Sea. *J. Trop. Oceanogr.* **2020**, *40*, 34–43. [CrossRef]
63. Zhao, Z.; Sun, Z.; Qiu, N.; Zhao, M.; Zhang, J.; Li, F.; Lin, J.; Lee, E. The paleo-lithospheric structure and rifting-magmatic processes of the northern South China Sea passive margin. *Gondwana Res.* **2023**, *120*, 162–174. [CrossRef]
64. Zhou, D. Mesozoic strata and sedimentary environment in SW Taiwan basin of the South China Sea and peikang high of Western Taiwan. *J. Trop. Oceanogr.* **2002**, *21*, 50–57. Available online: https://www.semanticscholar.org/paper/MESOZOIC-TRATA-AND-SEDIMENTARY-ENVIRONMENTIN-SW-OF-Di/4664810c06970041a510369dd5e303215fdab5be (accessed on 29 March 2024). (In Chinese with English abstract).
65. He, J.; Xia, B.; Wang, Z.; Sun, D. Petroleum geologic characteristics and exploration base of taixinan basin in eastern area of continental shelf in northern of the South China Sea. *Nat. Gas. Geosci.* **2006**, *17*, 345–350. Available online: https://www.semanticscholar.org/paper/PETROLEUM-GEOLOGICCHARACTERISTICS-AND-EXPLORATION-Jia-xiong/%20c1b68568f6cdd9a8c19c1eded725c54b84ed2597 (accessed on 20 January 2024). (In Chinese with English abstract).
66. Shao, L.; You, H.; Hao, H.; Wu, G.; Qiao, P.; Lei, Y. Petrology and depositional environments of Mesozoic strata in the northeastern South China Sea. *Geol. Rev.* **2007**, *53*, 164–169. Available online: https://en.cnki.com.cn/Article_en/CJFDTOTAL-DZLP200702004.htm (accessed on 22 March 2024).
67. Wu, G.; Wang, R.; Hao, H.; Shao, L. Microfossil evidence for development of marine Mesozoic in the north of South China Sea. *Mar. Geol. Quat. Geol.* **2007**, *27*, 79–85. Available online: http://jhydz.com.cn/en/article/id/080d7e39-a946-49be-b226-f2c5953be43b (accessed on 12 January 2024). (In Chinese with English abstract).
68. Zhong, G.; Wu, S.; Feng, C. Sedimentary model of Mesozoic in the northern South China Sea. *J. Trop. Oceanogr.* **2011**, *30*, 43–48. Available online: https://en.cnki.com.cn/Article_en/CJFDTOTAL-RDHY201101006.htm (accessed on 20 March 2024). (In Chinese with English abstract).
69. Zhong, G.; Feng, C.; Wang, Y.; Chen, S.; Sun, M.; Yi, H. Fault bounded models of oil–gas and gas–hydrate accumulation in the Chaoshan depression, the South China Sea. *Front. Earth Sci.* **2022**, *10*, 965898. [CrossRef]
70. Zhang, L.; Geng, A.; Wang, L.; Liao, Y.; Xu, G.; Wei, Z. Assessment of Mesozoic source rocks at the Margin of South China Continent. *Mar. Geol. Quat. Geol.* **2012**, *32*, 99–108. Available online: http://jhydz.com.cn/en/article/doi/10.3724/SP.J.1140.2012.01099 (accessed on 10 January 2024). [CrossRef]
71. Hao, H.; Lin, H.; Yang, M.; Xie, H.; Chen, J. The Mesozoic in Chaoshan depression: A new domain of petroleum exploration. *China Offshore Oil Gas* **2001**, *15*, 157–163. Available online: https://en.cnki.com.cn/Article_en/CJFDTOTAL-ZHSD200103000.htm (accessed on 15 January 2024). (In Chinese).
72. Duan, J.; Mi, H. Seismic facies and sedimentary facies study of Mesozoic in Chaoshan sag. *Resour. Ind.* **2012**, *14*, 100–105. Available online: http://www.resourcesindustries.net.cn/EN/abstract/abstract2169.shtml (accessed on 19 January 2024). (In Chinese with English abstract).
73. Han, X.; Yang, K.; Huang, Y. Origin and nature of cold seep in northeastern Dongsha area, the South China Sea: Evidence from chimney-like seep carbonates. *Chin. Sci. Bull.* **2013**, *58*, 1865–1873. [CrossRef]
74. Hao, H.; Wang, R.; Zhang, X.; Xie, H.; Chen, Z. Mesozoic marine sediment identification and distribution in the eastern Pearl River Mouth Basin. *China Offshore Oil Gas* **2004**, *16*, 84–88. Available online: https://www.semanticscholar.org/paper/Mesozoic-marine-sediment-identification-and-in-the-Petroleum/59300365ddad9383f3ac62aa620adfe674c94d52 (accessed on 27 January 2024).
75. Hao, H.; Shi, H.; Zhang, X.; Wang, T.; Tang, S. Mesozoic sediments and their petroleum geology conditions in Chaoshan sag: A discussion based on drilling results from the exploratory well LF35-1-1. *China Offshore Oil Gas* **2009**, *21*, 151–156. Available online: https://en.cnki.com.cn/Article_en/CJFDTOTAL-ZHSD200903001.htm (accessed on 20 February 2024).
76. He, J.; Zhong, C.; Yao, Y.; Yan, P.; Wang, Y.; Wan, Z.; Guan, J.; Zhang, J. The exploration and production test of gas hydrate and its research progress and exploration prospect in the northern South China Sea. *Mar. Geol. Front.* **2020**, *36*, 1–14.
77. Ji, Z.; Zhao, H.; Wang, H.; Li, C. Mesozoic petroleum system in Chaoshan depression. *Pet. Geol. Eng.* **2014**, *28*, 9–15. Available online: http://www.sydzygc.cn/article/id/f8de6c5b-37a5-4db6-8eda-58a9c202bf29 (accessed on 20 April 2024). (In Chinese with English abstract).
78. Liang, J.; Zhang, G.; Lu, J.; Su, P.; Sha, Z.; Gong, Y.; Su, X. Accumulation characteristics and genetic models of natural gas hydrate reservoirs in the NE slope of the South China Sea. *Nat. Gas Ind.* **2016**, *36*, 157–162.
79. Sha, Z.; Xu, Z.; Fu, S.; Liang, J.; Zhang, W.; Su, P.; Lu, H.; Lu, J. Gas sources and its implications for hydrate accumulation in the eastern Pearl River Mouth Basin. *Mar. Geol. Quat. Geol.* **2019**, *39*, 116–125.

80. Su, P.; Liang, J.; Fu, S.; Lv, W.G.; Gong, Y. Geological background and accumulation models of gas hydrate reservoir in northern South China Sea. *Geol. China* **2017**, *44*, 415–427.
81. Su, P.; Sha, Z.; Chang, S.; Liang, J.; Fu, S. Geological models of gas hydrate formation in the eastern sea area of the Pearl River Mouth Basin. *Nat. Gas Ind.* **2014**, *34*, 162–168.
82. Teng, T.; Su, X.; Liu, H.; Cui, H.; Chen, F.; Cheng, S.; Yang, S.; Wang, H.; Liang, J.; Su, P. Archaeal diversity in sediments of core 973-5 from deep-sea cold seep, Dongsha area in South China Sea. *Geoscience* **2020**, *34*, 104–116.
83. Wang, P.; Xia, K.; Huang, C. The Mesozoic Marine sediment distribution and Geology-Geophysics characteristic at the North-Eastern of South China Continent. *J. Trop. Oceanogr.* **2020**, *19*, 28–35. Available online: https://kns.cnki.net/kcms/detail/detail.aspx?dbcode=CJFD&dbname=CJFD2000&filename=RDHY200004004&uniplatform=NZKPT&v=SSOOogLH9I4fPQOJAHkYL7YYdSD0XwrleHteIOcywdqBscyno6lCU8nVfB-vZ_PQ (accessed on 22 April 2024).
84. Yang, S.; Tong, Z.; He, Q.; Hao, J. Mesozoic hydrocarbon generation history and igneous intrusion impacts in Chaoshan depression; South China Sea: A case of LF35-1-1 well. *China Offshore Oil Gas* **2008**, *20*, 152–156. Available online: https://en.cnki.com.cn/Article_en/CJFDTOTAL-ZHSD200803002.htm (accessed on 27 April 2024).
85. Zhang, J.; Sun, Z.; Zhang, S. Analysis of Mesozoic tectonic deformation in the Chaoshan depression of Pearl River Mouth Basin. *J. Trop. Oceanogr.* **2014**, *33*, 41–49. Available online: http://www.jto.ac.cn/EN/abstract/abstract578.shtml (accessed on 29 April 2024).
86. Zhang, Q.; Zhang, H.; Jiang, X.; Xu, Y.; Zhou, H. Application of elastic parameters inversion and attribute fusion technology in the "sweet spot" prediction. *Nat. Gas Geosci.* **2017**, *28*, 582–588. Available online: http://www.nggs.ac.cn/EN/abstract/abstract4347.shtml (accessed on 10 May 2024). (In Chinese with English abstract).
87. Zhang, W.; Liang, J.; Lu, J.; Wei, J.; Su, P.; Fang, Y.; Guo, Y.; Yang, S.; Zhang, G. Accumulation features and mechanisms of high saturation natural gas hydrate in Shenhu area, northern South China Sea. *Pet. Explor. Dev.* **2017**, *44*, 670–680. [CrossRef]
88. Zhao, Z.; Zhong, G.; Sun, M.; Feng, C.; Tu, G.; Yi, H. Hydrocarbon Accumulation Analysis Based on Quasi-3D Seismic Data in the Turbulent Area of the Northern South China Sea. *J. Mar. Sci. Eng.* **2023**, *11*, 338. [CrossRef]
89. Zhong, G.; Su, P.; Feng, C.; Chen, S.; Sun, M.; Yi, H.; Wang, Y.; Yu, J.; Zhao, J.; Zhao, Z. Mesozoic hydrocarbon accumulation model in the Northern South China Sea. *IOP Conf. Ser. Earth Environ. Sci.* **2022**, 1087. [CrossRef]
90. Dai, Q. Analysis of Magnetic Field of China Seas and Adjacent Regions. *Mar. Geol. Quat. Geol.* **1997**, *17*, 63–72.
91. Xia, K.; Huanq, C.; Huanq, Z. Upper Triassic-Cretaceous sediment distribution and hydrocarbon potential in South China Sea and its adjacent areas. *China Offshore Oil Gas* **2004**, *16*, 73–83.
92. Wu, S.; Liu, W. Tectonics of subduction zone in the East Asia continental margin. *Earth Sci. Front.* **2004**, *11*, 15–22.
93. Zhong, G.; Wu, N.; Lin, Z.; Yao, Y.; Yi, H. Characteristics of faults on the northeastern continental slope of the South China Sea and their controls on basin evolution. *Geol. China* **2008**, *35*, 456–462.
94. Lin, J.; Xu, M.; Zhou, Z. Ocean drilling investigation of the global subduction processes. *Adv. Earth Sci.* **2017**, *32*, 1253–1266, (In Chinese with English abstract).
95. Zhao, M.; Wang, Q.; Yang, F. Seismic survey in Huatung basin and its tectonic significance. *Earth Sci.* **2021**, *46*, 359–368, (In Chinese with English abstract).
96. Zhang, J.; Qiu, X.; Zhao, M. Abnormal data retrieval of three-dimensional OBS survey at the Bashi Channel area of the South China Sea. *Chin. J. Geophys.* **2018**, *61*, 1529–1538, (In Chinese with English abstract).
97. Qiu, y.; Liu, C.; Peng, G.; Huang, L.; Liang, C.; Li, H.; Wu, Z.; Yang, L. Structural characteristics and development model of fluid diapirs within the structural transition zone, northern South China Sea. *Geol. J.* **2024**, *59*, 1906–1923.
98. Feng, C.; Zhong, G.; Sun, M. Analysis of thermogenic gas sources of natural gas hydrates in the Dongsha waters. *Front. Earth Sci.* **2022**, *1*, 10. Available online: https://www.frontiersin.org/articles/10.3389/feart.2022.873615/full (accessed on 17 May 2024).
99. Farahani, M.V.; Hassanpouryouzband, A.; Yang, J.; Tohidi, B. Development of a coupled geophysical–geothermal scheme for quantification of hydrates in gas hydrate-bearing permafrost sediments. *Phys. Chem. Chem. Phys.* **2021**, *23*, 24249–24264. [CrossRef]
100. Hassanpouryouzband, A.; Joonaki, E.; Farahani, M.V.; Takeya, S.; Ruppel, C.; Yang, J.; English, N.J.; Schicks, J.M.; Edlmann, K.; Mehrabian, H. Gas hydrates in sustainable chemistry. *Chem. Soc. Rev.* **2020**, *49*, 5225–5309.
101. Kvamme, B.; Zhao, J.Z.; Wei, N.; Sun, W.T.; Saeidi, N.; Pei, J.; Kuznetsova, T. Hydrate production philosophy and thermodynamic calculations. *Energies* **2020**, *13*, 672. [CrossRef]
102. Kvamme, B.; Vasilev, A. Danube Fan and Nyegga-the largest contrast European gas hydrate deposits for CO_2 storing and CH_4 and H_2 production. *Int. J. Greenh. Gas Control* **2023**, *130*, 104014. [CrossRef]

Journal of
Marine Science and Engineering

MDPI

Article

Shallow Gas Distribution Influenced by the Interface of Sedimentary Facies in the Southwest of the Qiongdongnan Basin

Taotao Yang [1], Xiaohan Li [1], Jiapeng Jin [2,*], Jianwei Chen [1], Zhi Gong [3], Li Zhao [4], Wenlong Wang [5], Bo Liu [6], Jinzi Hu [3], Wenlu Wang [3] and Xiujuan Wang [3,*]

1. China National Petroleum Corporation Limited Southern Petroleum Exploration and Development Co., Ltd., Haikou 570206, China
2. Laoshan Laboratory, Qingdao 266236, China
3. Frontiers Science Center for Deep Ocean Multispheres and Earth System, Key Lab of Submarine Geosciences and Prospecting Techniques, MOE and College of Marine Geosciences, Ocean University of China, Qingdao 266100, China
4. Research Institute of Unconventional Oil & Gas and Renewable Energy (RIUE), China University of Petroleum (East China), Qingdao 266580, China; 15650148317@163.com
5. Yantai Center of Coastal Zone Geological Survey, China Geological Survey, Yantai 264000, China; wenlongwang2020@163.com
6. School of Earth Science and Engineering, Hebei University of Engineering, Handan 056038, China

* Correspondence: jpjin@qnlm.ac (J.J.); wangxiujuan@ouc.edu.cn (X.W.)

Abstract: Shallow gas, with huge resources, has been confirmed using three dimensional (3D) seismic data and more than 20 drilling sites in the deep water of the LS36 gas field, the Qiongdongnan Basin, the South China Sea. The interface of sedimentary facies in the southern boundary of the basin controls the distribution within the basin of clastic sediments coming from the north and west of the land uplifted. In this study, seismic data and geophysical attributes were used to investigate the controlling effect of the interface of sedimentary facies on the distribution of shallow gas within the basin. Our study shows that the shallow gas is mainly distributed in the Quaternary Ledong Formation in the southwest of the Qiongdongnan Basin, which was observed from acoustic impedance, amplitude versus offset (AVO), and seismic interpretations. The channelized submarine fans that onlap the interface of the sedimentary facies are distributed in a vertically stacked manner and are the main reservoirs for the shallow gas. Therefore, these sedimentary studies show that the sand-rich sediments are distributed along the interface of the sedimentary facies from the southwest to the northeast and are limited to the shallow gas within the basin. The Central Canyon provides an important deep gas source, while the flank of the canyon, gas chimney, and normal faults related to basement uplift provide pathways for vertical and lateral gas migration to form the shallow gas. This study shows that shallow gas may be widely distributed in other marginal sea basins, and sedimentary systems should be further studied in the future.

Keywords: shallow gas; interface of sedimentary facies; AVO; submarine fan; Qiongdongnan Basin

1. Introduction

Deep-water shallow gas generally refers to the natural gas that is present in water depths greater than the shelf-break zone (200–300 m) and accumulates in reservoirs in unconsolidated shallow strata (within 500 or 800 m below the seafloor) [1,2]. It is characterized by shallow burial, a variety of gas sources, diverse reservoir–caprock combinations, and a wide distribution. Due to exploitation issues such as high costs and low resource abundance, previous studies have regarded shallow gas as an engineering and geological hazard and that oil/gas drilling needs to avoid it [3–5]. In 2024, the China National Offshore Oil Corporation (CNOOC) discovered that a shallow field with a reserve of tens of billions of cubic meters, named as the LS36-1 gas field, was confirmed in the Lingshui Sag, the Qiongdongnan Basin, the South China Sea [6–8]. Shallow gas in deep-water areas has higher pressure and lower temperature; thus, it has high resource abundance and great potential. Previous studies have shown that shallow gas is present in the near seabed strata, is associated with gas hydrates, and is prone to leakage on the seabed [9–14]. It is closely related to engineering and scientific research such as that on seabed geological disasters, ocean acidification, and gas hydrate exploitation.

Research on deep-water shallow gas in oceans is relatively rare. Offshore gas fields with a burial depth of less than 500 m are mainly distributed in areas such as the Bay of Bengal [15], the North Sea [16], the Mediterranean Sea [17], offshore from Turkey [18], the Krishna–Godavari Basin in India [19], and the Japan Sea [4]. The most representative deep-water shallow gas fields are the Dhirubhai 22 and Dhirubhai 29 gas fields in the Bay of Bengal, and the water depth ranges from 1400 to 1832 m. The gas reservoir is located at 268 and 300 m below the seafloor, the strata age spans from the Mesozoic to the Cenozoic period, and the trap types are structural traps and lithologic–structural traps [20]. In comparison, the reservoir of the LS36-1 gas field in the Qiongdongnan Basin is younger, belonging to Quaternary strata. Analyzing its reservoir and fluid migration system can provide theoretical guidance for the shallow gas exploration in international sea areas.

Numerous middle and deep gas fields have been discovered in the Qiongdongnan Basin, including the Deep-Sea No. 1 gas field in the Central Canyon and the deep BD21-1 gas field, which offer abundant gas source conditions for shallow gas accumulation [21]. The west of the Qiongdongnan Basin is near to the Vietnam Uplift and the Hainan Uplift. Multiple fluvial systems form small deltas in the shallow-water area, providing a favorable terrigenous sedimentary system for the deep-water area (Figure 1). Additionally, carbonate reefs are located in the southern uplifts of the Xisha Islands, making it a favorable area for endogenic clastic sedimentation and carbonate turbidite sedimentation. Moreover, the platform edge reefs of the Xisha Uplift may provide endogenic clastics and carbonate turbidite deposits for the southern uplift of the Qiongdongnan Basin [22]. The interface of sedimentary facies' transition has developed between the northern sag and the southern uplift of the Qiongdongnan Basin (Figure 2), controlling the differences in sedimentary distribution patterns on both sides since the Pliocene period [23]. However, systematic research on shallow gas accumulation and distribution is lacking in the southwest of the Qiongdongnan Basin.

In this study, we analyze the reservoir characteristics and the geophysical properties related to shallow gas in the Qiongdongnan Basin using two dimensional (2D) and three dimensional (3D) seismic data to show the spatial distribution of shallow gas. Through comprehensive geophysical inversion and seismic interpretation, we summarize the main geological controlling factors of the shallow gas distribution in the southwest of the Qiongdongnan Basin and provide a geological model reference for deep-water shallow gas exploration.

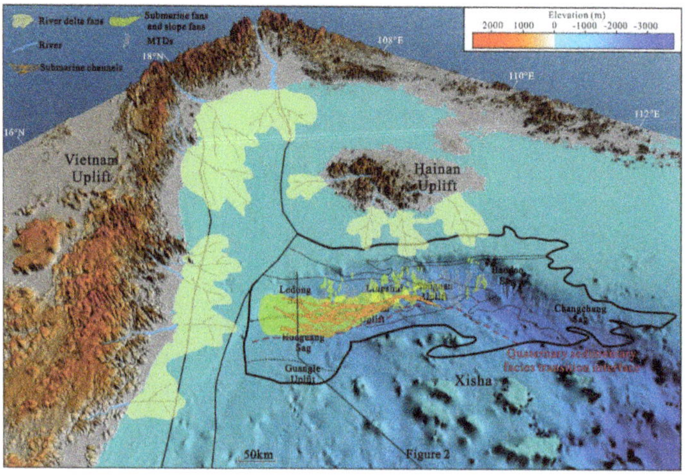

Figure 1. The current topographic map of the northwestern continental margin of the South China Sea showing the Qiongdongnan Basin and its internal tectonic regions. The Vietnam and Hainan Uplifts supply terrigenous clastics for sedimentation, and submarine fans, slope fans, and MTDs are distributed in the Qiongdongnan Basin. The channels are revised from [24] and the sand distribution in the northern slope is revised from [7].

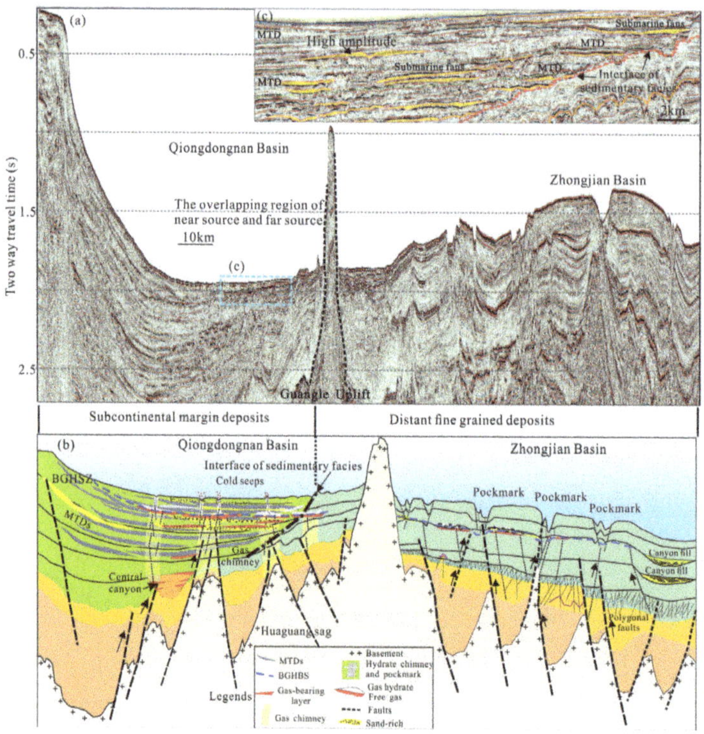

Figure 2. Seismic profiles (**a**) and interpretation profiles (**b**) across the Qiongdongnan Basin and the Zhongjian Basin showing that a large amount of clastic sediment is deposited in the northern sag of the interface of sedimentary facies transition within the Qiongdongnan Basin. (**c**) is the enlarged seismic profile, which shows a large number of submarine fan deposits in the north of the interface of sedimentary facies transition. The location of the seismic profile is shown in Figure 1.

2. Geological Setting

The South China Sea (SCS) is situated at the junction of the Eurasian, Pacific, and Indo-Australian tectonic plates, and its tectonic evolution has been shaped by processes such as rifting, subduction, collision, strike–slip movement, and mantle plume activity [25–28], resulting in a complex geological structure. Cenozoic rifting has led to the formation of a series of rift basins. The Qiongdongnan Basin is located in the northwest margin, the SCS, encompassing an area of ~80,000 km^2, with water depths ranging from 80 to 3000 m [8]. The evolution has progressed through three distinct stages: the faulting stage, the fault–depression transition stage, and the depression stage. Numerous NE-SW trending faults [29] subdivide the basin into several depressions, such as the Ledong depression and the Lingshui depression, as well as many uplifts. This geological configuration illustrates a tectonic pattern characterized by "north–south into belts and east–west into blocks" in the plane view. In profiles, the stratigraphy of the basin is delineated by breakup nonconformity (~23 Ma), leading to a structural arrangement described as "rifts in the lower part and depressions in the upper part" [30,31]. The Lingtou Formation and Yacheng Formation, which were deposited during the rifting period, host hydrocarbon source rocks associated with lacustrine and land–sea transition environments, respectively [32]. The Lingshui Formation, formed during the rifting–depression transitional stage, is characterized by deltaic and braided river sand deposits, which serve as effective hydrocarbon reservoirs within the basin. Additionally, the Sanya Formation, Meishan Formation, Huangliu Formation, Yinggehai Formation, and Ledong Formation all developed during the depression period and display gravity-flow deposits, submarine fan sands, and mudstone deposits [8]. Among these, the submarine fan sands in the Ledong Formation provide favorable reservoirs for the accumulation of shallow gas, while the mass transport deposits (MTDs) and mudstone deposits serve as favorable caprocks.

Sedimentation in the basin is distinguished by a prominent double-source system. During the rifting period, sediment sources were primarily derived from the Xisha uplift, the Indo-China Peninsula, and the South China margin, while in the depression period, sediment contributions predominantly came from the Indo-China Peninsula. In the late Quaternary period, these north–south sedimentary sources converged in the deep-water regions of the basin, resulting in a notable boundary of sedimentary facie change [23]. Since the Late Miocene period, due to paleo-topographic, the central sag area controlled the sedimentation process of clastic materials within the basin and received sediments from the uplifts in the north, west, and south [33]. In the Qiongdongnan Basin, fluid-related structures have been well discussed, including polygonal faults [34], gas chimneys [35], hydrothermal vents [36], cold seeps [37], mud volcanoes [38], and magmatism [39–41]. These structures highlight that there are abundant hydrocarbon-rich fluids within the basin, which migrate upward along the faults and gas chimneys above the structural uplifts, connecting the deep gas sources with the shallow reservoirs (Figure 2). Faults and gas chimneys serve as the primary vertical migration pathways for fluids [35,38], which contrast significantly with the fluid migration system observed in the Zhongjian Basin (Figure 2). In the Zhongjian Basin, the vertical migration system is primarily governed by a combination of basement faults and polygonal faults [42], with gas and gas hydrates chiefly stored in marginal platform fans or fine-grained sediments [43]. Fluid migrates to the seafloor, leading to the formation of pockmarks on the seafloor [43,44]. Therefore, there is a quite different sedimentary environment for the Qiongdongnan Basin and Zhongjian Basin, which influence the shallow gas, gas hydrate, and deep gas distribution (Figure 2).

3. Data and Method

Two-dimensional seismic data acquired in 2005 were reprocessed by the China National Petroleum Corporation (CNPC) in 2020 and 2023 to study gas hydrates and shallow gas in the southwest of the Qiongdongnan Basin. The reprocessing of the seismic data included improving the signal-to-noise ratio, preserving low frequencies, and broadening frequency bandwidths. The stack velocity obtained during reprocessing was used to invert acoustic impedance based on constrained sparse spike inversion (CSSI) due to the limitation of a well log in the study area. CSSI can effectively identify and quantitatively evaluate gas hydrate-bearing sediments and underlying free gas-bearing sediments, which has been applied in previous studies [7,18,45,46]. The wavelet is extracted from seismic data in shallow sediments. The low-frequency velocity model used in CSSI has been combined with stack velocity from seismic processing and the velocity trend at Site LS36-1c [8]. The inversion method has been widely used to delineate gas hydrate layers [45–47].

For gas hydrate- and free gas-bearing layers showing different responses, P-wave and S-wave velocities show obvious increases for gas hydrate-bearing layers, while P-wave velocity decreases greatly and S-wave velocity shows a slight increase when free gas is present in the sediment. Moreover, gas-bearing layers also show absorption attenuation anomalies for high frequencies. The amplitude versus offset (AVO) shows different changes which can be used to identify hydrocarbon gas. The velocity field from the seismic data processing was introduced to obtain more realistic and objective AVO intercept and gradient attributes, minimizing the impact during the conversion process from the offset to the incident angle. The AVO attributes for the intercept (I) and gradient (G) were obtained to show shallow gas and gas hydrates [48]. The intercept profile reflects the intensity of P-wave reflection at a normal incidence, and the gradient profile reflects the rate of change in the reflection amplitude with the incident angle. For gas hydrate layers, the AVO of BSR shows a Class III type anomaly [48]. Shallow gas, faults, channels, and polygonal faults were identified from the coherence, root mean square (RMS) amplitude, and instantaneous frequency attributes. The shallow gas had a high amplitude, was laterally discontinuous, and had enhanced reflections observed from seismic amplitude anomalies. Therefore, the attributes of the seismic data, AVO, and inversion were combined to delineate shallow gas in the southwest of the Qiongdongnan Basin.

4. Results

4.1. Shallow Gas Identification and Distribution

Drilling for gas hydrates and shallow gas has been carried out in the LS36-1 gas field [7,8,24,35,49]. The results show that in the well logs, the shallow gas layer is characterized by high resistivity, a significantly decreased P-wave velocity, and a slightly increased S-wave velocity. In the seismic profiles, it corresponds to a bright spot or enhanced reflections with reverse-polarity reflections compared to the seafloor reflection [7,8,24]. Therefore, shallow gas could be identified based on 2D and 3D seismic profiles and the anomalies of acoustic impedance, frequency, and AVO. The arbitrary seismic profile from LS36-1 in the south part of the Qiongdongnan Basin is used to show the shallow gas properties (Figure 3). Numerous enhanced reflections are widely distributed in the channel–levee system or submarine fan area, which are nearly parallel. There are multiple MTDs showing weak, chaotic, and discontinuous seismic reflections in the seismic profiles, which erode the lower sediments and form different-sized grooves (Figure 3). Many small channels are also formed which also erode the lower formation and are filled with clay-rich sediments. The high amplitude reflections are partially related to the base of the MTDs.

The shallow gas is interpreted from the seismic anomalies showing the fan-shaped distribution (Figure 4). The shallow gas is mainly distributed in Ledong Formation with three layers. The shallow gas identified from the seismic data in the southwest is mainly distributed in the prosperous and degenerating stages of the submarine fans, as can be seen with the LS36-1 shallow gas. Most of the shallow gas is located in the Ledong Sag and Huaguang Sag. In the HaiMa cold seeps, shallow gas is also widely distributed. The distribution of the shallow gas shows two directions and then converges to LS36-1.

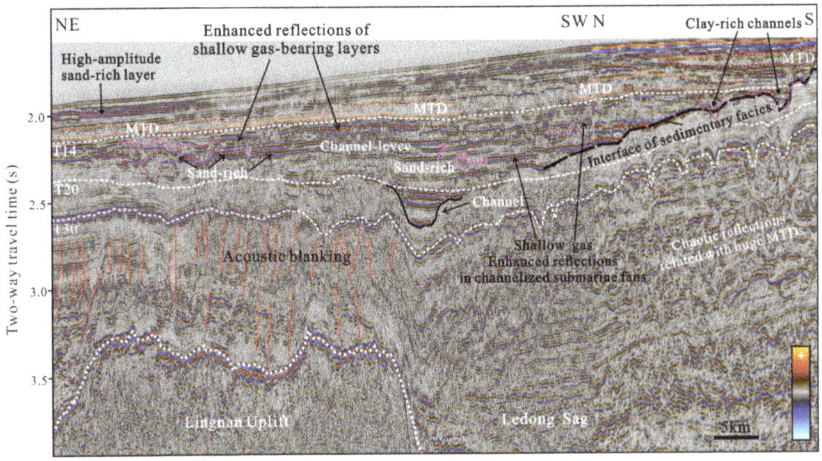

Figure 3. Seismic profiles showing the sedimentary facies' transition interface and internal shallow gas, submarine fans, channels, and MTDs. The location of the seismic profile is shown in Figure 4.

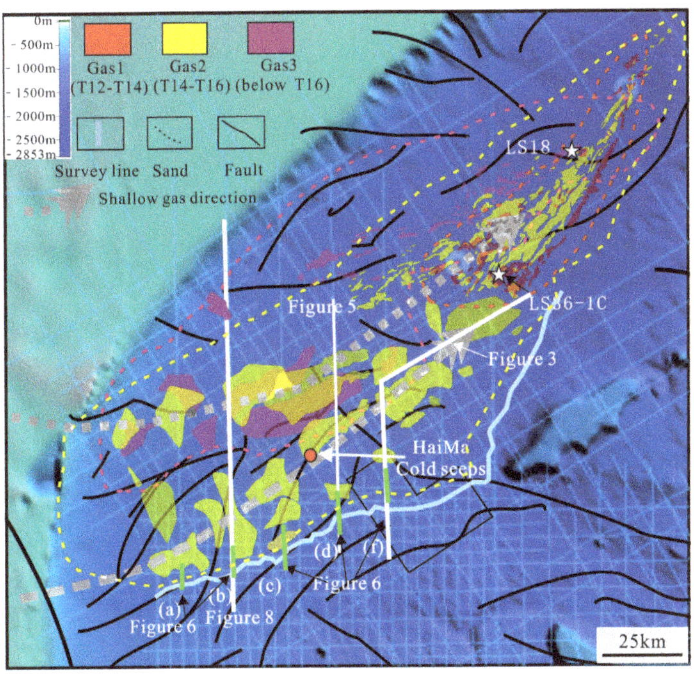

Figure 4. Shallow gas distributions in three layers and a sand-rich reservoir identified from 2D/3D seismic data showing the submarine fan-like distribution. The shallow gas in the LS 36-1 is revised from [7,8]. The thick white line represents the location of the whole seismic profiles, such as Figures 3, 5, and 8, while the green line indicates the scope of the seismic profiles used in Figure 7.

4.2. Geophysical Properties of Shallow Gas

For the sand-rich gas-bearing layers, the measured P-wave velocity from logging while drilling quickly decreases, with a value as low as 800 m/s, and the density also decreases, which forms a negative reflection coefficient interface and enhanced reflection in the seismic profiles [8]. For the gas hydrate-bearing layers, the P-wave velocity is about 2200 m/s and the seismic profile shows a positive reflection. Therefore, the characteristics of seismic reflection can be used to judge shallow gas, gas hydrate, and water-saturated sediments. The data from logging while drilling at Site LS36-1 show a low P-wave velocity, a higher S-wave velocity, and low P-wave-to-S-wave velocity ratios for the gas-bearing layers, and the AVO shows Class III–IV anomalies. Shallow gas reservoirs belong to unconsolidated high-porosity strata, and the top boundary of the reservoir exhibits typical Class III AVO anomalies. AVO is attributed with a negative intercept and a negative gradient. When the shallow gas reservoir has a gas hydrate cap, the velocity is relatively high, and the top boundary of the reservoir sometimes shows typical Class IV AVO anomalies, which are similar to the negative intercepts of Class III AVO anomalies.

The acoustic impedance profiles were inverted using the reprocessed seismic data based on CSSI (Figure 5) to identify shallow gas. The seismic profile shows an obvious change from negative to positive reflection for the same layer (Figure 5b), while the instantaneous frequency from the seismic profile also shows low frequency for the same layer where pull-down reflections occur (Figure 5c). The inverted P-wave velocities clearly show high and low values at interfaces which are obviously higher than those of the bases of mass transport deposits (Figure 5d). Gas chimneys showing chaotic and pull-down reflections with local enhanced reflections are observed below this anomalous zone, which indicate fluid migration along the gas chimneys (Figure 5a).

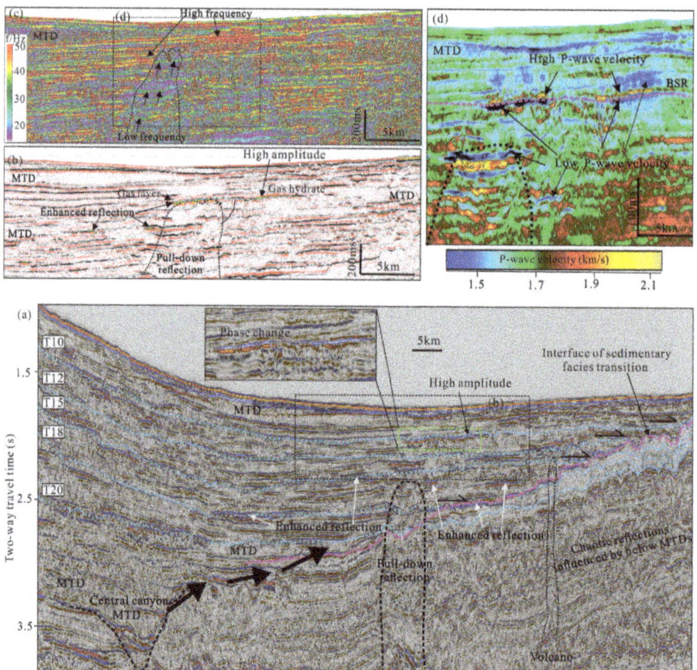

Figure 5. (**a**) Seismic profiles showing enhanced reflections related to submarine fan systems and shallow gas. (**b**) An enlarged seismic profile showing positive and negative amplitude anomalies and

pull-down reflections. (**c**) The instantaneous frequency profile shows low frequency for the gas-bearing layers and high frequency for the gas hydrate-bearing layers. (**d**) The inverted P-wave velocity from constrained sparse spike inversion (CSSI) showing the anomalies. The location of the seismic profile is shown in Figure 4.

The seismic anomalies and acoustic impedance show the occurrences of shallow gas and gas hydrate (Figure 5). However, the shallow gas with enhanced reflections shows multiple layers which are terminated at the sedimentary interface boundary. To confirm and delineate the anomalies, the intercept profile (P), gradient profile (G), and product of the intercept and gradient profile (P × G) of the AVO were extracted from the gathered data to show the reservoir properties and pore fluids (Figure 6). The P profile is near to the zero-offset profile showing the normal incidence reflection coefficient which can be used to identify hydrocarbon gas. The G profile can show the amplitude variation with the offset. For the gas hydrate-bearing layers near the BSR, the P profile shows a positive and strong value (Figure 6 upper), while the G and P × G profiles show negative values (Figure 6 lower). For shallow gas in free-gas layers, the P × G profile is sensitive to fluids showing strong and positive values in the deeper layers (Figure 6 lower).

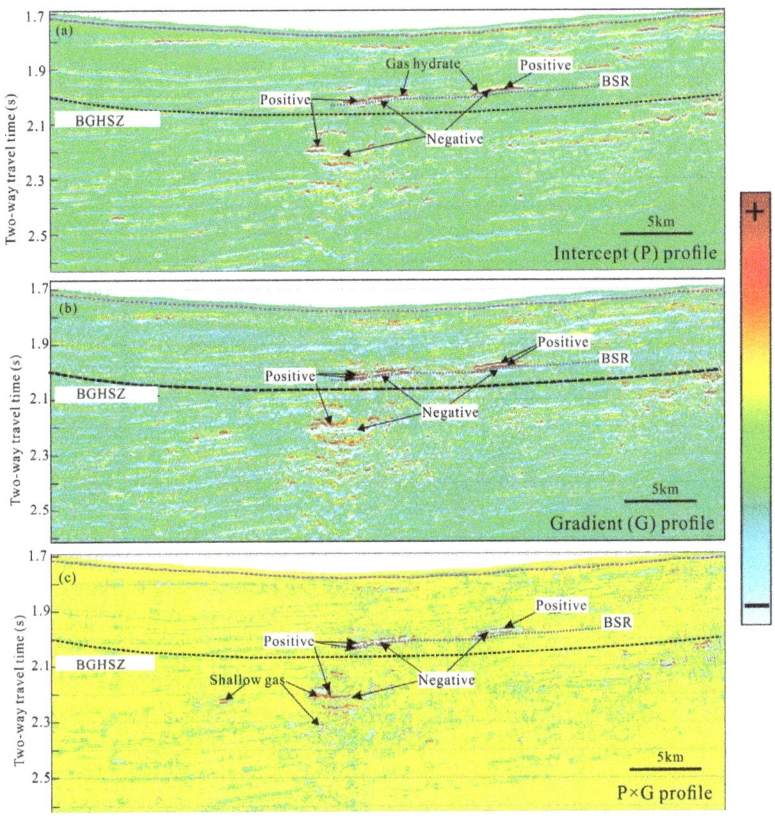

Figure 6. AVO attribute profiles show intercept profile (P), gradient profile (G), and product of intercept and gradient profile (P × G) for seismic profile of Figure 5b.

4.3. Sedimentary Facies' Identification

In the southern Qiongdongnan Basin, there is an interface of sedimentary facies' transition trending nearly east–west with a dip angle of ~1.1°. This interface initiates along the southern sidewall of the Central Canyon and coincides with the T30 interface, with an initial development age of ~5.5 Ma (Figures 2 and 3). It overlays the southern uplift, separating the northern sag from the southern Guangle uplift. The interface of the sedimentary facies' transition exhibits diachronous characteristics, extending from the T30 interface near the seabed and separating the Ledong Formation and Yinggehai Formation (Figure 3).

The seismic profile interpretations and the lateral variations in the stratigraphic thickness indicate that, due to its higher topography and greater distance from the sediment sources, the southern uplift mainly receives fine-grained terrestrial clastic and carbonate sedimentations, which are characterized by low deposition rates, near-parallel and continuous seismic facies, and a strong vertical inheritance of topography. In contrast, the northern sag receives a significant amount of terrestrial clastic materials, with depositional systems dominated by Ledong Formation submarine fans sourced from the western uplift, slope fans, and MTDs sourced from the northern shelf uplift, exhibiting high deposition rates (Figure 1) [7]. The submarine fan exhibits lenticular and strong amplitude reflections. The channel is marked by an erosional surface that cuts into the strata, and its interior can be filled with either strong-amplitude or weak-amplitude reflections. The MTDs develop distinct strong-amplitude top and bottom interfaces, with blanking and chaotic characteristics inside. Multiple well-connected muddy channels are visible above the submarine fans. MTDs and submarine fans are usually interbedded deposits. The interiors of channels can be filled with sandstone or mudstone sediments. Locally, the sedimentary facies' transition interface also shows the occurrence of channels (Figure 7). Within the submarine fans, there are negative-polarity and strong-amplitude reflections, which are locally segmented by channels, which are the identifications of shallow gas.

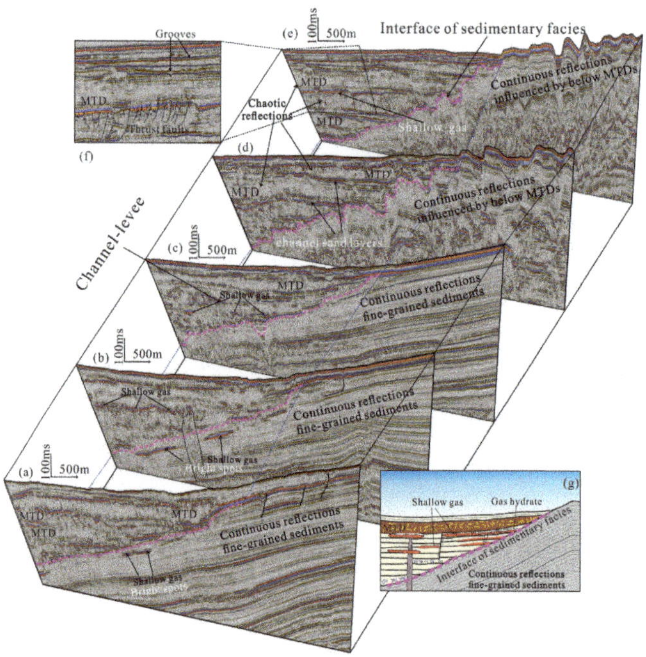

Figure 7. Continuous seismic profiles showing the spatial distribution of the sedimentary facies' transition interface, which separates the continuous fine-grained sediments in the south from the coarse-grained sediments in the north of the Qiongdongnan Basin. The seismic profiles from (**a**–**g**) are distributed from west to east, and the location of the seismic profiles are shown in Figure 4.

5. Discussion

5.1. The Formation of the Sedimentary Facies' Transition Interface

From the perspective of macroscopic "source–pathway–sink" analysis, the river systems in the Vietnam and Hainan Uplifts form delta fan systems in the shallow-water continental shelf (Figure 1). When the sea level drops and the sediment supply increases, terrigenous clastics enter the continental slope and the deep-water sag area through the shelf break [50]. Since the Pliocene period, shelf breaks have emerged in the northern and western area of the Qiongdongnan Basin, marking the onset of a stage of rapid subsidence [51,52]. With an abundant sediment source in the north and west, the continental shelf has advanced rapidly seaward. The channels that developed in the continental shelf and slope areas during the Pliocene and Miocene periods indicate an enhanced ability to transport sediments seaward (Figures 3 and 8) [24]. In this geological setting, a large number of interbedded submarine fans, slope fans, and MTDs can be observed in the north of the sedimentary facies' transition interface (Figures 3, 5, 6 and 8). Topography is a prerequisite for sedimentation, determining the location of the sedimentation center and the distribution and morphology of sedimentary systems [53]. Due to the relatively high paleo-topography in the southern uplift, the sag area is continuously filled with sediments, and the submarine fans are distributed along the sedimentary facies' conversion interface and gradually converge from west to east in plain view (Figures 3 and 8). Apart from submarine fans, MTDs and bathyal sediments are also preferentially deposited in the sag, onlapping onto the southern uplift area, thus forming the sedimentary facies' transition interface. Therefore, the sedimentary facies' transition interface is caused by the differential deposition of sediments in the northern sagging subsidence area and the southern uplift area under the control of topography.

5.2. Influence on Shallow Gas and Gas Hydrate Distribution

The geochemical indicators of the cores from the shallow gas drilling Sites 01 and 03 in the Qiongdongnan Basin indicate that the gas source of the shallow gas in the study area is a mixture of biogenic gas and thermogenic gas. Thermogenic gas usually originates from the continental coal source of the deep Yacheng Formation and Lingshui Formation, while biogenic gas comes from the marine source rocks of the Huangliu Formation and Yinggehai Formation [8]. The fluid migration system serves as a passage for biogenic gas and thermogenic gas from source rocks to reservoirs. The first function of the sedimentary facies' transition interface in shallow gas accumulation is to act as a fluid migration pathway, which laterally transports hydrocarbon gasses trapped within the Central Canyon, as well as those from deep sources and shallow strata into shallow sandy reservoirs (Figure 5). Faults and gas chimney structures, on the other hand, are important conditions for the vertical migration of fluids (Figures 5, 7 and 8).

The sedimentary facies' transition interface may form lithologic traps for shallow gas with reservoirs such as submarine fans. Seismic interpretations and attribute studies show that the shallow gas in the Qiongdongnan Basin is mainly enriched in coarse-grained reservoirs such as submarine fans and slope fans [7,8]. Compared to the Qiongdongnan Sea, the reservoirs of the Dhirubhai 22 and 29 gas fields in the Bay of Bengal are of an older age, belonging to the Mesozoic to the Cenozoic strata. Similarly, lithologic traps and structural traps also play important roles [20]. The coarse-grained reservoirs, due to their continuous onlap onto and pinching out at the sedimentary facies' transition interface, form a lithologic trap (Figure 8). In addition, a large number of channels develop above the sedimentary facies' transition interface, which serve as important transport pathway for sandy sediments [54]. When sandy sediments are deposited in the channels, these channels can also function as reservoirs for shallow gas and gas hydrates. When the shallow gas reservoir is located above the base of the gas hydrate stability zone, the shallow gas forms gas hydrates. Sandy gas hydrate reservoirs composed of submarine fans have been discovered in the LS36-1 gas reservoir area of the Qiongdongnan Basin [24]. However, when the channels are filled with muddy sediments, the shallow gas is separated by the channels and, together with MTDs, the muddy filled channels act as a caprock for the shallow gas (Figure 8).

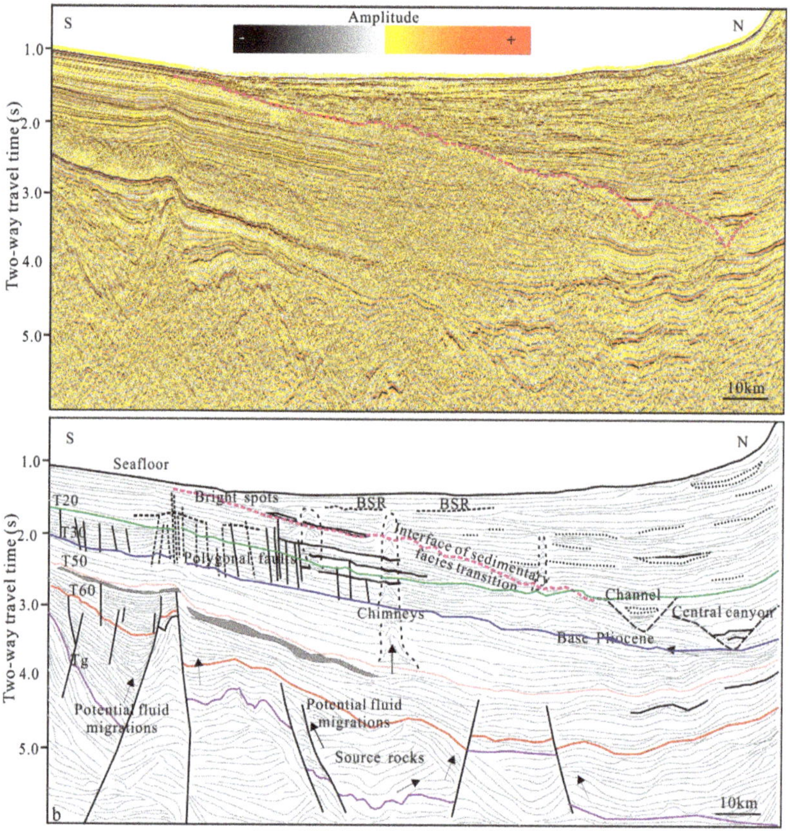

Figure 8. A shallow gas accumulation model in the southwest of the Qiongdonganan Basin showing submarine fan-shaped distributions under the control of the sedimentary facies' transition interface. The location of the seismic profile is shown in Figure 4.

6. Conclusions

The interface of the sedimentary facies' transition, formed since the Pliocene period, divides the northern sag area and the southern uplift area of the Qiongdongnan Basin, which causes the sediments to change from rapid deposition to slow deposition. The seismic facies show obviously different reflections. The enhanced reflections, low acoustic impedance, low frequency, and AVO attributes indicate the widely developed shallow gas in the high-deposition zone and are significantly controlled by the interface of the sedimentary facies' transition. The difference in paleo-topography between the northern sag and the southern uplift leads to different sedimentations, thus forming an interface of sedimentary facies' transition between the two areas. During the process of shallow gas reservoir formation, the sedimentary facies' transition interface can act as a lateral migration channel for fluids, form lithologic traps together with submarine fans, and trap the shallow gas within local channels. This work emphasizes the widely distributed shallow gas along the sedimentary facies between different sources in the Qiongdongnan Basin. There has been relatively little research on shallow gas in international oceans. However, the discovery of the shallow gas field LS36-1 in the Qiongdongnan area indicates that it may be widespread in international marine waters. Nevertheless, it has not attracted the attention of the industrial community yet, which highlights the significance of this study.

Author Contributions: All authors contributed equally. All authors have read and agreed to the published version of the manuscript.

Funding: This work is supported by the International Science and Technology Cooperation Program of China (2023YFE0119900), DE Research on key technologies of deep-water oil and gas exploration and evaluation (2024ZZ56), and the exploitation mechanism and technology of natural gas hydrates (024DJ96).

Institutional Review Board Statement: Not applicable.

Informed Consent Statement: Not applicable.

Data Availability Statement: The data presented in this study are available on request from the corresponding author due to (specify the reason for the restriction).

Conflicts of Interest: Authors Taotao Yang, Xiaohan Li and Jianwei Chen were employed by the company China National Petroleum Corporation Limited Southern, Petroleum Exploration and Development Co., Ltd. The remaining authors declare that the research was conducted in the absence of any commercial or financial relationships that could be construed as a potential conflict of interest.

References

1. Davis, A.-M. Shallow gas: An overview. *Cont. Shelf Res.* **1992**, *12*, 1077–1079. [CrossRef]
2. Fleischer, P.; Orsi, T.-H.; Richardson, M.-D.; Anderson, A.-L. Distribution of free gas in marine sediments: A global overview. *Geo Mar. Lett.* **2001**, *21*, 103–122.
3. Fonnesua, M.; Palermo, D.; Galbiati, M.; Marchesini, M.; Bonaminia, E.; Bendias, D. A new world-class deep-water play-type, deposited by the syndepositional interaction of turbidity flows and bottom currents: The giant Eocene Coral Field in northern Mozambique. *Mar. Pet. Geol.* **2020**, *111*, 179–201. [CrossRef]
4. Horozal, S.; Chae, S.J.; Kim, D.-H.; Seo, J.-M.; Lee, S.-M.; Han, H.-S.; Cukur, D.; Kong, G.-S. Seismic evidence of shallow gas in sediments on the southeastern continental shelf of Korea, East Sea (Japan Sea). *Mar. Pet. Geol.* **2021**, *133*, 105291. [CrossRef]
5. Hasan, M.N.; Pepper, A.; Mann, P. Basin-scale estimates of thermal stress and expelled petroleum from Mesozoic-Cenozoic potential source rocks, southern Gulf of Mexico. *Mar. Pet. Geol.* **2023**, *148*, 105995. [CrossRef]
6. Liu, K.; Cheng, P.; Tian, H.; Song, P.; Hu, W.-Y. Development model of shallow lithologic traps and natural gas accumulation mechanisms in marine deep-water areas: A case study in the Qiongdongnan basin, South China Sea. *Mar. Pet. Geol.* **2023**, *151*, 106211. [CrossRef]
7. Pei, J.; Liu, E.; Song, P.; Yan, D.; Luo, W.; Zhan, J.; Wang, Z.; Li, G.; Uysal, I.T.; Yang, P. Influencing Factors and Model of Shallow Gas Enrichment in the Quaternary Sediments of the Qiongdongnan Basin, South China Sea. *J. Mar. Sci. Eng.* **2024**, *12*, 1928. [CrossRef]
8. Xu, C.-G.; Wu, K.-Q.; Pei, J.-X.; Hu, L. Enrichment mechanisms and accumulation model of ultra-deep water and ultra-shallow gas: A case study of Lingshui 36-1 gas field in Qiongdongnan Basin, South China Sea. *Pet. Explor. Dev.* **2024**, *51*, 1–13, (In Chinese with English Abstract).
9. Schroot, B.-M.; Schüttenhelm, R.-T.-E. Shallow gas and gas seepage: Expressions on seismic and otheracoustic data from the Netherlands North Sea. *J. Geochem. Explor.* **2003**, *78*, 305–309. [CrossRef]
10. Körber, J.-H.; Sahling, H.; Pape, T.; dos Santos Ferreira, C.; MacDonald, I.; Bohrmann, G. Natural oil seepage at kobuleti ridge, eastern Black Sea. *Mar. Pet. Geol.* **2014**, *50*, 68–82. [CrossRef]
11. Egger, M.; Riedinger, N.; Mogollon, J.-M.; Jørgensen, B.-B. Global diffusive fluxes of methane in marine sediments. *Nat. Geosci.* **2018**, *11*, 421–425. [CrossRef]
12. Mazumdar, A.; Peketi, A.; Dewangan, P.; Badesab, F.; Ramprasad, T.; Ramana, M.V.; Patil, D.J.; Dayal, A. Shallow gas charged sediments off the Indian west coast: Genesis and distribution. *Mar. Geol.* **2009**, *267*, 71–85. [CrossRef]

13. Portnova, A.; Cooka, A.E.; Sawyera, D.E.; Yang, C.; Hillman, J.I.T.; Waitec, W.F. Clustered BSRs: Evidence for gas hydrate-bearing turbidite complexes in folded regions, example from the Perdido Fold Belt, northern Gulf of Mexico. *Earth Planet Sci. Lett.* **2019**, *528*, 115843. [CrossRef]
14. Choi, J.-Y.; Kang, N.-K.; Hwang, I.-G.; Lee, D.-H. Geochemical haracteristics and origins of hydrocarbon gases in the shallow gas field in the Pohang Basin, Korea. *Geosci. J.* **2022**, *26*, 349–365. [CrossRef]
15. Monteleone, V.; Marín-Moreno, H.; Bayrakci, G.; Best, A.; Shaon, F.; Hossain, M.-M.; Karim, A.-A.; Alam, M.-K. Seismic characterization and modelling of the gas hydrate system in the northern Bay of Bengal, offshore Bangladesh. *Mar. Pet. Geol.* **2022**, *141*, 105690. [CrossRef]
16. Cox, D.-R.; Huuse, M.; Newton, A.-M.-W.; Sarkar, A.-D.; Knutz, P.-C. Shallow gas and gas hydrate occurrences on the northwest Greenland shelf margin. *Mar. Geol.* **2021**, *432*, 106382. [CrossRef]
17. Lazar, M.; Schattner, U.; Reshef, M. The great escape: An intra-Messinian gas system in the eastern Mediterranean. *Geophys. Res. Lett.* **2012**, *39*, L20309. [CrossRef]
18. Cukur, D.; Krastel, S.; Tomonaga, Y.; Çağatay, M.-N.; Meydan, A.-F. The PaleoVan Science Team. Seismic evidence of shallow gas from Lake Van, eastern Turkey. *Mar. Pet. Geol.* **2013**, *48*, 341–353. [CrossRef]
19. Dubey, K.-M.; Chaubey, A.-K. Origin and distribution of shallow gas-charged sediment on the inner continental shelf of central west coast of India. *Curr. Sci.* **2019**, *116*, 1410–1417. [CrossRef]
20. Malinverno, A.; Goldberg, D.-S. Testing short-range migration of microbial methane as a hydrate formation mechanism: Results from Andaman Sea and Kumano Basin drill sites and global implications. *Earth Planet Sci. Lett.* **2015**, *422*, 105–114. [CrossRef]
21. Gan, J.; Zhang, Y.-Z.; Liang, G.; Yang, X.-B.; Li, X.; Yang, J.-H.; Guo, X.-X. On accumulation process and dynamic mechanism of natural gas in the deep water area of Central Canyon, Qiongdongnan basin. *Acta Geol. Sin.* **2018**, *92*, 2359–2367, (In Chinese with English Abstract).
22. Wu, S.-G.; Sun, Q.L.; Wu, T.-Y.; Yuan, S.; Yao, G. Polygonal fault and oil-gas accumulation in deepwater area of Qiongdongnan Basin. *Acta Pet. Sin.* **2009**, *30*, 22–27.
23. Yuan, S.Q.; Wu, S.G.; Thomas, L.; Yao, G.S.; Lv, F.L.; Cao, F.; Wang, H.R.; Li, L. Fine-grained Pleistocene deepwater turbiditechannel system on the slope of Qiongdongnan Basin, northern South China Sea. *Mar. Pet. Geol.* **2009**, *26*, 1441–1451.
24. Meng, M.-M.; Liang, J.Q.; Kuang, Z.-G.; Ren, J.F.; He, Y.; Deng, W.; Gong, Y. Distribution Characteristics of Quaternary Channel Systems and Their Controlling Factors in the Qiongdongnan Basin, South China Sea. *Front. Earth Sci.* **2022**, *10*, 902517. [CrossRef]
25. Hua, Y.; Zhao, D.; Xu, Y.-G. Azimuthal Anisotropy Tomography of the Southeast Asia Subduction System. *J. Geophys. Res. Solid Earth* **2022**, *127*, e2021JB022854. [CrossRef]
26. Lester, R.; McIntosh, K.; Van Avendonk, H.-J.-A.; Lavier, L.; Liu, C.-S.; Wang, T.-K. Crustal accretion in the Manila trench accretionary wedge at the transition from subduction to mountain-building in Taiwan. *Earth Planet. Sci. Lett.* **2013**, *375*, 430–440. [CrossRef]
27. Replumaz, A.; Lacassin, R.; Tapponnier, P.; Leloup, P.-H. Large river offsets and Plio-Quaternary dextral slip rate on the Red River fault (Yunnan, China). *J. Geophys. Res. Solid Earth* **2001**, *106*, 819–836. [CrossRef]
28. Sun, Z.; Zhou, D.; Zhong, Z.; Zeng, Z.; Wu, S. Experimental evidence for the dynamics of the formation of the Yinggehai basin, NW South China Sea. *Tectonophysics* **2003**, *372*, 41–58. [CrossRef]
29. Yang, G.; Yin, H.; Gan, J.; Wang, W.; Zhu, J.; Jia, D.; Xiong, X.; Xu, W. Explaining structural difference between the Eastern and Western Zones of the Qiongdongnan Basin, northern South China Sea: Insights from scaled physical models. *Tectonics* **2022**, *41*, e2021TC006899. [CrossRef]
30. Zhang, C.; Wang, Z.; Sun, Z.; Sun, Z.; Liu, J.; Wang, Z. Structural differences between the western and eastern Qiongdongnan Basin: Evidence of Indochina block extrusion and South China Sea seafloor spreading. *Mar. Geophys. Res.* **2013**, *34*, 309–323. [CrossRef]
31. Zhao, Z.; Sun, Z.; Liu, J.; Pérez-Gussinyé, M.; Zhuo, H. The continental extension discrepancy and anomalous subsidence pattern in the western Qiongdongnan Basin, South China Sea. *Earth Planet. Sci. Lett.* **2018**, *501*, 180–191. [CrossRef]
32. Shao, L.; Li, A.; Wu, G.; Li, Q.; Liu, C.; Qiao, P. Evolution of sedimentary environment and provenance in Qiongdongnan Basin in the northern South China Sea. *Acta Pet. Sin.* **2010**, *31*, 548–552, (In Chinese with English Abstract).
33. Lei, C.; Ren, J.-Y.; Pei, J.-X.; Lin, H.-T.; Tong, D.-J. Tectonic framework and multiple episode tectonic evolution in deepwater Qiongdongnan Basin, northern Continental Margin of South China Sea. *Earth Sci.-J. China Univ. Geosci.* **2011**, *36*, 151–162.
34. Han, J.; Leng, J.; Wang, Y. Characteristics and genesis of the polygonal fault system in southern slope of the Qiongdongnan Basin, South China Sea. *Mar. Pet. Geol.* **2016**, *70*, 163–174. [CrossRef]
35. Deng, W.; Liang, J.-Q.; Kuang, Z.-G.; Xie, Y.-F.; Yan, P. Estimation of Gas Hydrate Saturation Regarding the Hydrate Morphology in Hydrate-Bearing Sands in the Qiongdongnan Basin, South China Sea. *Pure Appl. Geophys.* **2023**, *180*, 2757–2773.
36. Geng, M.; Zhang, R.; Yang, S.; Guo, J.; Chen, Z.; Wan, Z. Focused Fluid Flow, Shallow Gas Hydrate, and Cold Seep in the Qiongdongnan Basin, Northwestern South China Sea. *Geofluids* **2021**, *2021*, 5594980. [CrossRef]

37. Wang, X.; Wu, S.; Kong, X.; Ma, B.; Li, W.; Wang, D.; Gao, J.; Chen, W. Subsurface fluid flow at an active cold seep area in the Qiongdongnan Basin, northern South China Sea. *J. Asian Earth Sci.* **2018**, *168*, 17–26. [CrossRef]
38. Zhang, W.; Liang, J.; Yang, X.; Su, P.; Wan, Z. The formation mechanism of mud diapirs and gas chimneys and their relationship with natural gas hydrates: Insights from the deep-water area of Qiongdongnan Basin, northern South China Sea. *Int. Geol. Rev.* **2018**, *62*, 789–810. [CrossRef]
39. Sun, Q.; Wang, C.; Xie, X. Sill swarms and hydrothermal vents in the Qiongdongnan Basin, northern South China Sea. *Geosystems Geoenvironment* **2022**, *1*, 100037. [CrossRef]
40. Wang, L.; Sun, Z.; Yang, J.; Sun, Z.; Zhu, J.; Zhuo, H.; Stock, J. Seismic characteristics and evolution of post-rift igneous complexes and hydrothermal vents in the Lingshui sag (Qiongdongnan basin), northwestern South China Sea. *Mar. Geol.* **2019**, *418*, 106043. [CrossRef]
41. Zhao, Y.; Tong, D.; Song, Y.; Yang, L.; Huang, C. Seismic reflection characteristics and evolution of intrusions in the Qiongdongnan Basin: Implications for the rifting of the South China Sea. *J. Earth Sci.* **2016**, *27*, 642–653. [CrossRef]
42. Yang, T.-T.; Li, L.; Lu, Y.-T.; Yu, K.-Q.; Yang, Z.-L.; Wang, X.-F.; Li, W. Gas hydrate accumulation associated with fluid escape structure in the western margin of South China Sea. *J. Oceanol. Limnol.* **2023**, *41*, 947–958. [CrossRef]
43. Zhang, G.-X.; Wang, X.-J.; Li, L.; Sun, L.-Y.; Guo, Y.-Q.; Lu, Y.-T.; Li, W.; Wang, Z.-Q.; Qian, J.; Yang, T.-T.; et al. Gas Hydrate Accumulation Related to Pockmarks and Faults in the Zhongjiannan Basin, South China Sea. *Front. Earth Sci.* **2022**, *10*, 902469. [CrossRef]
44. Wang, W.; Wang, X.; Lu, Y.; Li, S.; Jin, J.; Suo, Y.; Guo, Y.; Zhang, G.; Zhang, Z.; Zhou, J. Interaction between magmatism and polygonal faults revealed by three-dimensional seismic data in the Zhongjiannan Basin, South China Sea. *Mar. Pet. Geol.* **2024**, *163*, 106793. [CrossRef]
45. Lu, S.; McMechan, G.A. Estimation of gas hydrate and free gas saturation, concentration, and distribution from seismic data. *Geophysics* **2002**, *67*, 582–593. [CrossRef]
46. Wang, X.-J.; Lee, M.; Guo, Y.-Q.; Yang, S.-X.; Liang, J.-Q. Gas hydrate saturation from acoustic impedance and resistivity logs in the Shenhu area, South China Sea. *Mar. Pet. Geol.* **2011**, *28*, 1625–1633. [CrossRef]
47. Wang, X.-J.; Liu, B.; Jin, J.-P.; Lu, J.-A.; Zhou, J.-L.; Qian, J.; Wu, N.-Y. Increasing the accuracy of estimated porosity and saturation for gas hydrate reservoir by integrating geostatistical inversion and lithofacies constraints. *Mar. Pet. Geol.* **2020**, *115*, 104298. [CrossRef]
48. Wang, X.-C.; Pan, D.-Y. Application of AVO attribute inversion technology to gas hydrate identification in the Shenhu Area, South China Sea. *Mar. Pet. Geol.* **2017**, *80*, 23–31. [CrossRef]
49. Yu, H.; Wang, J.; Deng, W.; Kuang, Z.; Li, T.; Lei, Z. High-Resolution 3D Geological Modeling of Three-Phase Zone Coexisting Hydrate, Gas, and Brine. *J. Mar. Sci. Eng.* **2024**, *12*, 2171. [CrossRef]
50. Hao, Y.; Chen, P.; Wan, X. Late Tertiary Sequence Stratigraphy and Sea Level Changes in Yinggehai-Qiongdongnan Basin. *Geoscience* **2000**, *14*, 237–245.
51. Xie, X.-N.; Müller, R.-D.; Li, S.; Gong, Z.; Steinberger, B. Origin of Anomalous Subsidence along the Northern South China Sea Margin and its Relationship to Dynamic Topography. *Mar. Pet. Geol.* **2006**, *23*, 745–765. [CrossRef]
52. Yan, H.; Tian, X.; Xu, F.; Hu, B.; Liu, Z. Sediment Provenance of Offshore Mud Area of the Eastern Hainan Island in South China Seas in the Mid-Holocene. *Haiyang Xuebao* **2016**, *38*, 97–106, (In Chinese with English Abstract).
53. Armitage, D.A. High-Resolution Architectural Evolution of Depositional Elements in Deep-Marine Slope Environments: The Quaternary Niger Delta Slope, Quaternary Southwest Grand Banks Slope, Canada, and Cretaceous Tres Pasos Formation. Ph.D. Dissertation, Stanford University, Santiago, Chile, 2009.
54. Abreu, V.; Sullivan, M.; Pirmez, C.; Mohrig, D. Lateral Accretion Packages (LAPs): An Important Reservoir Element in Deep Water Sinuous Channels. *Mar. Pet. Geol.* **2003**, *20*, 631–648. [CrossRef]

Journal of
Marine Science and Engineering

MDPI

Article

High-Resolution 3D Geological Modeling of Three-Phase Zone Coexisting Hydrate, Gas, and Brine

Han Yu [1,*], Ju Wang [2], Wei Deng [1], Zenggui Kuang [1], Tingwei Li [1] and Zhangshu Lei [3]

[1] Guangzhou Marine Geological Survey, China Geological Survey, Guangzhou 510700, China; hahens@163.com (W.D.); kzg21001@163.com (Z.K.); litingwei2020@163.com (T.L.)
[2] School of Computer Science, Yangtze University, Jingzhou 434000, China; superjonwj@126.com
[3] School of Petroleum Engineering, China University of Petroleum (East China), Qingdao 266580, China; bz24010004@s.upc.edu.cn
* Correspondence: gmgs_yh4966@163.com

Abstract: Three-dimensional geological modeling is essential for simulating natural gas hydrate (NGH) productivity and formulating development strategies. Current approaches primarily concentrate on the single-phase modeling of either hydrate or free gas layers. However, an increasing number of instances suggest that the three-phase coexistence zone, which includes hydrate, gas, and water, is common and has become a focal point of international research, as this type of reservoir may present the most viable opportunities for exploitation. At present, there exists a significant gap in the research regarding modeling techniques for such reservoirs. This study undertakes a comprehensive modeling investigation of the three-phase zone reservoir situated in the sand layer of the Qiongdongnan Basin. By employing deterministic complex geological modeling techniques and integrating existing seismic and logging data, we have developed a three-phase coexistence zone model that precisely characterizes the interactions between geological structures and utilizes them as auxiliary constraints. This approach effectively mitigates the potential impact of complex geological conditions on model accuracy. Through a comprehensive analysis of 105 seismic profiles, we enhanced the model's accuracy, resulting in the creation of a three-phase coexistence zone model comprising 350,000 grids. A comparison between the modeling results and well data indicates a relatively small error margin, offering valuable insights for future development efforts. Furthermore, this method serves as a reference for modeling hydrates in marine environments characterized by three-phase coexistence on a global scale.

Keywords: natural gas hydrate (NGH); three-phase coexistence zone; deterministic complex geological modeling; Qiongdongnan Basin

1. Introduction

The persistent increase in global energy demand, coupled with the gradual exhaustion of conventional fossil fuel resources, has rendered the advancement of new clean energy technologies a central element of international energy policy. Among various alternative energy sources, methane hydrate is regarded as one of the most promising due to its substantial reserves and environmentally friendly combustion properties [1,2]. Methane hydrate, an unconventional energy source, typically occurs in deep sea sediments or permafrost regions. Its unique physical and chemical properties complicate the extraction process and present significant risks [3–6]. Modeling technology serves as a crucial tool for understanding and forecasting the natural gas hydrate extraction process, thereby facilitating the simulation and optimization of mining strategies [7]. By integrating seismic, logging, and geological data, this technology offers a comprehensive subsurface perspective that can enhance production efficiency and improve decision-making processes [8,9].

Hydrate modeling relies on an extensive array of seismic and logging data. How-ever, the scarcity of available data has resulted in a limited number of global reports on hydrate modeling, and the overall research quality remains relatively low. The predominant modeling approaches currently employed are largely derived from traditional oil and gas reservoir modeling techniques [10,11]. Wang et al. studied 74 crude oil and condensate gas samples from the Tarim, Junggar, Beibu Gulf, and Bohai Bay basins in China, analyzing the sedimentary environment and distribution characteristics of source rocks [12]. Jamil et al. investigated the effects of both earthquakes and non-earthquakes on sediment structure in deep sea sedimentary environments [13]. Additionally, Jamil et al. examined the development and distribution of Hybrid Event Beds in deep water sedimentary systems and their potential applications for deep sea energy exploration and development [14]. Argentino et al. analyzed the characteristics and identification of hydrates in sediments, providing important geochemical and geological evidence for the study of hydrates [15]. Ye et al. utilized geostatistical techniques to construct a three-dimensional geological model of hydrates within the W11-17 deposit situated in the Shenhu region of the South China Sea. Their research provided vertical predictions regarding the saturation and permeability of the reservoir, which effectively guided the design of horizontal wells. Similarly, Machiko et al. conducted an investigation in the eastern section of the Nankai Trough in Japan, employing two distinct methodologies to predict and validate the distribution of hydrate saturation [16]. Their results demonstrated that the amalgamation of well data and seismic information significantly enhances the precision of model predictions. It is noteworthy that the modeling approaches employed by these researchers predominantly focus on the single-phase modeling of either hydrate or gas. However, recent investigations into hydrate formations have demonstrated the widespread occurrence of the three-phase zone, which is defined by the coexistence of hydrate and natural gas within natural environments. This three-phase mixed layer has been identified in various geographical locations, including the Qiongdongnan area in the South China Sea [17], the Krishna-Godavari (K-G) Basin in India [18], the Black Sea in the western Atlantic [19], Hydrate Ridge off the coast of Oregon in the United States [20], Borneo in southern China [21], the West Kuranyi subduction zone in New Zealand [22], and a mine pit in Nigeria [23]. These reservoirs may represent promising targets for hydrate development due to their positioning at the boundary of stable conditions, which renders them more susceptible to depressurization and subsequent development. Qing et al. found that the hydrate reservoir is a complex system composed of natural gas hydrate, free gas, and water, based on their study of the Pearl River Mouth Basin. In the hydrate stability zone of the Shenhu District in the South China Sea, it has been confirmed that natural gas hydrates, free gases, and water coexist [24]. However, the three-phase hydrate system exhibits a complex superposition, complicating the characterization of its spatial morphology. At present, there is a deficiency of modeling studies that focus on the intricate spatial distribution of this system, and the existing modeling methodologies are insufficient for application to reservoirs exhibiting three-phase coexistence.

This study focuses on the three-phase coexistence zone within typical sandy reservoirs located in the Qiongdongnan sea area. Notably, drilling conducted at GMGS7 in Qiongdongnan in 2021 revealed the presence of hydrates and gas situated approximately 130 m beneath the seabed. The elevated temperatures within the central region of the gas layer have resulted in the predominant distribution of hydrates surrounding this gas layer, thereby facilitating the development of a three-phase coexistence zone between the gas and hydrate layers [3,17]. Utilizing logging data from nine wells alongside high-resolution 3D seismic data, this research employs a deterministic complex morphological modeling approach to construct a reservoir model of the three-phase zone. Optimization was achieved through rock physics modeling, which enhanced the detailed characterization of the internal properties of the reservoir. Additionally, the implementation of refined interpretation techniques further mitigated errors associated with geological structures. A comparative analysis of the modeling outcomes against well data indicates a minimal level of error.

2. Geological Setting

2.1. Geologic Background

The Qiongdongnan Basin is situated across the continental shelf and slope regions, adjacent to the Xisha Islands and the Zhongjiannan Basin to the south, the Shenhu Dark Sand Uplift and the edge of the Yinggehai Basin to the west, and the Hainan Island Uplift to the north [25,26]. The basin extends northeast in an overall direction of approximately NE60°. It is recognized not only as a significant potential region for natural gas hydrates but also as a crucial area for conventional oil and gas accumulation in China [27–29]. The structural characteristics of the Qiongdongnan Basin are illustrated in Figure 1. As illustrated in Figure 2, Due to their similar geological backgrounds and tectonic activities, the basin and the Pearl River Mouth Basin exhibit significant similarities in Cenozoic stratigraphic sedimentation. The basin has undergone continuous sedimentary processes, beginning with the Eocene continental Lingtou Formation, followed by the Lower Oligocene transitional Yacheng Formation and the Upper Oligocene shallow marine Lingshui Formation [30]. During the Neogene depression phase, the basin experienced continuous sedimentation variations from shore shallow marine facies to deep sea facies, with an increasing depth of seawater. The sedimentary sequence from this period include the Lower Miocene Sanya Formation, the Middle Miocene Meishan Formation, the Upper Miocene Huangliu Formation, and the Pliocene Yinggehai Formation [31].

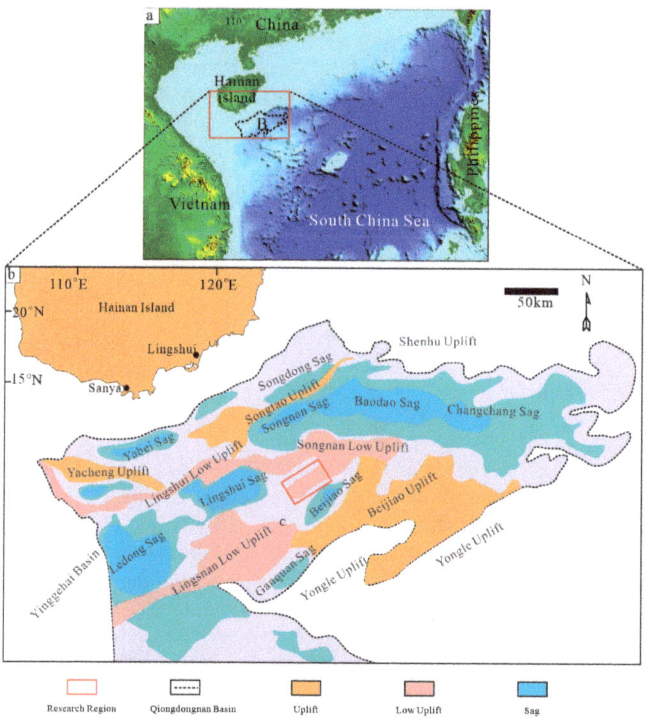

Figure 1. (**a**) The location of the Qiongdongnan basin in the south China Sea; (**b**) the tectonic units and location of the research area in the Qiongdongnan basin [32].

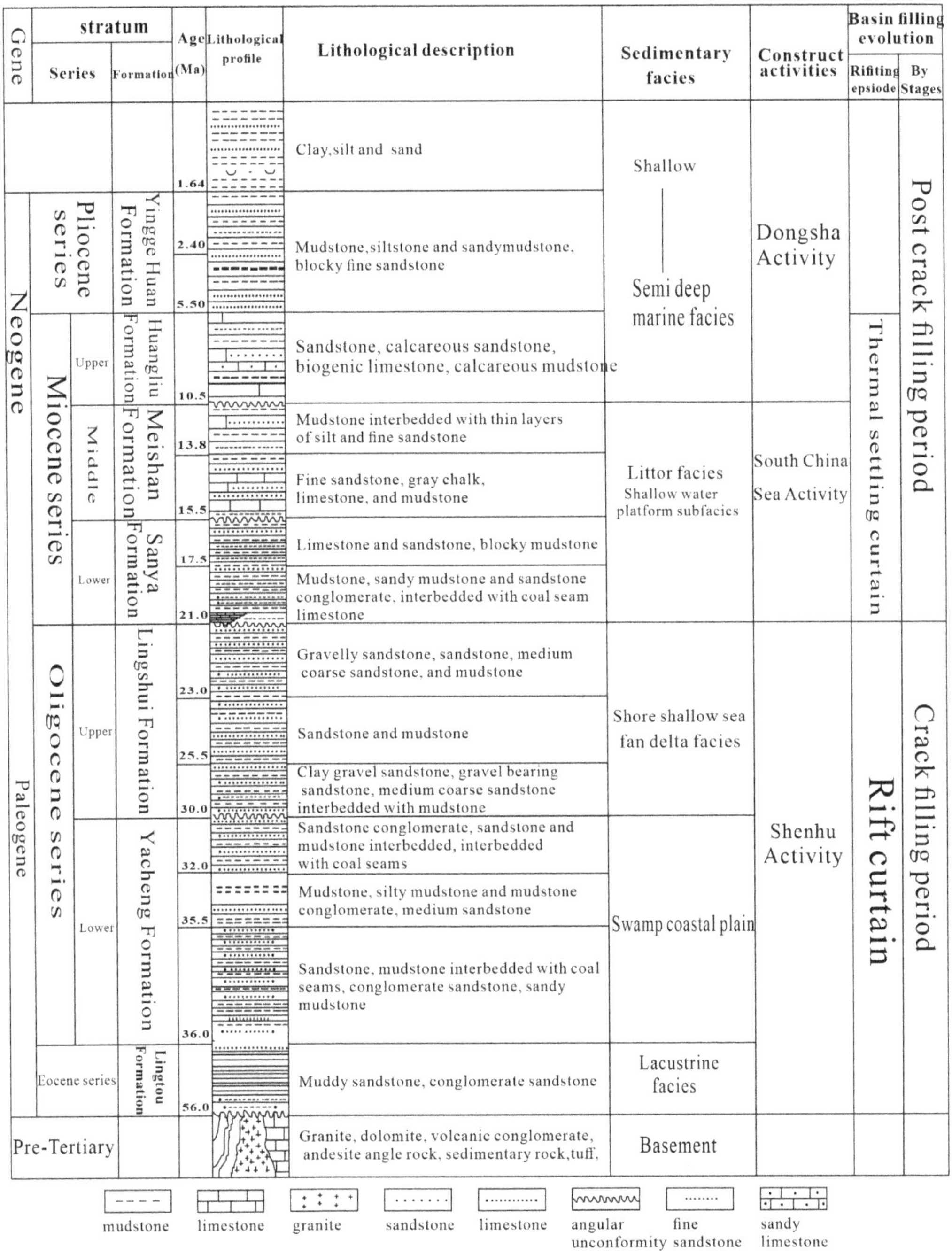

Figure 2. Schematic stratigraphic column of the Qingdongnan Basin [33].

The hydrate test area in the Qiongdongnan sea region is situated in the southeastern part of Hainan Island, on the northern slope of the South China Sea, and to the northwest of the Xisha Islands. This region is situated within the Qiongdongnan Basin, characterized by water depths ranging from 300 m to 2600 m, covering an approximate area of 3.28 × 108 km^2. The study area is positioned within the Songnan low uplift of the Qiongdongnan Basin, located approximately 646 km from Guangzhou and 177 km from Sanya, Hainan Province. The seabed exhibits a relatively flat profile, with water depths varying between 1600 and 1830 m, gradually increasing from west to east. The overall topography is elevated in the west and distribution towards the east, featuring a gentle slope (Figure 3).

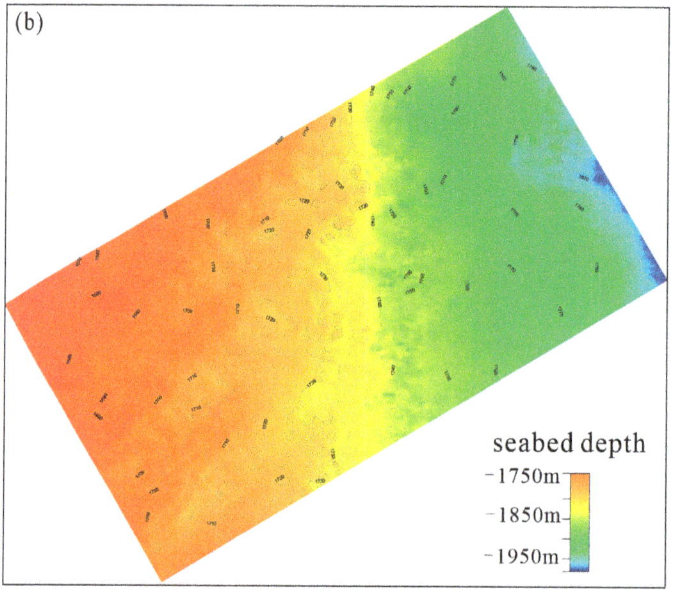

Figure 3. An overview of the seabed topographical features present in the study area.

Within this region, well W07 has demonstrated the presence of a sandstone hydrate accumulation zone. Notably, both hydrate and natural gas are found within the same sand layer, approximately 130 m beneath the seabed at well W07, which is characterized by the deposition of a natural river dike. The reservoir is situated in the Songnan low uplift, adjacent to the hydrocarbon-rich sub-depression in the eastern region of the Lingshui depression.

Previous studies have indicated that gas chimneys serve as primary conduits for fluid migration, linking deep gas sources to shallow hydrate layers, with the C2 content in the gas source reaching up to 20%. The gas chimney observed in the southeastern Qiongdongnan Basin is indicative of the upward migration of deep, high-temperature, and high-pressure fluids. This structure represents a novel type of geological structure primarily influenced by fractures formed under tectonic stress. The Qiongdongnan Basin exhibits favorable regional geological conditions for the formation of natural gas hydrates.

2.2. Distribution of Three-Phase Coexistence Zone

The study area contains a three-phase zone situated between a hydrate reservoir and a natural gas reservoir. The water depth in this region varies from 1745 to 1788 m. The depth of the simulated bottom simulated reflection (BSR) is observed to range from 1874 m to 1918 m, which corresponds to a depth of 97.5 m (mbsf) to 147.5 m (mbsf) below the seafloor. An analysis of the BSR indicates the presence of faults [34]. These faults extend approximately for 2.7 km, and in conjunction with the extinction lines, they create a geological trap with an estimated area of about 1.9 k^2 [35] (Figure 4).

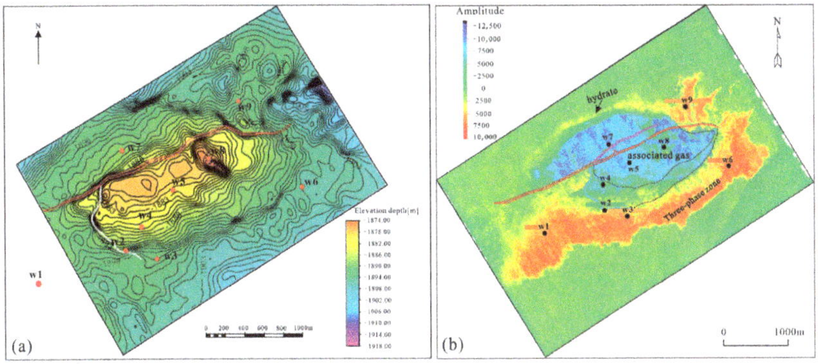

Figure 4. (**a**) Distribution map of bottom simulating reflection (BSR) for W11-17 deposit; (**b**) amplitude attribute map within the study area [17].

3. Material and Methods

3.1. Data

The seismic data utilized in this study was collected by the China National Offshore Oil Corporation (CNOOC) in 2017. This dataset includes offshore 3D seismic data, as well as extensive information gathered using Schlumberger's SonicScope and proVISIONN (Houston, TX, USA) while drilling logging tools. The data encompass gamma measurements, wellbore diameter, resistivity, high-definition resistivity images, various micro-detection data, density, longitudinal wave time differences, transverse wave time differences, T2 spectrum, and pore structure. A comprehensive evaluation was conducted.

3.2. Method

3.2.1. Logging Interpretation

By utilizing MicoScopeHD resistivity imaging for dip angle extraction, ELAN analysis, and acoustic longitudinal wave time difference extraction, we can determine the formation dip angle, as well as the development of fractures and micro faults. Conventional calculations of hydrate saturation and permeability are then performed. Following this, we extract acoustic and transverse wave time differences, conduct element energy spectrum extraction and Sigma processing, and analyze nuclear magnetic data to assess porosity and permeability. An accurate analysis of the ELAN lithology model is based on the element capture energy spectrum, and hydrate saturation is calculated using a combination of density and nuclear magnetic data. This approach not only allows for the identification of the mineral category and dry weight of the formation but also enables precise calculations of hydrate saturation and permeability, thereby enhancing the accuracy of the results.

3.2.2. Seismic Interpretation and Inversion

This study utilized the Geoframe software (2012) to interpret three-dimensional seismic data. Based on the characteristic that the top interface of hydrates produces a positive reflection while the top interface of gas layers generates a negative reflection, the top and bottom interfaces of two sets of sand layers and faults within the ore body were identified. By employing post-stack impedance and pre-stack AVO inversion, the distribution range of hydrates and free gases was clearly delineated.

3.2.3. Geologic Modeling

The geological characteristics of the seabed within the study area are as follows: in this region, only two wells, identified as W2 and W4, exhibit features that suggest the presence of a three-phase zone. In order to enhance the accuracy of the model representing the three-phase coexistence zone, this study employed a deterministic modeling approach that integrates the dense sampling characteristics of seismic and well data in both vertical and horizontal dimensions.

(1) The logging data from wells W1 to W9 have been subjected to correction processes. A band-pass filtering technique has been applied to reduce errors in the logging data, followed by environmental corrections. In sections lacking a density curve, a phase-controlled density prediction method has been utilized to estimate the density curve for subsequent calibration with seismic data [36].
(2) A deep time conversion of the logging data has been executed, wherein non-equidistant logging data in the time domain have been interpolated to generate an equidistant curve. Following this resampling, a time sampling reflection coefficient sequence that corresponds with the seismic data has been derived.
(3) A sand body distribution model has been developed through geostatistical methods, integrating lithological data derived from well information and seismic inversion.
(4) The morphological plane model of the three-phase coexistence zone has been established within the study area. Section interpretation has been employed to mitigate the influence of intricate geological conditions on model accuracy. A total of 105 sections have been examined throughout the three-phase zone, leading to the development of porosity and saturation models for hydrate and free gas.

4. Results

4.1. Well-Seismic Interpretation and Calibration of Three-Phase Coexistence Zone

In the reservoir evaluation map for well W2, the analyzed interval spans from 1899.1 to 1902.3 m. Within this interval, there is an observed increase in resistivity, a decrease in P-wave transit time, a significant reduction in nuclear magnetic porosity, a decline in neutron measurements, a slight decrease in density, an inversion in neutron density, and a reduction in sigma gamma values. Based on a comprehensive analysis, it is concluded that this layer represents a zone of coexisting hydrate and gas. The thickness of this layer is measured at 3.2 m, with an average effective porosity of 43.7%. The average hydrate saturation, as determined by the resistivity method, is calculated to be 54.6%, while the gas saturation is estimated at 16.0% (Figure 5a).

The detailed reservoir evaluation of the W4 well indicates that within the depth range of 1919.5 to 1924.3 m, there is an observed increase in resistivity, a slight decrease in nuclear magnetic resonance, a reduction in neutron measurements, and a minor increase in density. This comprehensive analysis suggests the presence of a layer characterized by the coexistence of hydrates and gas, with a total thickness of 4.8 m and an average effective porosity of 44.8%. Furthermore, the average saturation of hydrates and gas, as determined by the resistivity method, is calculated to be 72.5%, while the average permeability is assessed at 161.7 millidarcies (mD) (Figure 5b).

At present, the research domain in this area encompasses nine drilling wells. The results obtained from wave impedance inversion employing a traditional hard constraint are illustrated in Figure 6a [37].

As depicted in Figure 6b, the inversion results delineate a distinct three-phase coexistence zone of hydrates situated on the plane that demarcates hydrates from the associated gas. The hydrate layer manifests as a ring, positioned at the periphery of the associated gas. The yellow color denotes moderate hydrate saturation, whereas red indicates high hydrate saturation. In Region 1, the impedance slice transitions from red to yellow, signifying a decrease in hydrate quantity. Region 2 demonstrates a further reduction in impedance, while Region 3 is characterized by low impedance. The three-phase coexistence zone is located between curves 1 and 2, with its southwestern boundary being approximately 600 m wider than the eastern and northern boundaries, which are about 400 m narrower. Within the three-phase zone, hydrates are arranged in a wedge-like configuration, with gas situated beneath and hydrates positioned above.

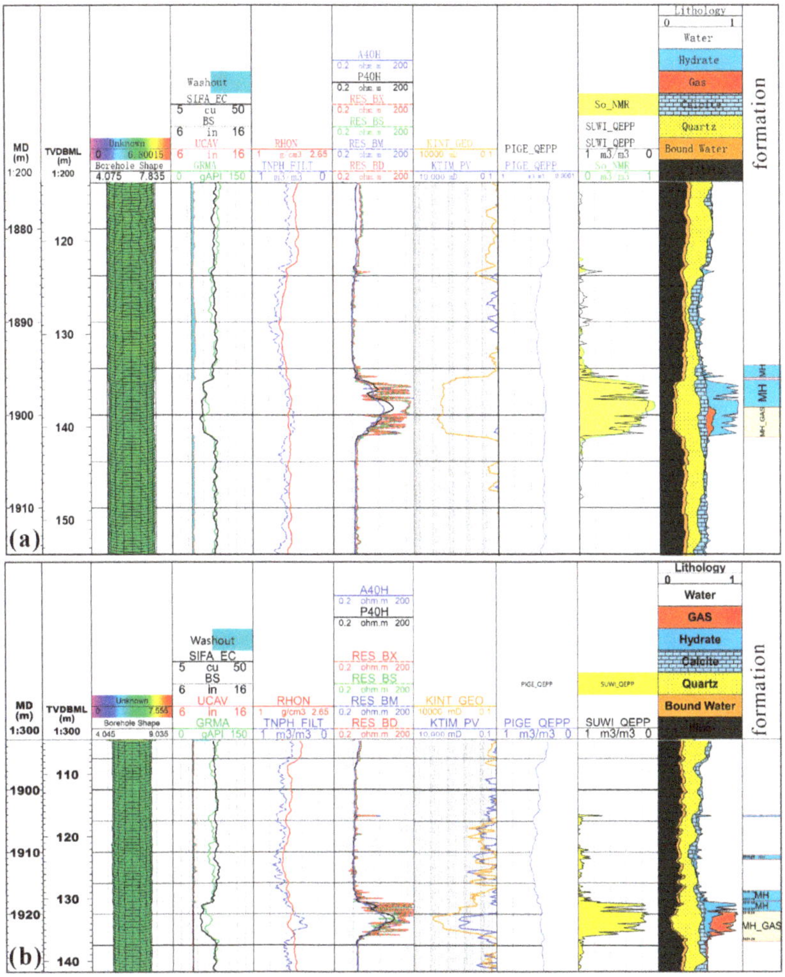

Figure 5. Comprehensive evaluation diagram of logging reservoir. Diagram (**a**) shows well W2, and diagram (**b**) shows well W4. Blue represents hydrate, red represents free gas, and yellow represents the three-phase zone.

Figure 7 illustrates the average velocity measured on the plane at the upper interface of the hydrate over a period of 15 milliseconds. The hydrate exhibits a pronounced high-value anomaly, with the hydrate layer attaining a maximum thickness of approximately 10 m and being relatively thin, predominantly concentrated in the southeastern and southern regions. The western section of the hydrate layer is comparatively thinner. An analysis of the lowest velocity over a 20 ms interval downward along the hydrate's upper interface reveals a low velocity anomaly, indicative of gas bearing anomalies. The fault appears to exert a significant influence on the gas layer, characterized by a thicker lower wall and a thinner hanging wall. The sand body exhibits a relatively uniform distribution of gas layers, although a reduction in gas layer thickness is observed within the three-phase zone.

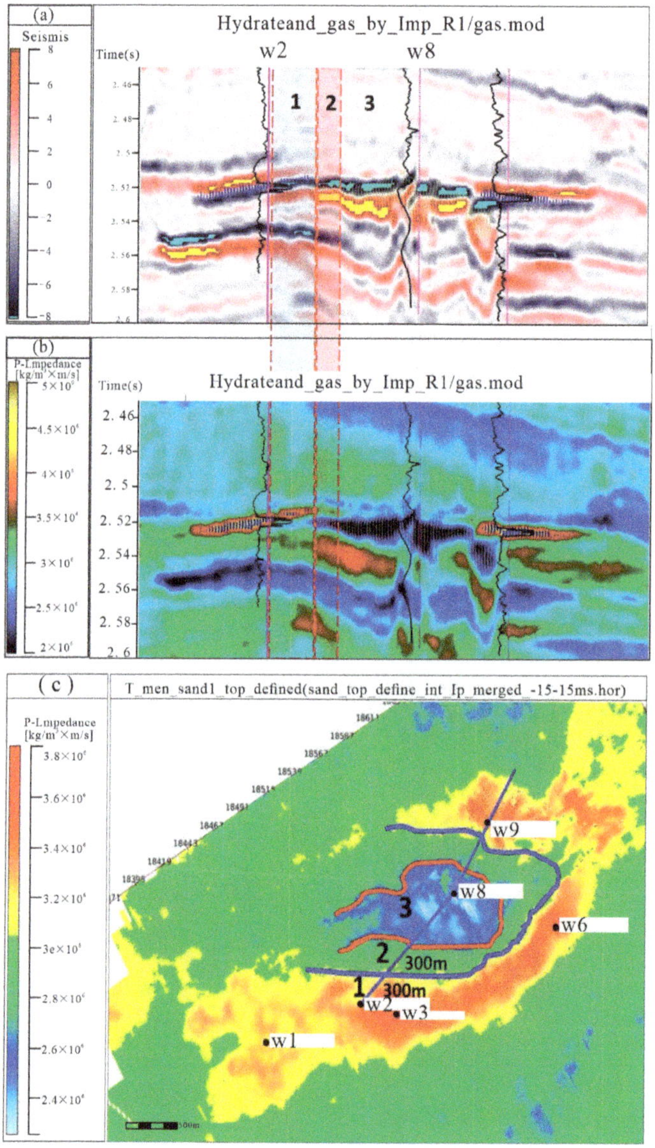

Figure 6. (**a**) Three-phase coexistence zone hydrate impedance profiles; (**b**) three-phase coexistence zone free gas impedance profiles; (**c**) 15 ms average wave impedance section above and below the hydrate top interface.

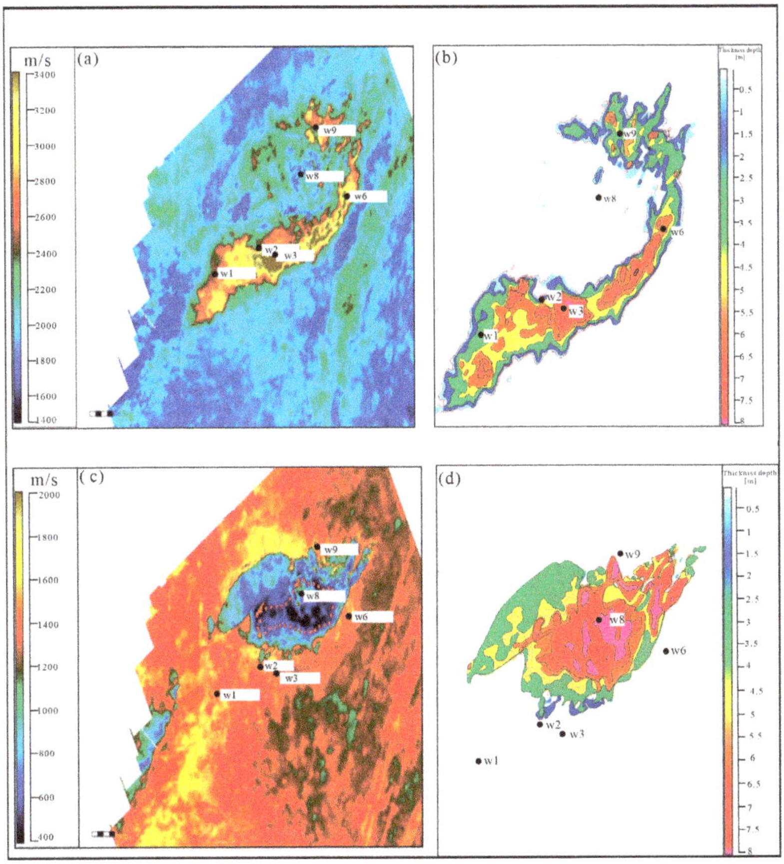

Figure 7. (**a**) Hydrate layer velocity slicing; (**b**) Thickness distribution of hydrate layer; (**c**) Associated gas layer velocity slicing; (**d**) Thickness distribution of associated gas layer.

4.2. Structure Model

A three-dimensional geological model of the three-phase coexistence zone was created. The three-phase zone model contains nine wells, with an average spacing of approximately 1 km between them. The three-phase zone model is spatially distributed in an elliptical shape oriented northeast-southwest, measuring about 4 km by 2.5 km (Figure 8a). The vertical logging data spans roughly 100 m. A detailed seismic data interpretation reveals six layers and one fault at the top and bottom interfaces of two sets of sand layers, as well as the upper and lower hosting rock interfaces. Hydrate and free gas are primarily in the first set of sand bodies. A significant fault trending east–west is present in the region (Figure 8b). The fault model illustrates the spatial characteristics of fault distribution, highlighting the displacement of strata and the variations among different fault blocks. Following the outlined procedures, the fault shape was repeatedly adjusted and validated to ensure it accurately reflects the actual conditions and effectively describes its distribution characteristics in three-dimensional space.

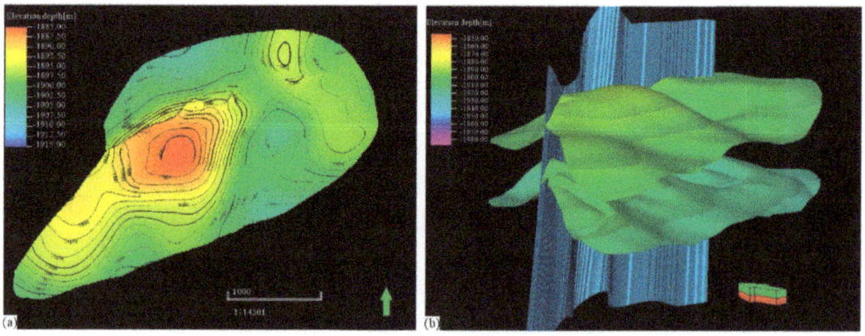

Figure 8. (**a**) Sand body planar structural features; (**b**) Sand body fault trend map.

4.3. Fine Interpretation

The sand body illustrated in Figure 8a primarily divides the strata into two minor zones, designated as Sand1 and Sand2. The three-phase zone, predominantly situated within the Sand1, exhibits an average thickness of approximately 8 m and is characterized by a structural elevated at the center and decreasing towards the peripheries. Utilizing an interactive interpretation approach that integrates both plane and section views, a total of 105 profiles are evaluated from left to right (Figures 9 and 10a) [38]. Four of these profiles have been selected for in-depth discussion. The W2 and W4 wells, which are situated within the three-phase zone, function as critical control points for this analysis. Each well is linked to a corresponding profile that is interpreted in a unique manner (Figure 10b).

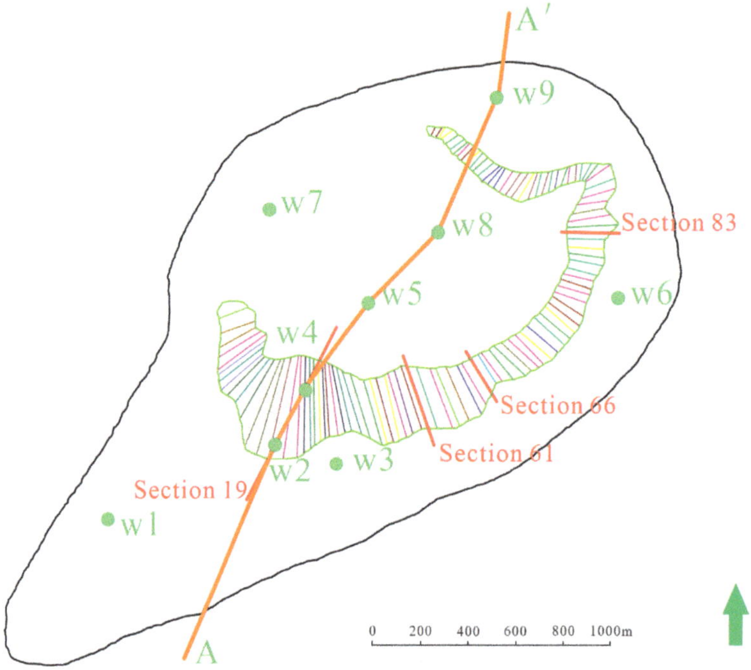

Figure 9. Distribution of three-phase zones and well locations within the study area.

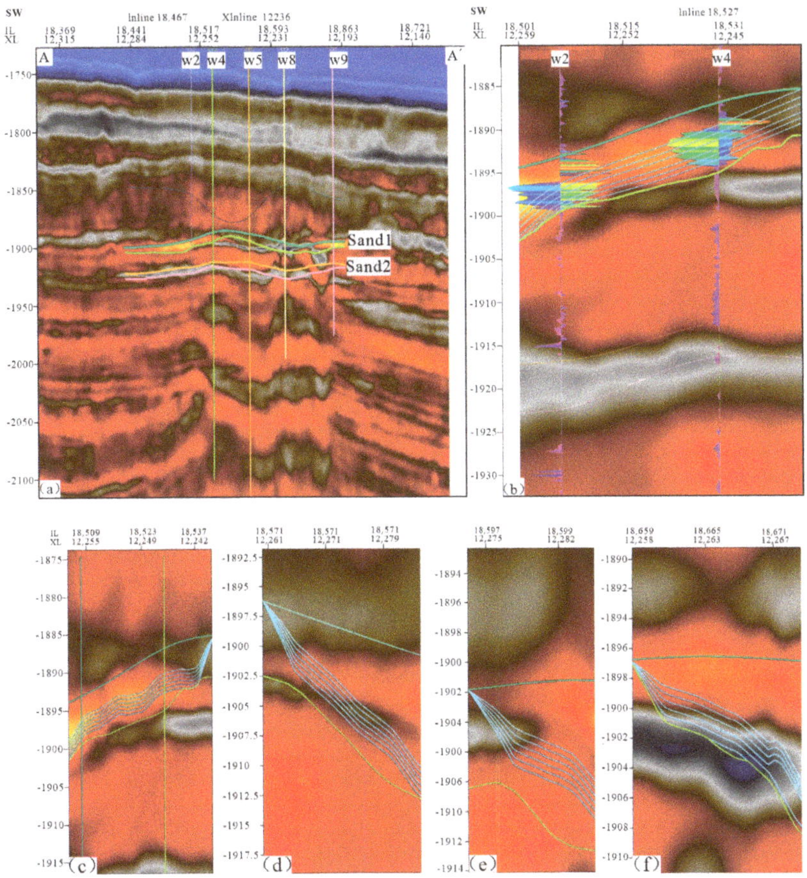

Figure 10. Seismic profile characteristics and layer interpretation. (**a**) A-A′ seismic profile; (**b**) Seismic profile of W2-W4 well; (**c**) Section 19; (**d**) Section 61; (**e**) Section 66; (**f**) Section 83.

Section 19 is oriented in a north–south direction, as illustrated in Figures 9 and 10c. This section exhibits five distinct layers of stratification within the three-phase zone, which is further categorized into five separate layers. The central layer is defined by an equal distribution model, with its lower boundary interfacing with free gas and its upper boundary in contact with hydrates.

A manual analysis of Section 61 and Section 66, trending from southeast to northwest, indicates that the front of this section displays a downward dip that is nearly parallel to the upper layer of the three-phase zone, a phenomenon attributed to geological constraints (Figure 10d,e).

Likewise, the upper boundary of Section 83, which trends from west to east, shows a downward slope along its trajectory until it reaches the boundary of the section (Figure 10f).

The internal strata of the three-phase coexistence zone are categorized into five discrete sublayers, arranged vertically from the base to the apex [25]. Figure 9 provides a detailed representation of the structural configuration of these sublayers within the three-phase zone. It is noteworthy that within these sublayers, there is a progressive increase in hydrate saturation from the lower to the upper sections (Figure 11a–c), whereas free gas saturation demonstrates a concomitant decrease (Figure 11d–f).

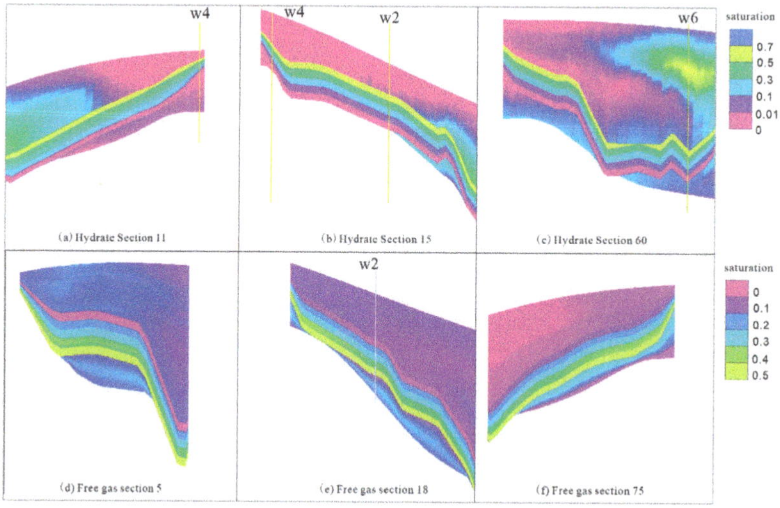

Figure 11. Three-phase coexistence zone structural sublayer distribution profile: hydrate (**a**–**c**) free gas (**d**–**f**).

4.4. Attribute Model

To construct models of porosity and saturation, as illustrated in Figure 12, it is essential to utilize variogram analysis based on well data. This process requires the careful selection of suitable parameters for range and azimuth, in addition to integrating seismic calibration results as a supplementary constraint.

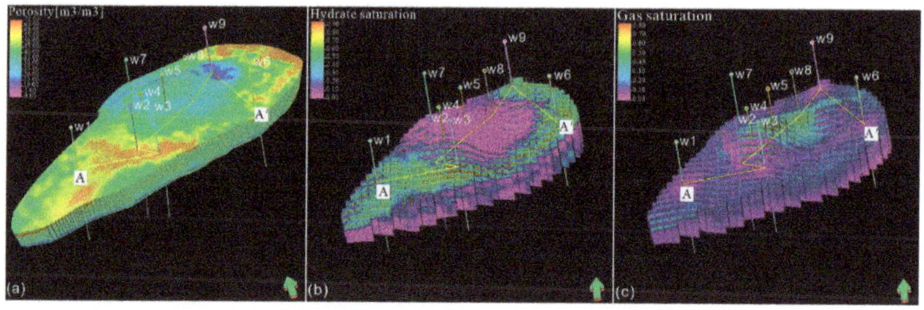

Figure 12. (**a**) Porosity model; (**b**) hydrate saturation model; (**c**) free gas saturation model.

The porosity model indicates that the spatial distribution of free gas within the study area is more limited compared to that of hydrate. Notably, there is an increase in porosity values in the upper section of Sand1, which may be attributed to the mass transport deposit (MTD) situated above Sand1, as illustrated in Figure 13a. In contrast, the porosity values within Sand1 exhibit relative uniformity, as depicted in the corresponding well profile in Figure 13a.

The hydrate saturation model reveals that areas of elevated hydrate saturation are predominantly located outside the three-phase zone, forming a "C" shaped distribution, as demonstrated in Figure 12b. The free gas saturation model indicates that the free gas present in the three-phase coexistence zone is characterized by high saturation levels, with a significant concentration of free gas found in the central region of the study area, as shown in Figure 12c. The A-A well profile presented in Figure 12c indicates that well 07 exhibits the highest free gas saturation, which may be correlated with its location at the center of the gas chimney.

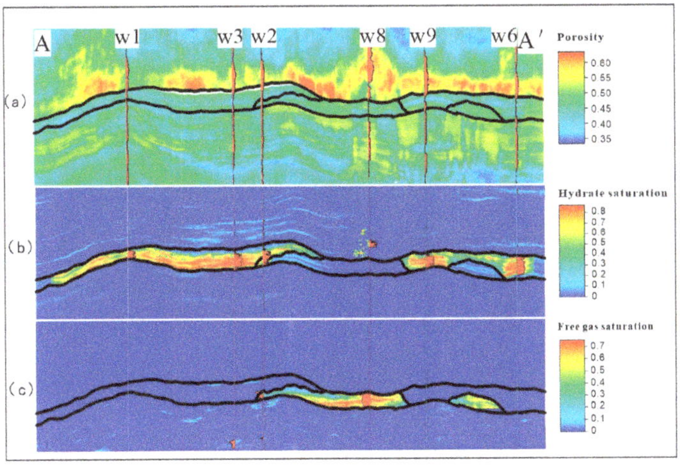

Figure 13. (**a**) Porosity model A-A′ connected well profile; (**b**) hydrate saturation model A-A′ connected well profile; (**c**) free gas saturation model A-A′ connected well profile.

5. Discussion

5.1. Error Analysis

The structural model is constructed using a methodology based on the corner grid system. This approach considers the spatial distribution of wells within the study area, the trends in geological structures, and the specific requirements associated with oilfield development. To improve the quality of the planar grid, a resolution of 10 m × 10 m is employed, with trend lines integrated as needed. In terms of vertical mesh generation, an average mesh accuracy of 0.5 m is attained, culminating in the creation of a total of 35,000 meshes.

5.1.1. Verification of Distribution Patterns

The three-dimensional distribution of hydrate demonstrates a notable alignment with the trends identified in wave impedance data. Specifically, the hydrate thickness is approximately 8 m in the western region, 5 m in the central region, and 3 m in the eastern region. Figure 14b depicts the three-dimensional structure of a minor layer within the three-phase zone, while Figure 14c,d provide three-dimensional structural maps from various perspectives within the three-phase coexistence zone of the study area.

5.1.2. Verification of Spatial Distribution of Parameters

Due to the limited number of control well points intersecting the three-phase zone in the hydrate study area, some discrepancies were observed in the distribution of model data, coarsening data, and original data. Nevertheless, the model effectively captures the regions of high porosity and high saturation, as evidenced by the seismic profile, indicating the general credibility of the model.

In comparing the models generated through three distinct modeling methodologies, values within the same model were selected for analysis. The comparison revealed that the saturation data derived from the deterministic complex morphological geological modeling method aligns closely with the variation range of the well data. In contrast, the porosity values obtained through Kriging interpolation and sequential Gaussian modeling exhibited more significant fluctuations. Figure 15. This analysis demonstrates that the three-dimensional multiscale model developed via deterministic complex morphological geological modeling is largely congruent with the prevailing geological understanding and actual logging data. Furthermore, the results substantiate the superiority of hierarchical modeling over models derived solely from a singular modeling approach.

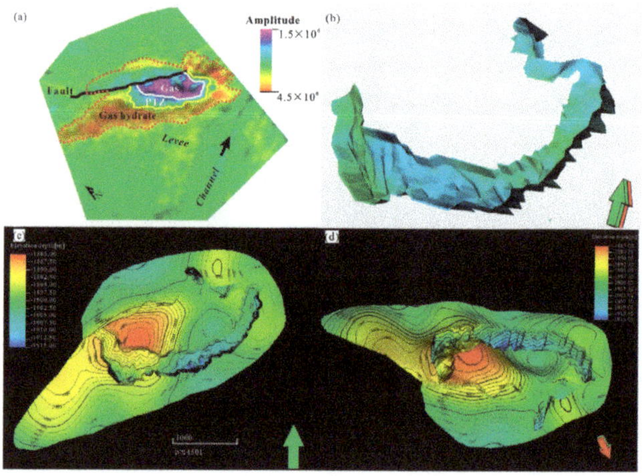

Figure 14. Comparison of hydrate forms. (**a**) Wave impedance slice map of the study area; (**b**) three-phase coexistence zone model small layer structure diagram; (**c**) top view of 3D model of three-phase coexistence area; (**d**) bottom view of 3D model of three-phase coexistence area.

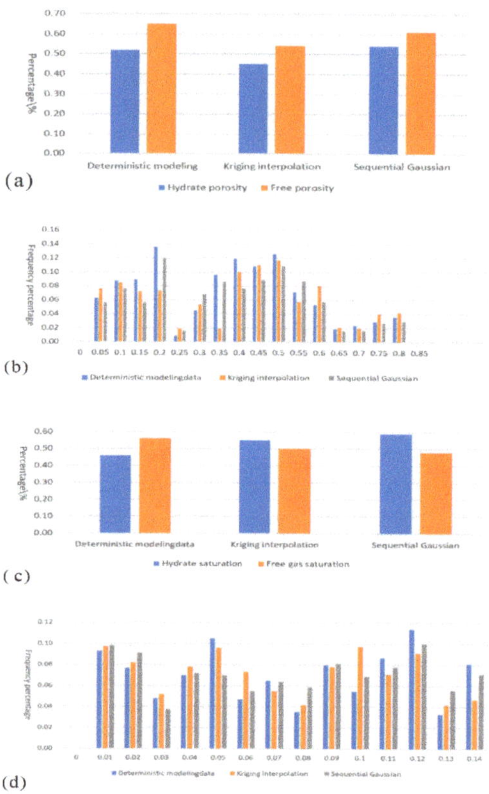

Figure 15. (**a**) Comparison of saturation data established by 3 modeling methods; (**b**) histograms of saturation probability distribution established by 3 modeling methods; (**c**) comparison of porosity data established by 3 modeling methods; (**d**) histograms of porosity probability distribution established by 3 modeling methods.

5.2. Geological Implications

The detailed depiction of the internal morphology and resource distribution patterns of the ore body provides a solid foundation for future production and development activities, such as the construction and advancement of horizontal wells. This modeling study has further clarified the reservoir formation pattern in the QDN area. Unlike the vertical distribution of hydrates and free gas observed in the Shenhu research area of the South China Sea, the hydrates and free gas in the QDN basin are distributed horizontally. As deep gas sources rise to high-permeability sand layers, the lateral migration speed increases, causing the free gas to gradually transform into hydrates as the pressure and temperature change. There is a three-phase coexistence zone of free gas and water between the two phases.

6. Conclusions

This article presents a high-precision model of three-phase coexistence zones by integrating detailed earthquake interpretation, deterministic complex shape modeling, and multiple data fusion techniques. Compared to previous studies, this approach has achieved significant advancements in data processing, spatial modeling, and result verification, offering new insights and methodologies for future hydrate development research. The geological modeling technique for three-phase coexistence zones proposed in this study demonstrates notable improvements and innovations over existing research. Unlike earlier studies that primarily concentrated on single-phase modeling—such as individual natural gas hydrates or free gas layers—the deterministic complex landform modeling method employed in this paper more accurately captures the spatial morphology of three-phase coexistence zones. This modeling framework not only synthesizes seismic and logging data but also incorporates detailed seismic interpretation and statistical geological methods, thereby addressing the limitations of traditional approaches in modeling complex three-phase systems.

Author Contributions: Conceptualization, H.Y. and J.W.; methodology, H.Y., J.W., and W.D; software, J.W. and Z.L; validation, H.Y., J.W., and Z.L.; formal analysis, W.D. and Z.L; investigation, H.Y. and J.W.; resources, Z.L., Z.K., and T.L.; data curation, W.D.; writing—original draft preparation, H.Y., J.W., and W.D.; writing—review and editing, Z.K.; visualization, J.W. and T.L; supervision, Z.K.; project administration, Z.K. and T.L.; funding acquisition, Z.K. All authors have read and agreed to the published version of the manuscript.

Funding: The research works are financially supported by National Key Research and Development Program of China (No. 2021YFC2800901). This work is supported by the National Natural Science Foundation of China (42376058) and the Guangdong Major Project of Basic and Applied Basic Research (2020B0301030003).

Data Availability Statement: In this research work presented in this paper, certain data has been obtained under strict confidentiality agreements and restrictions. The data in question is proprietary and sensitive in nature, and its disclosure is prohibited by the terms of the agreements with the data providers or due to legal and ethical obligations.

Conflicts of Interest: The authors declare no conflicts of interest.

References

1. Ye, J.-L.; Qin, X.-W.; Xie, W.-W.; Lu, H.-L.; Ma, B.-J.; Qiu, H.-J.; Liang, J.-Q.; Lu, J.-A.; Kuang, Z.-G.; Lu, C.; et al. The second natural gas hydrate production test in the South China Sea. *China Geol.* **2020**, *3*, 197–209. [CrossRef]
2. Zeng, X.; Yu, X.; Liang, J.; Kuang, Z.; Wang, J. Geological modeling and favorable area prediction of natural gas hydrate in Shenhu sea area. *Mar. Geol. Quat. Geol.* **2014**, *34*, 117–126.
3. Li, J.-F.; Ye, J.-L.; Qin, X.-W.; Qiu, H.-J.; Wu, N.-Y.; Lu, H.-L.; Xie, W.-W.; Lu, J.-A.; Peng, F.; Xu, Z.-Q.; et al. The first offshore natural gas hydrate production test in South China Sea. *China Geol.* **2018**, *1*, 5–16. [CrossRef]
4. Yang, S.; Liang, J.; Lu, J.; Qu, C.; Liu, B. New understanding of gas hydrate accumulation characteristics and main controlling factors in Shenhu area, northern South China Sea. *Geol. Front.* **2017**, *24*, 1–14. [CrossRef]
5. Wu, N.; Liang, J.; Wang, H.; Su, X.; Song, H.; Jiang, S.; Zhu, Y.; LU, Z. Research progress of marine gas hydrate accumulation system. *Mod. Geol.* **2008**, *3*, 356–362.

6. Dale, A.; Van Cappellen, P.; Aguilera, D.; Regnier, P. Methane efflux from marine sediments in passive and active margins: Estimations from bioenergetic reaction–transport simulations. *Earth Planet. Sci. Lett.* **2008**, *265*, 329–344. [CrossRef]
7. Liang, J.; Sha, Z.; Meng, M.; Wang, X.; Xu, M.; Fang, Y. Characteristics and distribution of high saturation natural gas hydrate reservoirs in the Shenhu Sea. *Geol. J.* **2024**, *98*, 2641–2653. [CrossRef]
8. Wang, X.; Wu, S.; Liu, X.; Yang, S.; Guo, Y.; Li, Q. Estimation of Natural Gas Hydrate Resources in Shenhu Sea Area Based on Logging and Seismic Data. *Adv. Geophys.* **2010**, *25*, 1288–1297.
9. Liang, J.; Wang, J.; Yang, C.; Kang, D.; Lu, J.; Liang, J. Geophysical characteristics of gas hydrate-bearing sediments in the eastern Pearl River Mouth Basin. *Gas Ind.* **2017**, *37*, 126–133.
10. Fan, J.; Liu, Y.; Zhang, X.; Liu, K.; Zheng, Y. Imulation and Comparison of Seismic Wave Fields in Natural Gas HydrateFormations. *Adv. Mar. Sci.* **2021**, *39*, 290–303.
11. Liu, X.; Yin, X.; Luan, X. Seismic rock physical modelling for gas hydrate-bearing sediments. *Sci. China Earth Sci.* **2018**, *61*, 1261–1278. [CrossRef]
12. Wang, G.; Cheng, B.; Wang, T.-G.; Simoneit, B.R.; Shi, S.; Wang, P. Monoterpanes as molecular indicators to diagnose depositional environments for source rocks of crude oils and condensates. *Org. Geochem.* **2014**, *72*, 59–68. [CrossRef]
13. Jamil, M.; Siddiqui, N.A.; Umar, M.; Usman, M.; Ahmed, N.; Rahman, A.H.A.; Zaidi, F.K. Aseismic and seismic impact on development of soft-sediment deformation structures in deep-marine sand-shaly Crocker fan in Sabah, NW Borneo. *J. King Saud Univ.-Sci.* **2021**, *33*, 101522. [CrossRef]
14. Jamil, M.; Siddiqui, N.A.; Ahmed, N.; Usman, M.; Umar, M.; Rahim, H.U.; Imran, Q.S. Facies Analysis and Sedimentary Architecture of Hybrid Event Beds in Submarine Lobes: Insights from the Crocker Fan, NW Borneo, Malaysia. *J. Mar. Sci. Eng.* **2021**, *9*, 1133. [CrossRef]
15. Argentino, C.; Conti, S.; Fioroni, C.; Fontana, D. Evidences for Paleo-Gas Hydrate Occurrence: What We Can Infer for the Miocene of the Northern Apennines (Italy). *Geosciences* **2019**, *9*, 134. [CrossRef]
16. Tamaki, M.; Suzuki, K.; Fujii, T.; Sato, A. Prediction and validation of gas hydrate saturation distribution in the eastern Nankai Trough, Japan: Geostatistical approach integrating well-log and 3D seismic data. *Interpretation* **2016**, *4*, SA83–SA94. [CrossRef]
17. Kuang, Z.; Cook, A.; Ren, J.; Deng, W.; Cao, Y.; Cai, H. A Flat-Lying Transitional Free Gas to Gas Hydrate System in a Sand Layer in the Qiongdongnan Basin of the South China Sea. *Geophys. Res. Lett.* **2023**, *50*, e2023GL105744. [CrossRef]
18. Zhou, J.; Wang, X.; Collett, T.S.; Li, S.; Kuang, Z.; Lu, Y.; Deng, W.; Yan, W.; Qian, J.; Jin, J. Characterization of a complex sand-rich gas hydrate reservoir system in the Indian marine continental margin with downhole log and seismic data. *Mar. Pet. Geol.* **2023**, *155*, 106370. [CrossRef]
19. Guerin, G.; Goldberg, D.; Meltser, A. Characterization of in situ elastic properties of gas hydrate-bearing sediments on the Blake Ridge. *J. Geophys. Res.* **1999**, *104*, 17781–17795. [CrossRef]
20. Milkov, A.V.; Dickens, G.R.; Claypool, G.E.; Lee, Y.J.; Borowski, W.S.; Torres, M.E.; Xu, W.; Tomaru, H.; Tréhu, A.M.; Schultheiss, P. Co-existence of gas hydrate, free gas, and brine within the regional gas hydrate stability zone at Hydrate Ridge (Oregon margin): Evidence from prolonged degassing of a pressurized core. *Earth Planet. Sci. Lett.* **2004**, *222*, 829–843. [CrossRef]
21. Paganoni, M.; Cartwright, J.A.; Foschi, M.; Shipp, R.C.; Van Rensbergen, P. Structure II gas hydrates found below the bottom-simulating reflector. *Geophys. Res. Lett.* **2016**, *43*, 5696–5706. [CrossRef]
22. Pecher, I.A.; Barnes, P.M.; Leah, I. *International Ocean Discovery Program Expedition 372 Preliminary Report Creeping Gas Hydrate Slides and Hikurangi Lwd Disclaimer*; Texan A&M University: College Station, TX, USA, 2018. [CrossRef]
23. Sultan, N.; Bohrmann, G.; Ruffine, L.; Pape, T.; Riboulot, V.; Colliat, J.; De Prunelé, A.; Dennielou, B.; Garziglia, S.; Himmler, T.; et al. Pockmark formation and evolution in deep water Nigeria: Rapid hydrate growth versus slow hydrate dissolution. *J. Geophys. Res. Solid Earth* **2014**, *119*, 2679–2694. [CrossRef]
24. Qin, X.-W.; Survey, M.O.N.R.C.G.; Lu, J.-A.; Lu, H.-L.; Qiu, H.-J.; Liang, J.-Q.; Kang, D.-J.; Zhan, L.-S.; Lu, H.-F.; Kuang, Z.-G.; et al. Coexistence of natural gas hydrate, free gas and water in the gas hydrate system in the Shenhu Area, South China Sea. *China Geol.* **2020**, *3*, 210–220. [CrossRef]
25. Fu, C.; Fan, X.; Yu, X.; Zhao, C.; He, Y.; Liang, J.; Su, P. Hierarchical modeling method and favorable zone prediction of hydrate reservoirs in Shenhu waters in continental slope of northern South China Sea. *China Offshore Oil Gas* **2019**, *31*, 83–92. [CrossRef]
26. Su, P.; He, J.; Liang, J.; Zhang, W. Natural gas hydrate migration and accumulation system and its controlling factors on northern deep water slope of the south China Sea. *Mar. Geol. Front.* **2017**, *33*, 7. [CrossRef]
27. Li, L.; Lei, X.; Zhang, X.; Sha, Z. Gas hydrate and associated free gas in the Dongsha Area of northern South China Sea. *Mar. Pet. Geol.* **2013**, *39*, 92–101. [CrossRef]
28. Li, L.; Lei, X.; Zhang, X.; Zhang, G. Heat flow derived from BSR and its implications for gas hydrate stability zone in Shenhu Area of northern South China Sea. *Mar. Geophys. Res.* **2012**, *33*, 77–87. [CrossRef]
29. Wang, X.; Hutchinson, D.R.; Wu, S.; Yang, S.; Guo, Y. Elevated gas hydrate saturation within silt and silty clay sediments in the Shenhu area, South China Sea. *J. Geophys. Res.* **2011**, *116*, B05102. [CrossRef]
30. Huang, B.; Xiao, X.; Li, X. Geochemistry and origins of natural gases in the Yinggehai and Qiongdongnan basins, offshore South China Sea. *Org. Geochem.* **2003**, *34*, 1009–1025. [CrossRef]
31. Yu, X.; Zhang, Z.; Su, X.; Chen, F.; Li, Y. Primary discussion on accumulation conditions for sedimentation of gas hydrate and its distribution in south China sea. *Earth Sci. Front.* **2004**, *11*, 311–315. [CrossRef]

32. Meng, M.; Liang, J.; Kuang, Z.; Ren, J.; He, Y.; Deng, W.; Gong, Y. Distribution characteristics of quaternary channel systems and their controlling factors in the Qiongdongnan Basin, South China Sea. *Front. Earth Sci.* **2022**, *10*, 902517. [CrossRef]
33. Xie, W.; Zhang, Y.; Sun, Z.; Jiang, J. Characteristics and Formation Mechanism of Faults in Qiongdongnan Basin. *Mar. Geol. Quat. Geol.* **2007**, *27*, 71–78.
34. Ge, J.; Zhao, X.; Fan, Q.; Fang, X.; Song, P.; Xiang, Z. Preliminary study on depositional and reservoir characteristics of high-saturation gas hydrate worldwide. *Mar. Geol. Front.* **2023**, *39*, 1–13. [CrossRef]
35. Yu, X.; Wang, J.; Liang, J.; Li, S.; Zeng, X.; Li, W. Depositional characteristics and accumulation model of gas hydrates in northern South China Sea. *Mar. Pet. Geol.* **2014**, *56*, 74–86. [CrossRef]
36. Bosch, M.; Carvajal, C.; Rodrigues, J.; Torres, A.; Aldana, M.; Sierra, J. Petrophysical seismic inversion conditioned to well-log data: Methods and application to a gas reservoir. *Geophysics* **2009**, *74*, O1–O15. [CrossRef]
37. Bashir, Y.; Siddiqui, N.A.; Morib, D.L.; Babasafari, A.A.; Ali, S.H.; Imran, Q.S.; Karaman, A. Cohesive approach for determining porosity and P-impedance in carbonate rocks using seismic attributes and inversion analysis. *J. Pet. Explor. Prod. Technol.* **2024**, *14*, 1173–1187. [CrossRef]
38. Yin, S.; Gao, Y.; Hu, Z.; Xiong, T.; Feng, W.; Zhao, J.; Cheng, L. Multiple-point geostatistical simulation of outcrop based on UAV oblique photographic data: A case study of Shihezi Formation in Pingtou township, Lvliang city, Shanxi. *Acta Pet. Sin.* **2021**, *42*, 198–216. [CrossRef]

Journal of
Marine Science and Engineering

MDPI

Article

Controls on Deep and Shallow Gas Hydrate Reservoirs in the Dongsha Area, South China Sea: Evidence from Sediment Properties

Chenyang Bai [1,2,3,*], Hongbin Wang [2,4,*], Qing Li [2,4], Yu Zhang [3] and Xiaolei Xu [3]

1 Key Laboratory of Polar Geology and Marine Mineral Resources, Ministry of Education, China University of Geosciences (Beijing), Beijing 100083, China
2 Laboratory for Marine Mineral Resources, Laoshan Laboratory, Qingdao 266237, China; qing.li@live.cn
3 School of Ocean Sciences, China University of Geosciences (Beijing), Beijing 100083, China; zyfn135@163.com (Y.Z.); littlexu0830@163.com (X.X.)
4 Qingdao Institute of Marine Geology, China Geological Survey, Qingdao 266237, China
* Correspondence: baicy@cugb.edu.cn (C.B.); oceanwang7106@163.com (H.W.)

Abstract: The Dongsha area, a key region in the northern South China Sea (SCS), features both diffusive deep and seepage shallow gas hydrate reservoirs. Utilizing sediment samples from gas hydrate reservoirs and adjacent layers at sites W08 and W16 in the Dongsha area, this study aims to uncover the sediment property differences between deep and shallow gas hydrate reservoirs and their impact on gas hydrate accumulation through grain size, X-ray diffraction, and specific surface area (SSA) analyses. The findings classify the study intervals into four distinct layers: shallow non-gas hydrate layer (shallow-NGHL), shallow gas hydrate reservoir (shallow-GHR), deep non-gas hydrate layer (deep-NGHL), and deep gas hydrate reservoir (deep-GHR). In the clayey silt sediment reservoirs, grain size has a minor influence on gas hydrate reservoirs. Both shallow and deep NGHLs, characterized by high smectite content and SSA, possess a complex structure that impedes gas and fluid migration and offers limited potential reservoir space. Consequently, both shallow and deep NGHLs function as sealing beds. The deep GHR, having low smectite content and SSA, exhibits a strong capacity for gas and fluid migration and greater potential reservoir space. As a result, sediment properties significantly influence the deep GHR. Seepage primarily controls the shallow GHR.

Keywords: gas hydrate; accumulation mechanism; gas hydrate reservoir; sediment properties; Dongsha area

1. Introduction

Gas hydrates are increasingly recognized as a promising clean energy alternative to traditional fossil fuels, due to their abundant reserves and high thermal efficiency [1–3]. Meanwhile, the environmental impacts and geohazards resulting from gas hydrate decomposition are increasingly the focus of research [4–7]. Gas hydrate reservoirs are located globally in permafrost and marine sediments, including the Malik in Canada [8], the Qilian Mountains in western China [9], the Alaminos Canyon [10], the Cascadia subduction zone [11], the Hikurangi Margin of New Zealand [12], the northern South China Sea (SCS) [13], the Nankai Trough [14], the Ulleung Basin [15], and the Krishna–Godavari Basin [16]. Most gas hydrates globally are found in seafloor sediments. However, the northern SCS's gas hydrates are found in clayey silt, a sediment significantly finer than the well-studied sandy gas hydrate reservoirs [13,17,18]. Most of offshore gas hydrates are located in fine-grained sediments [10,19]. Successful trial productions in the northern SCS during 2017 and 2020 demonstrated the feasibility of recovering and utilizing gas hydrate resources in fine-grained sediments [13,20,21]. Nevertheless, further research is required on the theory of gas hydrate accumulation in fine-grained sediment reservoirs.

The Dongsha area is a critical zone for gas hydrate exploration in the northern SCS. Geophysical techniques, including seismic and logging data analysis, have characterized gas hydrate reservoirs in the Dongsha area [22–25]. Several studies have synthesized diverse gas hydrate accumulation models on a macroscale [25–28]. In the Dongsha area, there are two types of gas hydrate reservoirs: deep (below approximately 60 m below the seafloor [mbsf]) and shallow (within approximately 25 mbsf). High-flux methane seepage predominantly influences shallow reservoirs, while low-flux methane diffusion primarily governs deep reservoirs [26]. At the macroscale, methane flux and migration pathways, including fractures and gas chimneys, dictate gas hydrate accumulation. However, the determinants and mechanisms behind gas hydrate accumulation at the microscale, especially in fine-grained reservoirs, are still not well understood. Prior research has demonstrated that sediment properties influence gas hydrate accumulation. Generally, an increase in coarse-grained components within sediments positively impacts gas hydrate accumulation [3,14,29–31]. However, research in the Shenhu area of the northern SCS revealed that sediment properties' influence on gas hydrates in fine-grained sediment reservoirs differs from their impact in sandy gas hydrate reservoirs. The controlling effect of grain size on gas hydrate accumulation is weaker than that of sedimentary processes or mineral composition [32,33]. The Dongsha area, proximate to the Shenhu region, exhibits a more complex gas hydrate occurrence compared with Shenhu [28]. Numerous studies have focused on the gas resources [34], distribution of gas hydrate reservoirs [17,24,35], and morphology of gas hydrates in the Dongsha area [17]. In the Dongsha area, the canyon sedimentary system and sediment waves play a crucial role in controlling gas hydrate accumulation on a macro scale [23,36]. Analysis of geophysical and geochemical data reveals that the process of gas hydrate accumulation in the Dongsha area is complex, involving multiple stages of formation and decomposition [25,37]. Consequently, sediment properties potentially exert significant influence on gas hydrate accumulation. However, detailed studies on the impact of sediment properties on gas hydrate control in the Dongsha area are scarce, and the variations in sediment properties between deep and shallow gas hydrate reservoirs have not been documented. Identifying potential differences in sediment properties between deep and shallow gas hydrate reservoirs, along with understanding the control mechanisms of sediment properties on gas hydrate accumulation, could be crucial for refining the microscale gas hydrate accumulation model.

In 2013, the Guangzhou Marine Geological Survey's second gas hydrate expedition (GMGS-2) began in the Dongsha area, leading to the discovery of high-quality gas hydrate layers at multiple locations [17,33,35]. During GMGS-2, cores were collected from the gas hydrate reservoir and adjacent sediment layers across multiple sites. The collected cores spanned both deep and shallow gas hydrate reservoirs in the Dongsha area [17]. Analyzing sediment properties differences, such as grain size, mineral composition, and biological content, between deep and shallow gas hydrate reservoirs, we identified key factors influencing gas hydrate enrichment and clarified the unique accumulation patterns of deep and shallow gas hydrates at the microscale.

2. Geological Setting

The SCS is one of the largest marginal seas in the western Pacific Ocean. Located in a transitional zone, the northern slope of the SCS connects quasi-passive and active continental margins [2]. This region features a complex geological structure, containing several large and medium-sized Cenozoic sedimentary basins with deposition thicknesses over 10,000 m [23]. In the northern SCS, sedimentation rates for late Pleistocene and Holocene strata range from 17.9 to 19.6 cm/ka and 9.6 to 14.6 cm/ka, respectively, at water depths of 200 to 3400 m and temperatures of 2 to 5 °C [23]. The rapid deposition rates, combined with favorable temperature and pressure conditions, promote the formation of gas hydrates [35].

The Dongsha area is in the eastern part of the Pearl River Mouth Basin, mainly located near the central uplift area of the Taixinan Basin, where the water depth gradually increases from 600 m to more than 2100 m (Figure 1a). The seabed topography of the GMGS2 in the Dongsha area is complex, with extensive occurrences of submarine ridges, canyons, and landslides. Faults and diapirs in the Dongsha area are highly active and can be divided into three types: inherent faults, minor normal faults that appear in the upper part of large gas chimneys, and strike–slip faults in deep water [35]. Gas chimneys have been widely observed in the Dongsha area, particularly on submarine ridges [17,23,24,35]. Faults and gas chimneys are important channels for upward fluid migration in the Dongsha area [17,23,25].

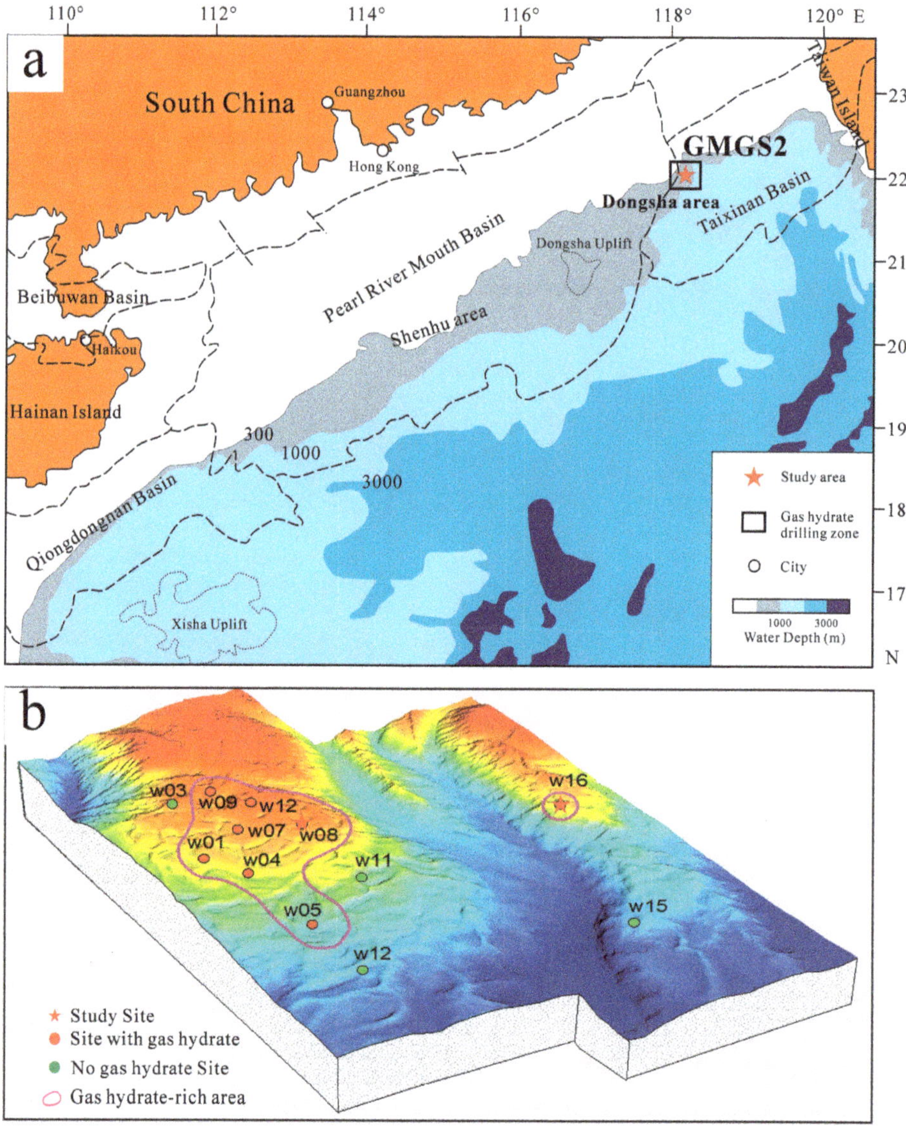

Figure 1. Geological map of the study area: (**a**) depicts the geological and water depth of the northern slope of the SCS (modified by [38]); (**b**) shows an enlarged view of the black box in (**a**), highlighting the GMGS2 drilling area (modified by [17]).

Logging data were obtained from 13 GMGS2 sites, with 8 indicating gas hydrate presence. Gas hydrate samples were collected through coring at sites W05, W07, W08, W09, and W16 (Figure 1b) [24,35]. In the Dongsha area, gas hydrates predominantly appear in massive, layered, vein-like, tubercular, and dispersed forms [17,23]. Gas hydrates in the study area were found in both deep (below ~60 mbsf) and shallow (within ~25 mbsf) reservoirs. The study area exhibited both dispersed and leaked gas hydrates. The gas hydrate-bearing layer reaches a maximum thickness of 32 m, with saturation levels exceeding 50% [17,35]. Sites W08 and W16 featured both deep and shallow gas hydrate reservoirs, in contrast with other sites in the study area, which had only one reservoir type [17]. Consequently, this study focused on sites W08 and W16 as key sites (Figure 1b).

3. Materials and Methods

3.1. Sites and Samples

The water depths at sites W08 and W16 were 801 m and 871 m, respectively. At site W08, shallow gas hydrates appeared 8–24 mbsf, mainly exhibiting massive, layered, vein-like, and tubercular forms. Deep gas hydrates appeared at 61–95 mbsf and were mainly massive (Figure 2a). Furthermore, some authigenic carbonate minerals have been found near the seabed and in the ~60 mbsf sediment layers [17]. At site W16, shallow gas hydrates appeared at 10–24 mbsf and mainly exhibited a tubercular form. Deep gas hydrates appeared at 182–206 mbsf and mainly exhibited dispersed and vein-like forms (Figure 2b) [17,38].

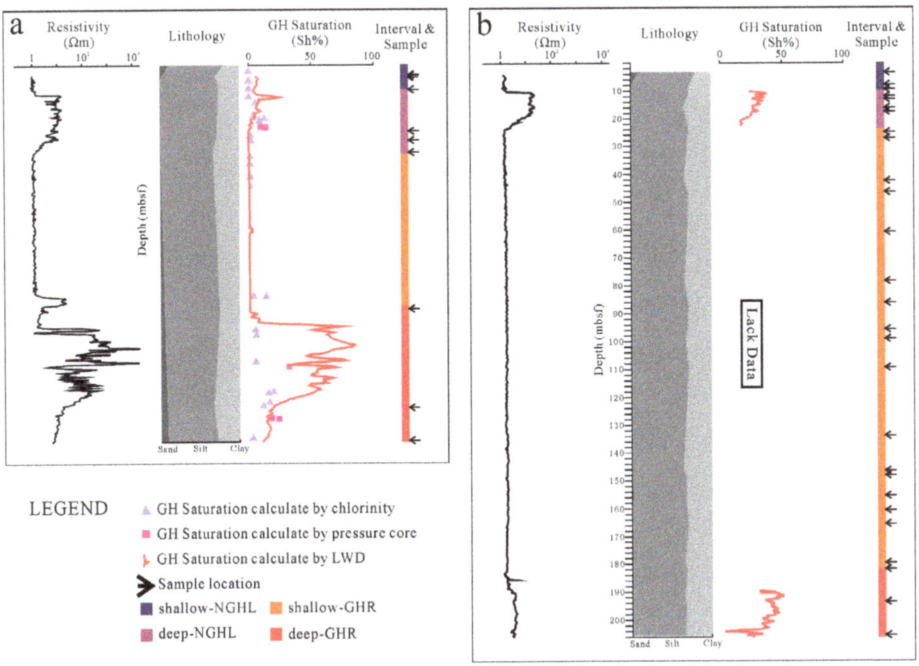

Figure 2. Column of sites W08 (**a**) and W16 (**b**). Due to the data limitations, gas hydrate saturation between 22 and 186 mbsf in site W16 are not obtained in this study. The left column displays the LWD resistivity values, represented by black curves. The middle columns represent lithology, plotted according to the fractions of sand, silt, and clay in the sediments. The right columns show GH (gas hydrate) saturation levels calculated using chlorinity (light purple triangles), pressure core (pink squares), and LWD resistivity values (red curves). The 'Interval & Sample' column indicates the study units and their respective sampling depths, marked by black arrows.

To delineate the differences more accurately between gas hydrate reservoirs and non-gas hydrate layers, this study extensively sampled sediments from both the hydrate-bearing layers and the adjacent non-hydrate layers. In the present study, 10 and 27 samples were collected from sites W08 and W16, respectively (Table 1 and Figure 2). A total of 37 samples were used for grain-size measurements, X-ray diffraction (XRD), and specific surface area (SSA) tests.

Table 1. The sample list of sites W08 and W16.

No.	Site	Sample	Depth (mbsf)	No.	Site	Sample	Depth (mbsf)
1	W08	W08-1	4.4	20	W16	W16-10	46.35
2		W08-2	4.5	21		W16-11	60.7
3		W08-3	5	22		W16-12	78.3
4		W08-4	8	23		W16-13	86.1
5		W08-5	18.2	24		W16-14	95.5
6		W08-6	20.5	25		W16-15	99
7		W08-7	23.4	26		W16-16	109.3
8		W08-8	61.6	27		W16-17	133.7
9		W08-9	85.7	28		W16-18	146.35
10		W08-10	93.7	29		W16-19	148
11	W16	W16-1	7.75	30		W16-20	155.3
12		W16-2	9.35	31		W16-21	160.5
13		W16-3	11.8	32		W16-22	165.35
14		W16-4	12.85	33		W16-23	179.5
15		W16-5	16	34		W16-24	181.65
16		W16-6	17.35	35		W16-25	193.35
17		W16-7	25	36		W16-26	193.41
18		W16-8	27	37		W16-27	205.3
19		W16-9	42.3				

3.2. Grain-Size Measurements

Grain-size measurement methods followed the protocols outlined by [33,39]. A total of 37 samples were evenly divided into two groups for grain-size analysis. Each group underwent a distinct preprocessing method: either A-type or B-type. In the A-type preprocessing method, approximately 1 g of sediment was weighed and exposed to 10 mL of 15% hydrogen peroxide (H_2O_2) for 24 h to eliminate organic matter. Subsequently, 10 mL of 20% hydrochloric acid (HCl) was added for another 24 h to remove inorganic carbon. The sample was then treated with 10 mL of 2-M sodium carbonate (Na_2CO_3) in a hot bath at 85 °C for 5 h to dissolve biological opal. Afterward, the sample was rinsed thrice with distilled water to eliminate residual Na_2CO_3. Following the removal of the supernatant, 10 mL of 5.1% sodium hexametaphosphate ($Na_4P_2O_7$ $10H_2O$) was introduced as a dispersing agent before conducting the grain-size analysis. Conversely, the B-type preprocessing method retains all components within the sample. Therefore, it involves only the final step of the A-type method, which is the addition of 10 mL of 5.1% $Na_4P_2O_7$ $10H_2O$. Grain-size data derived from various preprocessing methods possess distinct geological implications [33]. The A-type preprocessing method, which removes organic matter, inorganic carbon, and biological opal, allows the grain-size data to reflect the hydrodynamic conditions more accurately during deposition. This is because biological components and organic matter primarily originate from the sedimentation and transformation of organisms' post-mortem, rather than from hydrodynamic transport [18,33,40]. However, the B-type preprocessing method, which preserves all components within the sample, enables the grain-size data to represent the conditions conducive more effectively to gas hydrate formation.

After preprocessing, grain sizes were measured using a Marvin Mastersizer 3000 laser diffraction particle size analyzer. The calculation of sorting, kurtosis, and skewness was performed using the McManus method [41]. The analysis was conducted at the Beijing International Center for Gas Hydrates, Peking University, and the Key Laboratory of

Polar Geology and Marine Mineral Resources, Ministry of Education, China University of Geosciences (Beijing).

3.3. X-ray Diffraction

XRD data for both whole rock and clay fractions were obtained using a D8 Advance instrument (Bruker Corp., Karlsruhe, Germany). The XRD analysis was performed in accordance with the methodologies below.

Sediment samples, with a dry weight exceeding 5 g, were dried at 45 °C for 24 h in a drying cabinet prior to analysis. For clay-size fraction XRD analysis, samples were ground to 200 mesh and both organic and calcareous components were removed. Fractions smaller than 2 μm were separated using centrifugation, in accordance with Stock's Law. Subsequently, the clay component was combined with pure water to prepare aliquots for air-drying and ethylene glycol treatment prior to measurement [42,43]. The diffractometer, operating with a Cu K-alpha radiation source (λ = 0.154 nm) at 40 kV and 30 mA, was utilized. Scanning was conducted at a rate of 2° per minute over a range of 3° to 65°. Quantification of the results was achieved using Rockquan 2016 and ClayQuan 2016 software. The XRD analysis was carried out at the Micro Structure Analytical Laboratory, Peking University.

3.4. Specific Surface Area

SSA measurements were conducted with Quadrasorb EVO and Quantachrome Nova Station. The SSA was determined according to the method described by [33], utilizing the Brunauer–Emmet–Teller (BET) equation [44] for calculation. The SSA analysis was performed at the Institute of Analysis and Testing, Beijing Academy of Science and Technology.

4. Results

4.1. Lithology

The lithologies of sites W08 and W16 are the same, dominated by clayey silt (Figure 2). The grain-size data showed that the sand fraction in the 10 sediment samples from site W08 ranged from 0.11 to 9.60%, with an average of 2.30%. The silt fraction ranged from 63.75 to 73.18%, averaging 68.80%. The clay fraction ranged from 24.81 to 32.77%, averaging 28.90%. The grain-size data showed that the sand fraction in the 27 sediment samples from site W16 ranged from 0 to 2.71%, with an average of 0.95%. The silt fraction ranged from 65.36 to 73.79%, averaging 69.26%. The clay fraction ranged from 35.30 to 34.22%, averaging 29.80%.

4.2. Vertical Distribution of Gas Hydrate

According to previous research and cruise reports, the gas hydrate saturation of shallow and deep reservoirs at site W08 is estimated to be 10–14% and up to 33%, respectively, by the degassing of the pressure core (Figure 2a) [17]. Logging-while-drilling (LWD) data showed that the gas hydrate saturations of shallow and deep reservoirs at site W16 are ~42% and up to 50%, respectively (Figure 2b) [17,45].

4.3. Grain-Size Parameters

4.3.1. A-Type Preprocessing Method

Grain-size analysis using the A-type preprocessing method revealed that the median particle size of the 10 sediment samples from site W08 varied between 6.55 and 7.19 Φ (10.67 and 6.85 μm), averaging 6.84 Φ (8.73 μm). Sediment sorting values ranged from 1.22 to 1.48, with a mean of 1.37. Kurtosis values for the sediments spanned from 0.03 to 0.17, averaging 0.06. Sediment skewness varied from 0.92 to 1.17, with an average value of 1.02 (Figure 3a).

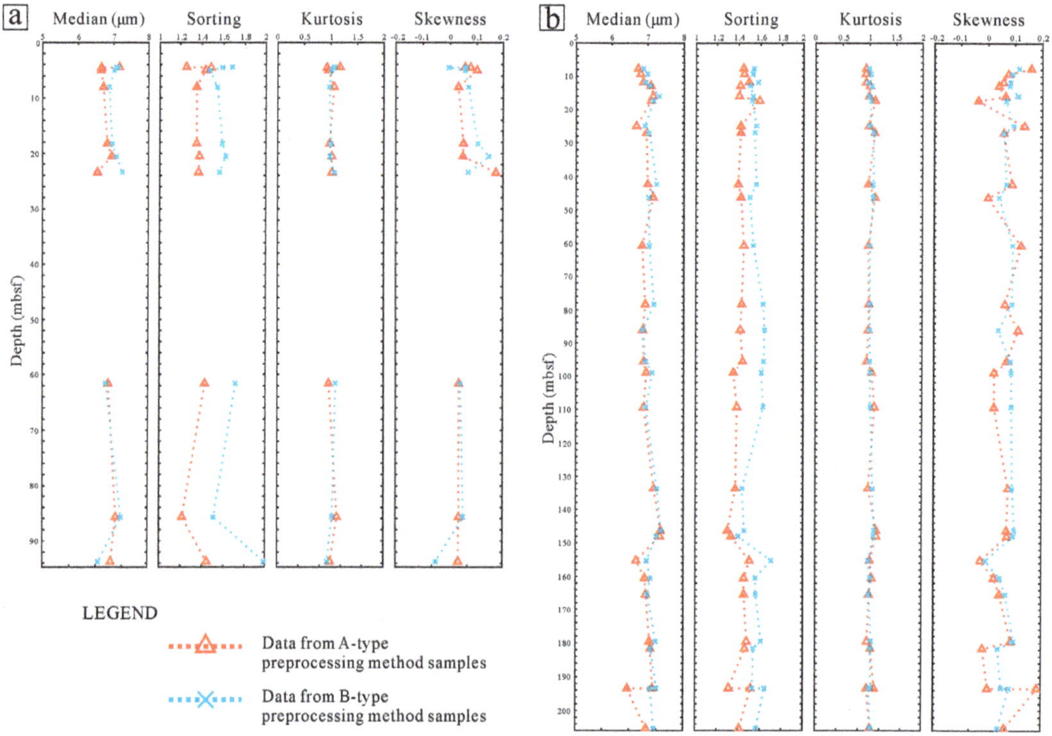

Figure 3. Column of grain-size parameters (median grain size, sorting, kurtosis, and skewness) of sites W08 (**a**) and W16 (**b**). Red triangles and dashed lines represent grain-size parameters from A-type preprocessing method samples and their vertical trends. Blue "X" marks and dashed lines indicate grain-size parameters from B-type preprocessing method samples and their vertical trends.

Analysis using the A-type preprocessing method indicated that the median particle size of the 27 sediment samples from site W16 ranged from 6.45 to 7.37 Φ (11.44 and 6.05 μm), with an average of 6.97 Φ (7.98 μm). Sorting values for the sediments ranged from 1.30 to 1.59, with a mean of 1.43. Kurtosis of the sediments was between −0.04 and 0.18, averaging 0.06. Skewness for the sediments was in the range of 0.93 to 1.13, with an average value of 1.02 (Figure 3b).

4.3.2. B-Type Preprocessing Method

Grain-size analysis, utilizing the B-type preprocessing method, revealed that the median particle size of the 10 sediment samples from site W08 varied between 6.57 and 7.25 Φ (10.53 and 6.57 μm), averaging 6.99 Φ (7.87 μm). Sediment sorting values ranged from 1.47 to 1.99, with a mean of 1.63. Kurtosis values for the sediments spanned from −0.05 to 0.15, averaging 0.05. Sediment skewness varied between 0.94 and 1.09, with an average value of 1.02 (Figure 3a).

Analysis of grain size, conducted with the B-type preprocessing method, indicated that the median particle size for the 27 sediment samples from site W16 was between 6.87 and 7.37 Φ (8.55 and 6.05 μm), with an average of 7.09 Φ (7.34 μm). Sorting values for the sediments were between 1.40 and 1.70, with a mean of 1.56. Kurtosis for the sediments ranged from −0.01 to 0.11, averaging 0.07. Skewness of the sediments varied from 0.97 to 1.09, with an average value of 1.02 (Figure 3b).

4.4. Mineral Components

Whole-rock XRD analysis of the 37 samples from sites W08 and W16 revealed a predominance of quartz, feldspar, carbonates, and clay minerals, complemented by minor amounts of pyrite and gypsum (Figure 4). Quartz content in the 10 samples from site W08 varied between 23% and 41%, averaging 36.6%. Feldspar content ranged from 12% to 20%, with an average of 15.8%. Carbonate content ranged from 6% to 40%, averaging 11.5%. Clay mineral content varied from 20% to 36%, with an average of 31.7% (Figure 4a). In the 27 samples from site W16, quartz content ranged from 3% to 42%, averaging 37.8%. Feldspar content in these samples ranged from 15% to 22%, with an average of 16.8%. Carbonate content in these samples varied from 9% to 13%, averaging 10.5%. Clay mineral content in these samples ranged from 28% to 38%, with an average of 32.7% (Figure 4b).

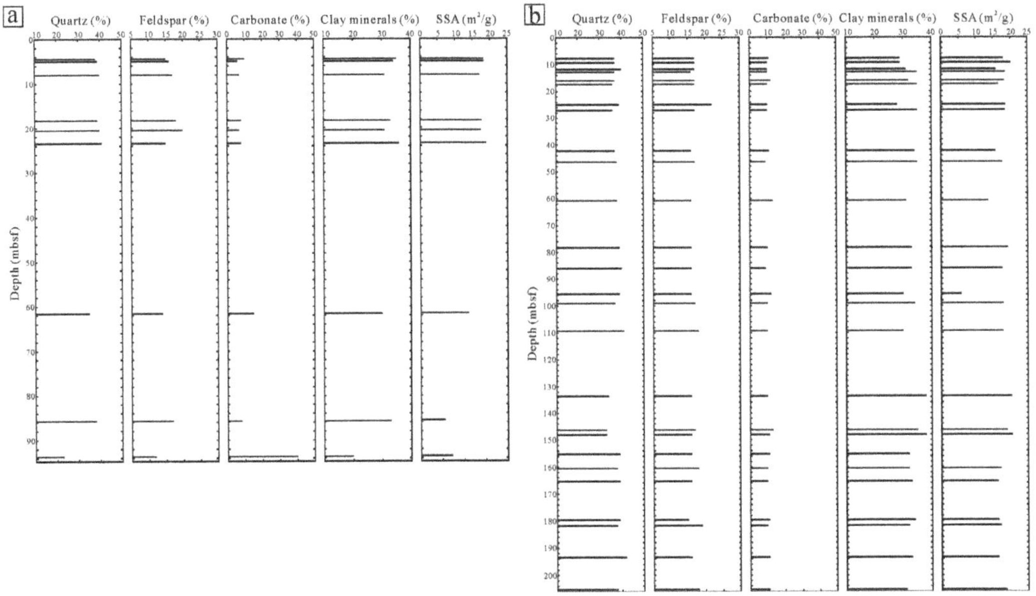

Figure 4. Column of whole rock minerals components and SSA from sites W08 (**a**) and W16 (**b**). Black lines show the proportions of minerals (quartz, feldspar, carbonate, or clay) in the whole rock, alongside SSA values.

XRD analysis of the clay fraction revealed that the 37 samples from sites W08 and W16 predominantly contained smectite, illite, kaolinite, and chlorite (Figure 5). In the 10 samples from site W08, smectite content varied between 29% and 42%, averaging 36.5%. Illite content ranged from 48% to 54%, with a mean of 50.7%. Kaolinite content was between 4% and 6%, with an average of 5.1%. Kaolinite content was between 4% and 6%, with an average of 5.1%. For the 27 samples from site W16, smectite content ranged between 32% and 46%, with an average of 39.5%. Illite content in these samples ranged from 47% to 54%, with an average of 48.9%. Kaolinite content varied between 3% and 7%, averaging 4.8%. Chlorite content was found to be between 4% and 11%, with an average of 6.8% (Figure 5b).

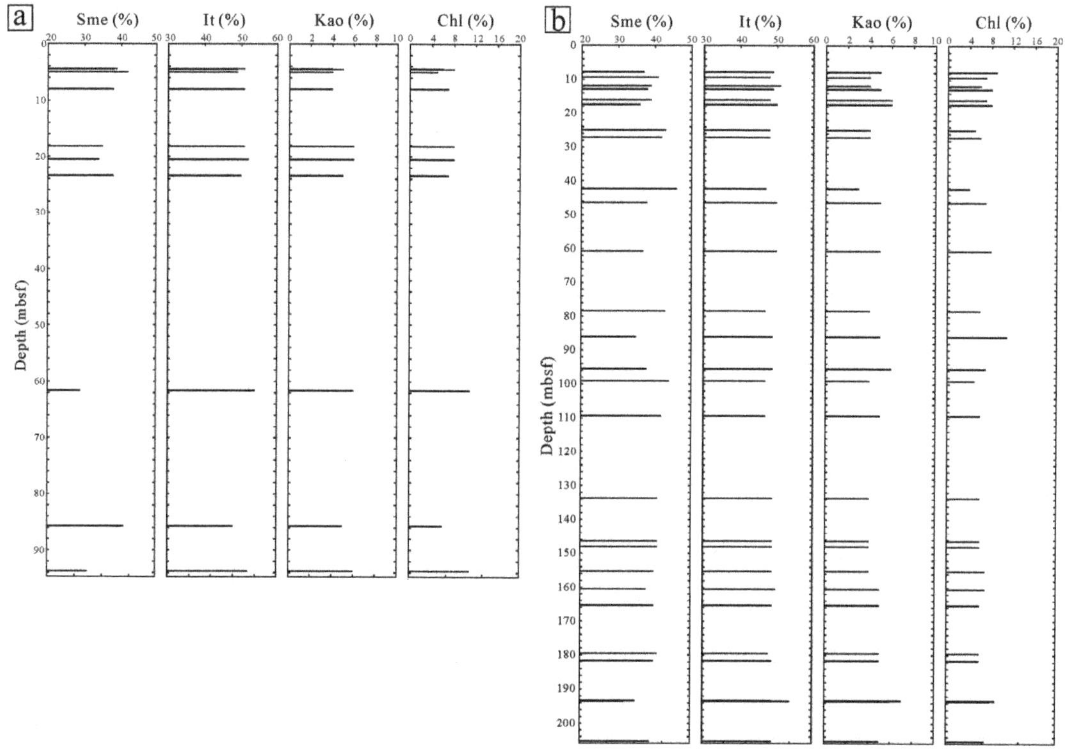

Figure 5. Column of clay minerals components from sites W08 (**a**) and W16 (**b**). Black lines denote the percentages of clay minerals (smectite, illite, kaolinite, chlorite) in the total clay mineral content. Sme, It, Kao, and Chl are short for smectite, illite, kaolinite, and chlorite, respectively.

4.5. Specific Surface Area

Clay mineral content in these samples ranged from 28% to 38%, with an average of 32.7% (Figure 4b). The SSA of the 10 samples collected from site W08 varied between 6.94 and 18.82 m^2/g, averaging 15.23 m^2/g. The SSA of the 27 samples from site W16 ranged from 14.89 to 20.44 m^2/g, with a mean value of 17.61 m^2/g.

5. Discussion

5.1. Characteristics of Deep and Shallow Gas Hydrate Reservoirs

Gas hydrate enrichment levels are commonly assessed through resistivity logs, Cl-anomalies in pore water, and pressure core degassing [33,46–48]. Resistivity logs serve as the most straightforward and reliable indicators for identifying gas hydrate-rich layers, which typically exhibit high resistivity values [46,48,49]. At site W08, the intervals between 0–8 and 24–61 mbsf displayed low resistivity values, indicating an absence of gas hydrates during drilling [17]. Significant increases in resistivity were observed at 8–24 and 61–95 mbsf at site W08, accompanied by Cl^- anomalies in pore water and degassing in pressure core samples, suggesting gas hydrate presence in these intervals (Figure 2a) [35]. The distribution patterns of gas hydrates at site W16 mirror those observed at site W08. At site W16, intervals of 0–10 mbsf and 24–182 mbsf exhibited low resistivity without any gas hydrate detection. Significant increases in resistivity at 10–24 and 182–206 mbsf at site W16 were noted, with gas hydrates identified during drilling operations (Figure 2b) [17,35].

Previous studies have found that gas hydrates in shallow layers mostly fill fractures in [17]. Generally, the seepage of high-flux methane controls the gas hydrate in shallow layers, and the fluid migration channel is mainly a fault or gas chimney. The gas sources are the decomposition of gas hydrates in deep layers and the direct upward seepage of deep biological and thermal gases [28]. The gas hydrates in the deep layers were mainly dispersed and in some massive forms. Dispersed gas hydrates are formed under the control of low-flux methane diffusion [17]. Research in the Shenhu area in the northern SCS indicates that massive gas hydrates in the deep layer may also be formed by low-flux methane diffusion [33,50]. Therefore, the research intervals of sites W08 and W16 can be categorized into four units: shallow and deep non-gas hydrate layers (shallow-NGHL and deep-NGHL) and shallow and deep gas hydrate reservoirs (shallow-GHR and deep-GHR) (Table 2 and Figure 2). The shallow and deep GHRs were controlled by seepage and diffusion, respectively, and the gas hydrate saturation of the deep GHRs was higher than that of the shallow GHRs (Figure 2).

Table 2. Classification of the study intervals at sites W08 and W16.

Intervals	Site	Depth Range (mbsf)	Maximum Gas Hydrate Saturation (%) *
Shallow non-gas hydrate layers (NGHL)	W08 W16	0~8 0~10	-
Shallow gas hydrate reservoirs (GHR)	W08 W16	8~24 10~24	14% 42%
Deep non-gas hydrate layers (NGHL)	W08 W16	24~61 24~182	-
Deep gas hydrate reservoirs (GHR)	W08 W16	61~95 182~206	33% 50%

* The data of gas hydrate saturation were obtained from [17,35].

5.2. Influence of the Lithology and Sedimentary Process

The grain-size data obtained through the B-type preprocessing method reflect the sediments' original geometric parameters during gas hydrate accumulation [33]. These data facilitate the lithological analysis of the study intervals at sites W08 and W16.

In the study interval, the sediments mainly consisted of clayey silt. There were no differences in lithology between the different units (Figure 6). The silt fraction was dominant, followed by the clay fraction. The sand fraction was rare (Figure 7). The differences in sand, silt, and clay fractions among the different units in the study intervals were not significant (Figure 7). The average sand fraction in the deep GHR was slightly higher, which was caused by the abnormally high sandy components in some samples. According to [17,51], there were many bioclasts at ~90 mbsf at site W08 and the grain size of these bioclasts is generally large, causing an increase in the sandy fraction.

The absence of lithological differentiation suggests that the sedimentary processes within the study intervals were likely straightforward and stable. Grain-size data, obtained through the A-type preprocessing method, are utilized to characterize sedimentary processes. The A-type preprocessing method eliminates organic matter, inorganic carbon, and biogenic opal. Minerals in the sediments, post-preprocessing and excluding some authigenic minerals, are transported and deposited through sedimentary processes. The grain size of these minerals aids in reconstructing the sedimentary processes [18,39,52].

There was a significant positive correlation between the median grain size and silt fraction, which indicated that the silt fraction was the primary particle size of the sediments transported during the sedimentary process (Figure 8a). The median grain size did not correlate with the sorting, skewness, or kurtosis (Figure 8b–d). There were no differences between the units in the study intervals, suggesting that the differences in sedimentary processes between the different units were very weak (Figure 8b–d).

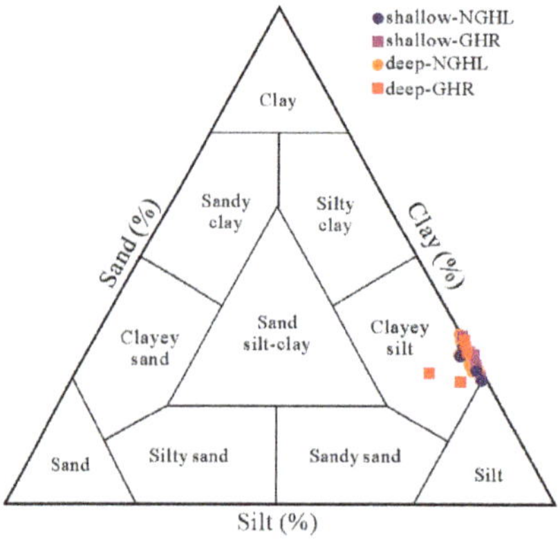

Figure 6. Ternary diagram showing lithological classification of sediments from sites W08 and W16.

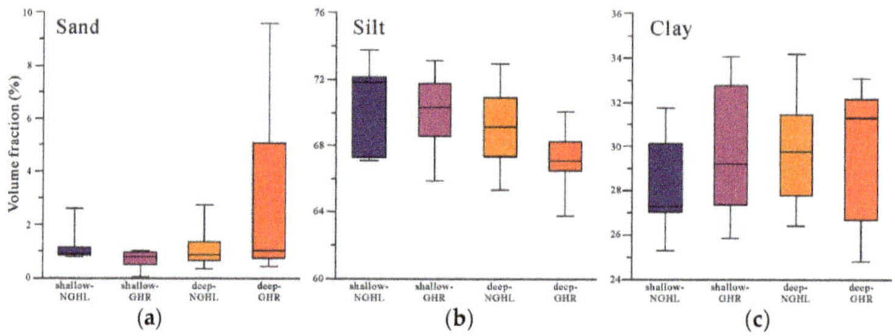

Figure 7. Box plot of the sand, silt, and clay fraction of different unit categories. (**a**–**c**) show the statistical data of sand, silt, and clay fractions, respectively.

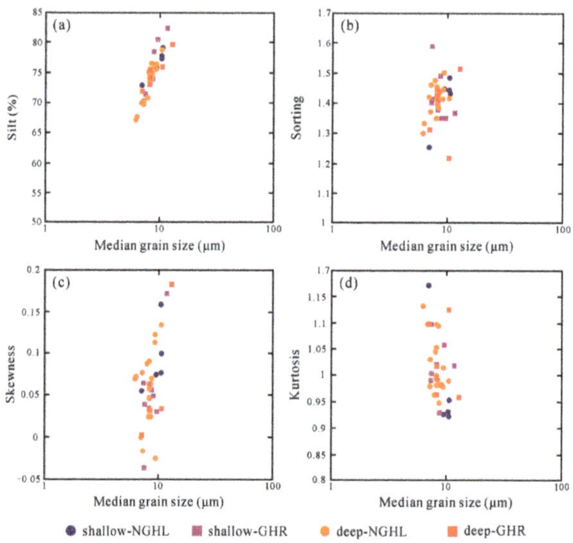

Figure 8. Dispersion diagrams showing the relationship between median grain size and parameters. (**a**) shows the relationship between silt fraction and median grain size; (**b**–**d**) show the relationship between sorting, skewness, and kurtosis and median grain size, respectively.

As depicted in Figure 9, sediment data distribution within the study intervals aligns with the C-M baseline, suggesting gravity-driven processes dominated the study area. Research indicates that fine-grained turbidite layers result from sediment sliding along paleochannels [18,53–55]. Numerous studies demonstrate that fine-grained turbidite layers facilitate gas hydrate accumulation, a finding consistent across the Dongsha area [18,33,55,56]. However, the sedimentary processes of both deep and shallow GHRs, as well as between GHRs and NGHLs, showed no significant differences, suggesting the presence of microscale factors influencing gas hydrate enrichment in the study area.

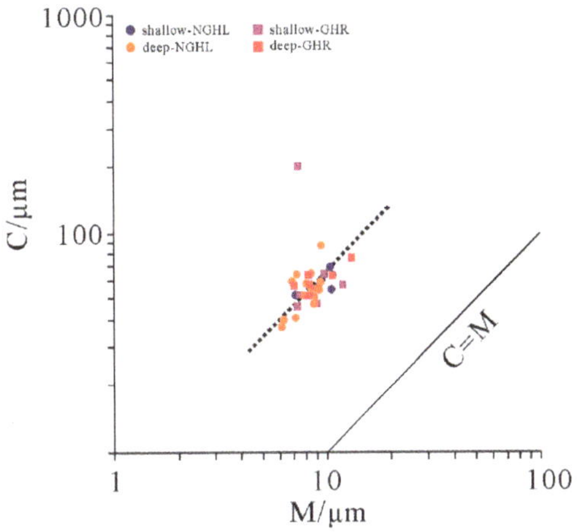

Figure 9. C-M diagram of sediments from sites W08 and W16.

5.3. Influence of Grain Size

Grain-size data obtained through B-type preprocessing are instrumental in analyzing gas hydrate accumulation. Statistical analysis of grain-size parameters revealed no significant differences in median particle size, skewness, and kurtosis across the study intervals (Figure 10a,c,d). The deep GHR exhibited a slightly coarser median particle size (Figure 10a) and poorer sorting (Figure 10b) compared with other units, aligning with its higher sand fraction. The most notable difference between group A and B samples' grain-size parameters was in sorting, especially within the deep-GHR (Figures 3 and 11a). The key distinction in preprocessing between group A and B samples lies in the preservation of organic matter, inorganic carbon, and biogenic opal. A clear positive correlation was observed between sorting differences and carbonate mineral content. The abundance of foraminiferal fossils in the Dongsha area suggests that foraminifera are the primary contributors to the poor sorting observed in the deep GHR. Prior research has shown that foraminifera, characterized by their chambered structure, are typically classified within the sandy fraction. Consequently, the high abundance of foraminifera is positively correlated with gas hydrate accumulation [17,33,40], potentially elucidating the enrichment of gas hydrates in the deep GHR.

In general, the difference in grain-size parameters between the GHRs and NGHLs is not significant, especially because the significance of grain-size parameters for shallow GHRs is weak. The sorting of the deep GHR is poor, reflecting the effect of foraminifera on the gas hydrate reservoir.

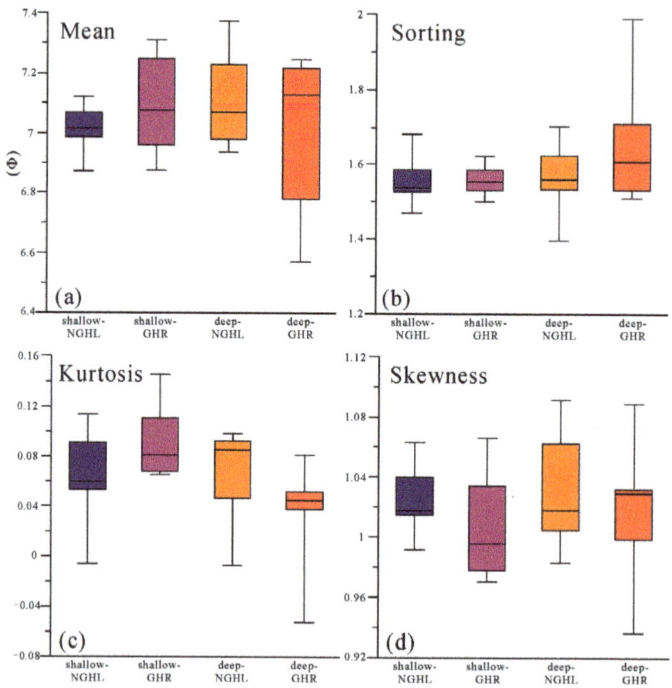

Figure 10. Box plot of the grain-size parameters of different unit categories. (**a**–**d**) show the statistical data of mean grain size, sorting, kurtosis, and skewness, respectively.

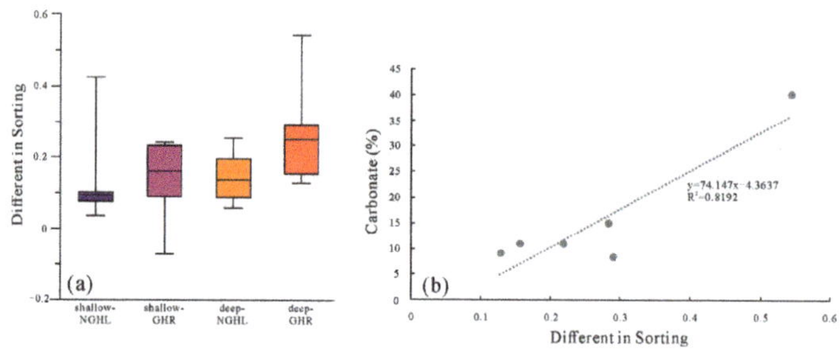

Figure 11. Box plot of difference in sorting (between A- and B-type preprocessing methods) (**a**) and the relationship between the carbonate minerals component and difference in sorting (**b**).

5.4. Influence of Mineral Components

Mineral components are another potential factor affecting gas hydrate accumulation. The quartz and feldspar contents were similar in different units of the study interval (Figure 12a,b). The carbonate mineral content was similar in the shallow-NGHL, shallow-GHR, and deep-NGHL and was relatively high in the deep-GHR (Figure 12c). This is because some samples in the deep-GHR have high foraminiferal abundance, which is consistent with the results of SEM, grain-size measurement, and lithology analysis. The clay mineral contents in the shallow-NGHL, shallow-GHR, and deep-NGHL were significantly higher than those in the deep-GHR (Figure 12d).

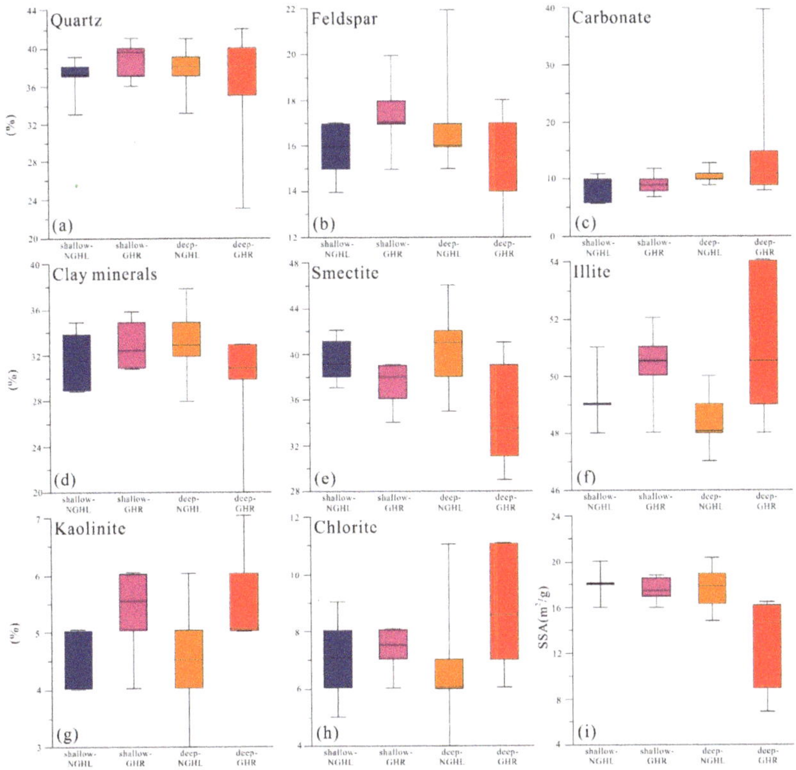

Figure 12. Box plot of the mineral components and SSA of different unit categories. (**a**–**d**) show the statistical data of quartz, felspar, carbonate minerals, and clay minerals contents in the whole rock, respectively; (**e**–**h**) show the statistical data of smectite, illite, kaolinite, and chlorite contents in the clay minerals, respectively; (**i**) shows the statistical data of SSA.

To further reveal the influence of clay minerals on gas hydrate accumulation, XRD analysis of the clay fraction was performed. During the study period, smectite and illite were the dominant types of clay minerals, accounting for more than 80% of all clay minerals. Thus, changes in the smectite and illite contents were the focus of this study. The smectite content in the shallow- and deep-GHRs was significantly lower than that in the shallow- and deep-NGHLs (Figure 12e), whereas the opposite was true for the illite content (Figure 12f). In the GHR, the smectite in the deep-GHR is lower than that in the shallow-GHR (Figure 12e), and the content of illite showed the opposite trend (Figure 12f). The kaolinite and chlorite contents were lower and similar in the different units of the study interval (Figure 12g,h). The difference in mineral components indicates that clay minerals, especially smectite, control the gas hydrate enrichment. Previous studies have indicated that the complex structure of smectite reduces its permeability and restricts fluid migration. The strong adsorption of free gas restricts gas migration [57–59]. To characterize the complexity of the sediment structure, we analyzed the SSA data of the samples. The results showed that the SSA of the deep-GHR was significantly lower than that of the shallow-NGHL, shallow-GHR, and deep-NGHL (Figure 12i). The SSA of the shallow-GHR is like that of the shallow- and deep-NGHLs (Figure 12i) because, although the shallow-GHR has less smectite content, the overall clay mineral content is not low (Figure 5). Further analysis of the correlation between SSA and mineral content showed that the correlation between the felsic mineral content (quartz and feldspar) and SSA was not significant (Figure 13a). In contrast, the content of all clay minerals and the content of smectite in the clay minerals were positively correlated with the SSA (Figure 13b,c). The smectite content in the whole rock was significantly and positively correlated with the SSA (Figure 13d). The correlation analysis indicated that clay minerals, especially smectite, controlled the SSA.

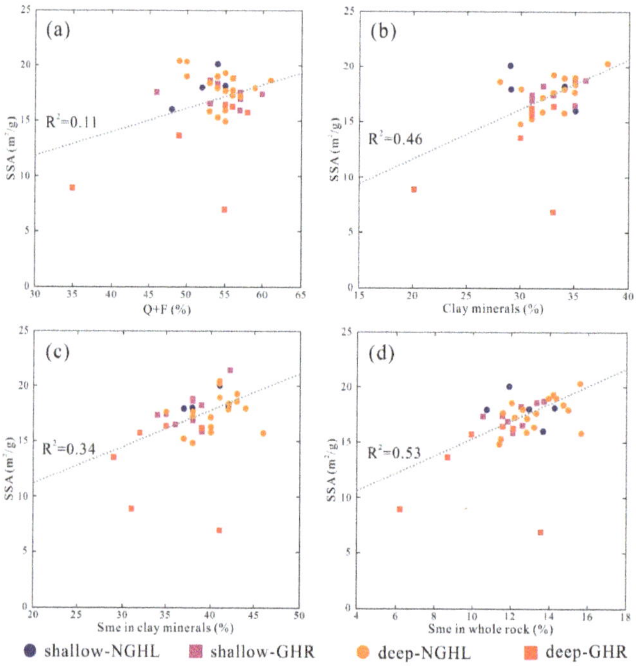

Figure 13. Dispersion diagrams showing the relationship between minerals components and SSA. (**a**) is the correlation between the total content of quartz and feldspar and SSA; (**b**) is the correlation between the content of clay minerals and SSA; (**c**) is the correlation between the content of smectite in clay minerals and SSA; (**d**) is the correlation between the content of smectite in whole rock and SSA.

SSA serves as a robust indicator for assessing the deep-GHR, reflecting changes in sediment mineral composition. Specifically, smectite and other clay minerals hinder gas hydrate accumulation. A reduced smectite content in deep GHRs promotes gas hydrate accumulation. Clay minerals in shallow GHRs exert less influence on gas hydrate accumulation compared with those in deep GHRs. However, the abundant clay minerals and smectite in NGHLs above both deep and shallow GHRs could serve as a seal, aiding in the preservation of underlying gas hydrates.

Additionally, some studies suggest that the stronger binding of water in the smectite decreases the formation efficiency of gas hydrates [60,61], contributing to the poor accumulation in smectite-rich layers. Furthermore, the complex microstructure of the I/S, as indicated by the mineral composition and SSA of sediments, influences the enrichment of gas hydrates. The complex microstructure of the smectite strongly adsorbs gas and fluids, and the vertical overlap between smectite-rich and low smectite content layers likely plays a crucial role in the distribution of high-quality gas hydrate reservoirs.

5.5. Constraints of Sediment Properties on Deep and Shallow Gas Hydrate Reservoirs

The sediment property data in the study interval indicate that mineral components may be key factors controlling gas hydrate accumulation in fine-grained sediments. In the Dongsha area, gas and fluid migrate along the gas chimney and fault to the gas hydrate stabilization zone (Figure 14a). In the deep-GHR, the smectite content was low, and the foraminiferal abundance was high. The sediment layer of the deep GHR had high permeability and a larger intergranular pore space, which improved the reservoir space and migration of gas and fluid conditions (Figure 14d). The deep NGHL had high smectite content. The sediment layer of the deep NGHL has a complex structure, poor permeability, and strong adsorption of gas, which makes it difficult for gas and fluid to migrate (Figure 14c). Thus, the vertical superposition of the deep-GHR and NGHL forms a "lithological trap" conducive to gas hydrate accumulation. When environmental changes lead to an increased methane flux from gas hydrate decomposition or gas chimneys and faults exist, gas and fluid can break through the sealing of the deep NGHL and continue to migrate upward [17,38,62]. With the release of pressure, the gas and fluid were blocked again when they migrated to the shallow NGHL and formed a shallow GHR (Figure 14b) [17,35]. Breakthroughs in deep-NGHL sealing require a pathway or overpressure release to impale the overlying layer. Thus, the shallow GHR is more clearly controlled by tectonism, in which gas hydrates are mostly found in fractures with vein-like and tubercular forms (Figure 14b). The control of the sediment properties in the shallow GHR was weaker than that in the deep GHR.

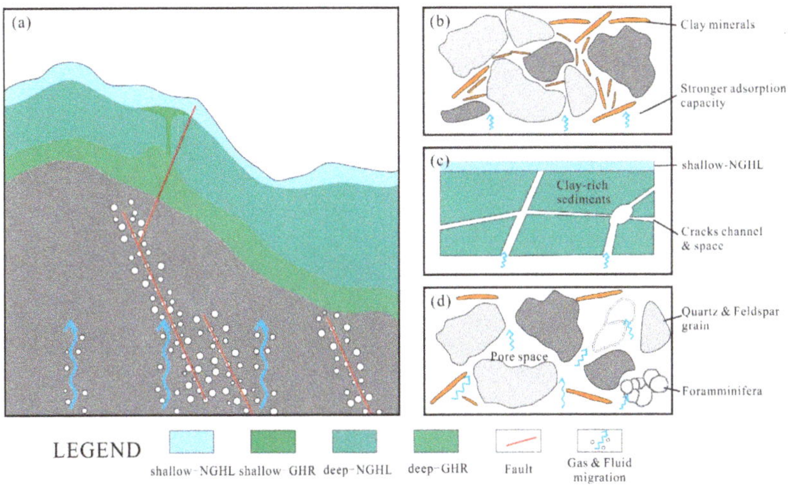

Figure 14. Gas hydrate enrichment patterns in the deep and shallow layer. The geomorphic of the Dongsha area is drawn according to [35], and the vertical scale of the deposition layers is enlarged. (**a**) gas hydrate enrichment patterns in macroscale; (**b**) gas and fluid migration in the NGHL; (**c**) gas hydrate enrichment in the shallow GHR; (**d**) gas hydrate enrichment in the deep GHR.

6. Conclusions

Gas hydrate reservoirs in the Dongsha region are classified into deep and shallow types. The deep reservoirs are primarily influenced by diffusion, whereas the shallow ones are affected by seepage. Both types are located within fine-grained sediments.

Grain-size parameters exert a minor influence on gas hydrate accumulation, whereas mineral composition, especially an increase in clay minerals like smectite, significantly impacts gas hydrate accumulation in fine-grained sediments, generally hindering it. The specific surface area of sediments, primarily determined by the content of clay minerals such as smectite, plays a crucial role in gas hydrate accumulation in deep diffusion reservoirs; a lower specific surface area is indicative of more favorable conditions.

Sediment properties exert limited control on gas hydrate accumulation in shallow seepage reservoirs. Tectonism primarily controls gas hydrate accumulation in shallow seepage reservoirs. However, the overlying reservoir with high smectite content acts as a seal, enhancing gas hydrate accumulation in shallow seepages. The deep diffusion gas hydrate reservoir, characterized by low smectite and high foraminiferal content, offers migration pathways and reservoir space. The overlying layer of the deep diffusion reservoir, rich in smectite, effectively impedes gas and fluid migration. The reservoir, with a high-smectite upper layer and a low-smectite lower layer, forms an optimal environment for gas hydrate accumulation.

Elucidating the influence of mineral composition, grain size, and sedimentary processes on gas hydrate accumulation aids in understanding the distribution patterns of gas hydrate reservoirs and supports further exploration and development efforts. Additionally, this analysis provides valuable insights into the laws governing gas hydrate accumulation in fine-grained sediments.

Author Contributions: Conceptualization, C.B. and H.W.; methodology, Y.Z. and X.X.; validation, Q.L.; formal analysis, Y.Z.; resources, H.W. and Q.L.; writing—original draft preparation, C.B.; writing—review and editing, H.W. and Q.L.; visualization, C.B.; project administration, H.W.; funding acquisition, H.W. and C.B. All authors have read and agreed to the published version of the manuscript.

Funding: This work was financially supported by the Laoshan Laboratory (No. LSKJ202203501) and the National Natural Science Foundation of China (No. 42376217).

Institutional Review Board Statement: Not applicable.

Informed Consent Statement: Not applicable.

Data Availability Statement: Datasets underpinning the study's results and findings are available via Figshare (DOI: http://doi.org/10.6084/m9.figshare.22281625, accessed on 16 March 2023).

Acknowledgments: We extend our gratitude to all contributors to the China National Gas Hydrate Program Expedition 2.

Conflicts of Interest: The authors declare no conflicts of interest.

References

1. Tréhu, A.; Ruppel, C.; Holland, M.; Dickens, G.; Torres, M.; Collett, T.; Goldberg, D.; Riedel, M.; Schultheiss, P. Gas Hydrates in Marine Sediments: Lessons from Scientific Ocean Drilling. *Oceanography* **2006**, *19*, 124–142. [CrossRef]
2. Wu, S.; Wang, X.; Wong, H.K.; Zhang, G. Low-Amplitude BSRs and Gas Hydrate Concentration on the Northern Margin of the South China Sea. *Mar. Geophys. Res.* **2007**, *28*, 127–138. [CrossRef]
3. Boswell, R.; Frye, M.; Shelander, D.; Shedd, W.; McConnell, D.R.; Cook, A. Architecture of Gas-Hydrate-Bearing Sands from Walker Ridge 313, Green Canyon 955, and Alaminos Canyon 21: Northern Deepwater Gulf of Mexico. *Mar. Pet. Geol.* **2012**, *34*, 134–149. [CrossRef]
4. Vasheghani Farahani, M.; Hassanpouryouzband, A.; Yang, J.; Tohidi, B. Development of a Coupled Geophysical–Geothermal Scheme for Quantification of Hydrates in Gas Hydrate-Bearing Permafrost Sediments. *Phys. Chem. Chem. Phys.* **2021**, *23*, 24249–24264. [CrossRef]
5. Vasheghani Farahani, M.; Hassanpouryouzband, A.; Yang, J.; Tohidi, B. Insights into the Climate-Driven Evolution of Gas Hydrate-Bearing Permafrost Sediments: Implications for Prediction of Environmental Impacts and Security of Energy in Cold Regions. *RSC Adv.* **2021**, *11*, 14334–14346. [CrossRef] [PubMed]
6. Ruppel, C. Permafrost-Associated Gas Hydrate: Is It Really Approximately 1% of the Global System? *J. Chem. Eng. Data* **2015**, *60*, 429–436. [CrossRef]
7. Ruppel, C.D.; Kessler, J.D. The Interaction of Climate Change and Methane Hydrates. *Rev. Geophys.* **2017**, *55*, 126–168. [CrossRef]
8. Schoderbek, D.; Farrell, H.; Howard, J.; Raterman, K.; Silpngarmlert, S.; Martin, K.; Smith, B.; Klein, P. *ConocoPhillips Gas Hydrate Production Test*; ConocoPhillips Co.: Houston, TX, USA, 2013; p. 1123878.
9. Zhu, Y.; Zhang, Y.; Wen, H.; Lu, Z.; Jia, Z.; Li, Y.; Li, Q.; Liu, C.; Wang, P.; Guo, X. Gas Hydrates in the Qilian Mountain Permafrost, Qinghai, Northwest China. *Acta Geol. Sin. Eng.* **2010**, *84*, 1–10. [CrossRef]
10. Boswell, R.; Shelander, D.; Lee, M.; Latham, T.; Collett, T.; Guerin, G.; Moridis, G.; Reagan, M.; Goldberg, D. Occurrence of Gas Hydrate in Oligocene Frio Sand: Alaminos Canyon Block 818: Northern Gulf of Mexico. *Mar. Pet. Geol.* **2009**, *26*, 1499–1512. [CrossRef]
11. Pohlman, J.W.; Kaneko, M.; Heuer, V.B.; Coffin, R.B.; Whiticar, M. Methane Sources and Production in the Northern Cascadia Margin Gas Hydrate System. *Earth Planet. Sci. Lett.* **2009**, *287*, 504–512. [CrossRef]
12. Kroeger, K.F.; Crutchley, G.J.; Kellett, R.; Barnes, P.M. A 3-D Model of Gas Generation, Migration, and Gas Hydrate Formation at a Young Convergent Margin (Hikurangi Margin, New Zealand). *Geochem. Geophys. Geosyst.* **2019**, *20*, 5126–5147. [CrossRef]
13. Li, J.; Ye, J.; Qin, X.; Qiu, H.; Wu, N.; Lu, H.; Xie, W.; Lu, J.; Peng, F.; Xu, Z.; et al. The First Offshore Natural Gas Hydrate Production Test in South China Sea. *China Geol.* **2018**, *1*, 5–16. [CrossRef]

14. Fujii, T.; Suzuki, K.; Takayama, T.; Tamaki, M.; Komatsu, Y.; Konno, Y.; Yoneda, J.; Yamamoto, K.; Nagao, J. Geological Setting and Characterization of a Methane Hydrate Reservoir Distributed at the First Offshore Production Test Site on the Daini-Atsumi Knoll in the Eastern Nankai Trough, Japan. *Mar. Pet. Geol.* **2015**, *66*, 310–322. [CrossRef]
15. Yi, B.Y.; Lee, G.H.; Kang, N.K.; Yoo, D.G.; Lee, J.Y. Deterministic Estimation of Gas-Hydrate Resource Volume in a Small Area of the Ulleung Basin, East Sea (Japan Sea) from Rock Physics Modeling and Pre-Stack Inversion. *Mar. Pet. Geol.* **2018**, *92*, 597–608. [CrossRef]
16. Boswell, R.; Myshakin, E.; Moridis, G.; Konno, Y.; Collett, T.S.; Reagan, M.; Ajayi, T.; Seol, Y. India National Gas Hydrate Program Expedition 02 Summary of Scientific Results: Numerical Simulation of Reservoir Response to Depressurization. *Mar. Pet. Geol.* **2019**, *108*, 154–166. [CrossRef]
17. Zhang, G.; Liang, J.; Lu, J.; Yang, S.; Zhang, M.; Holland, M.; Schultheiss, P.; Su, X.; Sha, Z.; Xu, H.; et al. Geological Features, Controlling Factors and Potential Prospects of the Gas Hydrate Occurrence in the East Part of the Pearl River Mouth Basin, South China Sea. *Mar. Pet. Geol.* **2015**, *67*, 356–367. [CrossRef]
18. Su, M.; Luo, K.; Fang, Y.; Kuang, Z.; Yang, C.; Liang, J.; Liang, C.; Chen, H.; Lin, Z.; Wang, C.; et al. Grain-Size Characteristics of Fine-Grained Sediments and Association with Gas Hydrate Saturation in Shenhu Area, Northern South China Sea. *Ore Geol. Rev.* **2021**, *129*, 103889. [CrossRef]
19. Boswell, R.; Collett, T.S. Current Perspectives on Gas Hydrate Resources. *Energy Environ. Sci.* **2011**, *4*, 1206–1215. [CrossRef]
20. Ye, J.; Qin, X.; Xie, W.; Lu, H.; Ma, J.; Qiu, H.; Liang, J.; Lu, J.; Kuang, G.; Lu, C.; et al. Main progress of the second gas hydrate trial production in the South China Sea. *Geol. China* **2020**, *47*, 557–568. (In Chinese with English Abstract)
21. Qin, X.; Lu, C.; Wang, P.; Liang, Q. Hydrate phase transition and seepage mechanism during natural gas hydrate production tests in the South China Sea: A review and prospect. *Geol. China* **2022**, *49*, 749–769. (In Chinese with English abstract) [CrossRef]
22. Wang, X.; Liu, B.; Qian, J.; Zhang, X.; Guo, Y.; Su, P.; Liang, J.; Jin, J.; Luan, Z.; Chen, D.; et al. Geophysical Evidence for Gas Hydrate Accumulation Related to Methane Seepage in the Taixinan Basin, South China Sea. *J. Asian Earth Sci.* **2018**, *168*, 27–37. [CrossRef]
23. Bai, C.; Zhang, G.; Lu, J.; Liang, J.; Yang, Z.; Yan, W.; Zhu, D.; Tian, Y. Deep-Water Sediment Waves as a Special Gas Hydrate Reservoirs in the Northeastern South China Sea. *Mar. Pet. Geol.* **2019**, *101*, 476–485. [CrossRef]
24. Liu, B.; Chen, J.; Pinheiro, L.M.; Yang, L.; Liu, S.; Guan, Y.; Song, H.; Wu, N.; Xu, H.; Yang, R. An Insight into Shallow Gas Hydrates in the Dongsha Area, South China Sea. *Acta Oceanol. Sin.* **2021**, *40*, 136–146. [CrossRef]
25. Guan, J.; Liang, Y.; Wang, S.; Wan, L.; Fan, S.; Su, P.; Zhang, W.; Liang, D. New Insight on the Stratigraphic-Diffusive Gas Hydrate System since the Pleistocene in the Dongsha Area of the Northeastern South China Sea. *J. Mar. Sci. Eng.* **2022**, *10*, 434. [CrossRef]
26. Liang, J.; Zhang, G.; Lu, J.; Su, P.; Sha, Z.; Gong, Y.; Su, X. Accumulation characteristics and genetic models of natural gas hydrate reservoirs in the NE slope of the South China Sea. *Nat. Gas Ind.* **2016**, *36*, 157–162. (In Chinese with English Abstract)
27. Wang, X.; Qian, J.; Collett, T.S.; Shi, H.; Yang, S.; Yan, C.; Li, Y.; Wang, Z.; Chen, D. Characterization of Gas Hydrate Distribution Using Conventional 3D Seismic Data in the Pearl River Mouth Basin, South China Sea. *Interpretation* **2016**, *4*, SA25–SA37. [CrossRef]
28. Zhang, W.; Liang, J.; Liang, Q.; Wei, J.; Wan, Z.; Feng, J.; Huang, W.; Zhao, J.; Meng, M.; Deng, W.; et al. Gas Hydrate Accumulation and Occurrence Associated with Cold Seep Systems in the Northern South China Sea: An Overview. *Geofluids* **2021**, *2021*, 5571150. [CrossRef]
29. Tsuji, Y.; Ishida, H.; Nakamizu, M.; Matsumoto, R.; Shimizu, S. Overview of the MITI Nankai Trough Wells: A Milestone in the Evaluation of Methane Hydrate Resources. *Resour. Geol.* **2004**, *54*, 3–10. [CrossRef]
30. Sun, S.-C.; Liu, C.-L.; Ye, Y.-G.; Liu, Y.-F. Phase Behavior of Methane Hydrate in Silica Sand. *J. Chem. Thermodyn.* **2014**, *69*, 118–124. [CrossRef]
31. Heeschen, K.U.; Schicks, J.M.; Oeltzschner, G. The Promoting Effect of Natural Sand on Methane Hydrate Formation: Grain Sizes and Mineral Composition. *Fuel* **2016**, *181*, 139–147. [CrossRef]
32. Yang, C.; Lu, K.; Liang, J.; Lin, Z.; Zhang, B.; Liu, F.; Su, M.; Fang, Y. Control effect of shallow-burial deepwater deposits on natural gas hydrate accumulation in the Shenhu sea area of the northern South China Sea. *Nat. Gas Ind.* **2020**, *40*, 68–76. (In Chinese with English Abstract)
33. Bai, C.; Su, P.; Su, X.; Cui, H.; Shang, W.; Han, S.; Zhang, G. Characterization of the Sediments in a Gas Hydrate Reservoir in the Northern South China Sea: Implications for Gas Hydrate Accumulation. *Mar. Geol.* **2022**, *453*, 106912. [CrossRef]
34. Su, P.; Wei, W.; Shi, C.; Li, J.; Han, W.; Xiao, Z.; Liang, J.; Wang, F.; Wan, Z. Gas Production Characteristics of Biogenic Gas Simulated in Hydrate-Developing Area of Dongsha Area, South China Sea. *Front. Mar. Sci.* **2023**, *10*, 1184641. [CrossRef]
35. Sha, Z.; Liang, J.; Zhang, G.; Yang, S.; Lu, J.; Zhang, Z.; McConnell, D.R.; Humphrey, G. A Seepage Gas Hydrate System in Northern South China Sea: Seismic and Well Log Interpretations. *Mar. Geol.* **2015**, *366*, 69–78. [CrossRef]
36. Wang, X.; Kneller, B.; Sun, Q. Sediment Waves Control Origins of Submarine Canyons. *Geology* **2023**, *51*, 310–314. [CrossRef]
37. Ruan, W.; Hu, C.; Li, Z.; Jia, Y. Effects of the Last Deglaciation Climate Warming on Hydrate Dissociation in the Northern South China Sea. *J. Mar. Syst.* **2024**, *242*, 103945. [CrossRef]
38. Zhao, J.; Wang, J.; Phillips, S.C.; Liang, J.; Su, P.; Lin, Q.; Chen, C.; Liu, J. Non-Evaporitic Gypsum Formed in Marine Sediments Due to Sulfate-Methane Transition Zone Fluctuations and Mass Transport Deposits in the Northern South China Sea. *Mar. Chem.* **2021**, *233*, 103988. [CrossRef]

39. Bista, D.; Kienast, S.S.; Hill, P.S.; Kienast, M. Sediment Sorting and Focusing in the Eastern Equatorial Pacific. *Mar. Geol.* **2016**, *382*, 151–161. [CrossRef]
40. Chen, F.; Su, X.; Lu, H.; Zhou, Y.; Zhuang, C. Relations between biogenic component (Foraminifera) and highly saturated gas hydrates distribution from Shenhu Area, Northern South China Sea. *Earth Sci.* **2013**, *38*, 907–915. (In Chinese with English Abstract)
41. Mcmanus, J. Grain Size Determination and Interpretation. In *Techniques in Sdimentology*; Blackwell: Hoboken, NJ, USA, 1988.
42. Moore, D.M.; Reynolds, R.C. *X-ray Diffraction and the Identification and Analysis of Clay Minerals*; Oxford University Press: Oxford, UK; New York, NY, USA, 1989; ISBN 978-0-19-505170-4.
43. Hillier, S. Quantitative Analysis of Clay and Other Minerals in Sandstones by X-ray Powder Diffraction (XRPD). In *Clay Mineral Cements in Sandstones*; Wiley-Blackwell: Hoboken, NJ, USA, 2003.
44. Brunauer, S.; Emmett, P.H.; Teller, E. Adsorption of Gases in Multimolecular Layers. *J. Am. Chem. Soc.* **1938**, *60*, 309–319. [CrossRef]
45. Wang, X.; Qian, J.; Lee, M. Methods for estimation of gas hydrate and free gas saturations and application to the northern slope of South China Sea. *Mar. Geol. Quat. Geol.* **2017**, *37*, 35–47, (In Chinese with English Abstract).
46. Archie, G.E. The Electrical Resistivity Log as an Aid in Determining Some Reservoir Characteristics. *Trans. AIME* **1942**, *146*, 54–62. [CrossRef]
47. Lu, S.; McMechan, G.A. Estimation of Gas Hydrate and Free Gas Saturation, Concentration, and Distribution from Seismic Data. *Geophysics* **2002**, *67*, 582–593. [CrossRef]
48. Zhang, W.; Liang, J.; Wei, J.; Lu, J.; Su, P.; Lin, L.; Huang, W.; Guo, Y.; Deng, W.; Yang, X.; et al. Geological and Geophysical Features of and Controls on Occurrence and Accumulation of Gas Hydrates in the First Offshore Gas-Hydrate Production Test Region in the Shenhu Area, Northern South China Sea. *Mar. Pet. Geol.* **2020**, *114*, 104191. [CrossRef]
49. Collett, T.S.; Lee, M.W.; Zyrianova, M.V.; Mrozewski, S.A.; Guerin, G.; Cook, A.E.; Goldberg, D.S. Gulf of Mexico Gas Hydrate Joint Industry Project Leg II Logging-While-Drilling Data Acquisition and Analysis. *Mar. Pet. Geol.* **2012**, *34*, 41–61. [CrossRef]
50. Li, J.; Lu, J.; Kang, D.; Ning, F.; Lu, H.; Kuang, Z.; Wang, D.; Liu, C.; Hu, G.; Wang, J.; et al. Lithological Characteristics and Hydrocarbon Gas Sources of Gas Hydrate-Bearing Sediments in the Shenhu Area, South China Sea: Implications from the W01B and W02B Sites. *Mar. Geol.* **2019**, *408*, 36–47. [CrossRef]
51. Zhuang, C.; Chen, F.; Cheng, S.; Lu, H.; Wu, C.; Cao, J.; Duan, X. Light Carbon Isotope Events of Foraminifera Attributed to Methane Release from Gas Hydrates on the Continental Slope, Northeastern South China Sea. *Sci. China Earth Sci.* **2016**, *59*, 1981–1995. [CrossRef]
52. Passega, R. Grain Size Representation by CM Patterns as a Geologic Tool. *J. Sediment. Res.* **1964**, *34*, 830–847. [CrossRef]
53. Li, X.S.; Zhou, Q.J.; Su, T.Y.; Liu, L.J.; Gao, S.; Zhou, S.W. Slope-Confined Submarine Canyons in the Baiyun Deep-Water Area, Northern South China Sea: Variation in Their Modern Morphology. *Mar. Geophys. Res.* **2016**, *37*, 95–112. [CrossRef]
54. Jin, J.; Wang, X.; Chen, D.; Guo, Y.; Su, P.; Liang, J.; Qian, J. Distribution of gas hydrate in Shenhu area: Identified with well log and seismic multi-attributes. *Mar. Geol. Quat. Geol.* **2017**, *37*, 122–130. (In Chinese with English Abstract)
55. Jiang, H.; Su, M.; Lei, X.; Kuang, Z.; Wu, N.; Liu, L.; Yang, R. Distribution of fine-grained turbidites on canyon ridges in the Shenhu area of northern South China Sea and its implications. *Mar. Geol. Quat. Geol.* **2018**, *38*, 52–62. (In Chinese with English Abstract)
56. Su, M.; Sha, Z.; Zhang, C.; Wang, H.; Wu, N.; Yang, R.; Liang, J.; Qiao, S.; Cong, X.; Liu, J. Types, Characteristics and Significances of Migrating Pathways of Gas-Bearing Fluids in the Shenhu Area, Northern Continental Slope of the South China Sea. *Acta Geol. Sin. Engl. Ed.* **2017**, *91*, 219–231. [CrossRef]
57. Ji, L.; Zhang, T.; Milliken, K.L.; Qu, J.; Zhang, X. Experimental Investigation of Main Controls to Methane Adsorption in Clay-Rich Rocks. *Appl. Geochem.* **2012**, *27*, 2533–2545. [CrossRef]
58. Zhang, Y.; Chen, T.; Zhang, Y.; Ren, W. Calculation Methods of Seepage Coefficient for Clay Based on the Permeation Mechanism. *Adv. Civ. Eng.* **2019**, *2019*, 6034526. [CrossRef]
59. Bai, C.; Su, P.; Su, X.; Guo, J.; Cui, H.; Han, S.; Zhang, G. Sediment Microstructure in Gas Hydrate Reservoirs and Its Association with Gas Hydrate Accumulation: A Case Study From the Northern South China Sea. *Front. Earth Sci.* **2022**, *10*, 876134. [CrossRef]
60. Sun, Y.; Jiang, S.; Li, S.; Wang, X.; Peng, S. Hydrate Formation from Clay Bound Water for CO_2 Storage. *Chem. Eng. J.* **2021**, *406*, 126872. [CrossRef]
61. Jiang, S. Study on the Influence of Clay Minerals on Hydrate Formation and Reformation. Ph.D. Thesis, Jilin University, Changchun, China, 2023.
62. Yang, J.; Wang, X.; Jin, J.; Li, Y.; Li, J.; Qian, J.; Shi, H.; Zhang, G. The Role of Fluid Migration in the Occurrence of Shallow Gas and Gas Hydrates in the South of the Pearl River Mouth Basin, South China Sea. *Interpretation* **2017**, *5*, SM1–SM11. [CrossRef]

Journal of
Marine Science and Engineering

MDPI

Article

Representative Dynamic Accumulation of Hydrate-Bearing Sediments in Gas Chimney System since 30 Kyr BP in the QiongDongNan Area, Northern South China Sea

Jinan Guan [1,2,3], Menghe Wang [1,2], Wei Zhang [4,5,*], Lihua Wan [2], Matthias Haeckel [3] and Qi Wu [6,*]

1 School of Energy Science and Engineering, University of Science and Technology of China, Hefei 230026, China; guanja@ms.giec.ac.cn (J.G.); wangmhe@mail.ustc.edu.cn (M.W.)
2 CAS Key Laboratory of Gas Hydrate, Guangdong Provincial Key Laboratory of New and Renewable Energy Research and Development, Guangzhou Institute of Energy Conversion, Chinese Academy of Sciences, Guangzhou 510640, China; wanlh@ms.giec.ac.cn
3 GEOMAR Helmholtz Centre for Ocean Research Kiel, 24148 Kiel, Germany; mhaeckel@geomar.de
4 Sanya Institute of South China Sea Geology, Guangzhou Marine Geological Survey, China Geological Survey, Sanya 572024, China
5 National Engineering Research Center for Gas Hydrate Exploration and Development, Guangzhou 510075, China
6 Guangxi Key Laboratory of Geomechanics and Geotechnical Engineering, College of Civil Engineering, Guilin University of Technology, Guilin 541004, China
* Correspondence: zwgmgs@foxmail.com (W.Z.); giecwuqi@foxmail.com (Q.W.)

Abstract: A stratigraphic complex composed of mass transport deposits (MTDs), where the gas occurrence allows for the formation of a gas chimney and pipe structure, is identified based on seismic interpretation in the QiongDongNan area of the northern South China Sea. During the Fifth Gas Hydrate Drilling Expedition of the Guangzhou Marine Geological Survey, this type of complex morphology that has close interaction with local gas hydrate (GH) distribution was eventually confirmed. A flow-reaction model is built to explore the spatial–temporal matching evolution process of massive GH reservoirs since 30 kyr before the present (BP). Five time snapshots, including 30, 20, 10, and 5 kyr BP, as well as the present, have been selected to exhibit key strata-evolving information. The results of in situ tensile estimation imply fracturing emergence occurs mostly at 5 kyr BP. Six other environmental scenarios and three cases of paleo-hydrate existence have been compared. The results almost coincide with field GH distribution below the bottom MTD from drilling reports, and state layer fracturing behaviors always feed and probably propagate in shallow sediments. It can be concluded that this complex system with 10% pre-existing hydrates results in the exact distribution and occurrence in local fine-grained silty clay layers adjacent to upper MTDs.

Keywords: QiongDongNan slope area; methane hydrate-bearing sediments; gas chimney; mass transport deposit; evolution dynamics; fracturing criterion

1. Introduction

Gas hydrate (GH) is an ice-like nonstoichiometric crystalline solid formed of water and methane. Natural GH can be found on the seabed, in ocean sediments, in deep lake sediments, as well as in the permafrost regions. The amount of carbon potentially trapped in natural hydrate-bearing deposits may be between 3×10^{15} to 1×10^{16} m^3, which makes them of major interest as a potential energy resource [1,2]. The presence and occurrence of GH in oceanic continental slope areas have attracted extensive attention and particular interest for the sake of energy potential, marine geohazards, and climate change issues [3,4].

A seismically imaged gas chimney is a type of stratigraphic sedimentary and hydraulic area where vertical seismic reflection profiles generally indicate that seafloor geological structures in this layer possess excellent transport capacity for upward gas-bearing fluids, and also lots of methane gas is trapped in conduit pores [5–7]. It ordinarily extends several strata sequences, provides an effective aisle for functionally connecting deep hydrocarbon sources with the shallow bottom boundary of the local gas hydrate stability zone (HSZ), and conveys these multiphase fluids into the seafloor surface through affiliated pipe structures on many occasions [8,9]. Hovland and Sommerville (1985) reported the identification of gas-charged sediments and the characteristics of two natural gas seepages in the North Sea, which suggested a gas–liquid mixture fluid seepage caused the pockmark development [10]. After that, the impacts of seabed pockmarks associated with gas chimneys on offshore construction and gas hydrate occurrence have been extensively investigated by Hovland and collaborators [11,12]. In passive continental marginal slope regions, diverse deposition activities along the down-slope direction promote shear stress to exceed the shear strength and subsequently cause sedimentary failure, thus generating another laterally extended complicated structure of mass transport deposits (MTDs) [13,14].

Plentiful biogenic methane produced from shallow organic matter and thermogenic methane produced by deep source rocks cross together through a main gas chimney area and enter into shallow seabed deposits, contributing to the formation of complex structural seepage GH reservoirs in a suitable thermodynamic environment [15–17]. If the fluent pipe structure is disturbed or even stabbed across by MTDs, the upward flow of methane-bearing fluids is blocked by a sudden decrease in layer permeability and is thus inclined to stop and remain in local layers to form abundant GHs with various occurrence states. Subsequently, pore pressure below gradually increases, and hydraulic failure is likely to occur. Therefore, gas chimneys and local MTD systems play a key role in the dynamic migration, occurrence, and accumulation of GH reservoirs [18–20]. These activities are presumed to have occurred in many gas chimney–hydrate geological systems from time to time with geologic and environmental impacts [21].

Combining the visible distribution states of GH from the drilling core samples, we discovered that the accumulation type of GH reservoirs in the QiongDongNan Basin (QDNB) area of the northwestern South China Sea (SCS) (Figure 1a), to some extent, is similar to other typical marine reservoirs, such as the Ulleung basin in South Korea and the Krishna-Godavari (KG) basin in India [22,23]. The GH system in the QiongDongNan (QDN) area is deemed to have the same distribution and accumulation traits described above. Previous gas and oil resource exploration surveys have identified large-scale gas chimney morphotypes in the target sediments by analyzing high-resolution seismic sections. The large range of low-frequency chaotic seismic reflection area, which is obviously different from the surrounding stratigraphy, indicates the presence of abundant gas. The boundary of this chimney scope was confined by the termination of the blanking reflection. The instantaneous frequency and some of the internal events can clearly reveal the fluid pathway of gas chimney and pipe structure. Bottom simulating reflector (BSR) characteristics usually appear at the top of these shapes, and fluid leakage channels show a pull-up trace of seismic indications reaching the seafloor [24–26]. During the Fifth Gas Hydrate Drilling Expedition of the Guangzhou Marine Geological Survey (GMGS5) expedition, visible GH crystals recovered from drilling cores mainly show various features such as massive, laminates, nodules and veins massive, layers, nodules, and veins (sites W07, 08, and 09). GHs cement into fractured clay is the dominant phase in fine-grain silty sediments and have a concentrated distribution [27,28].

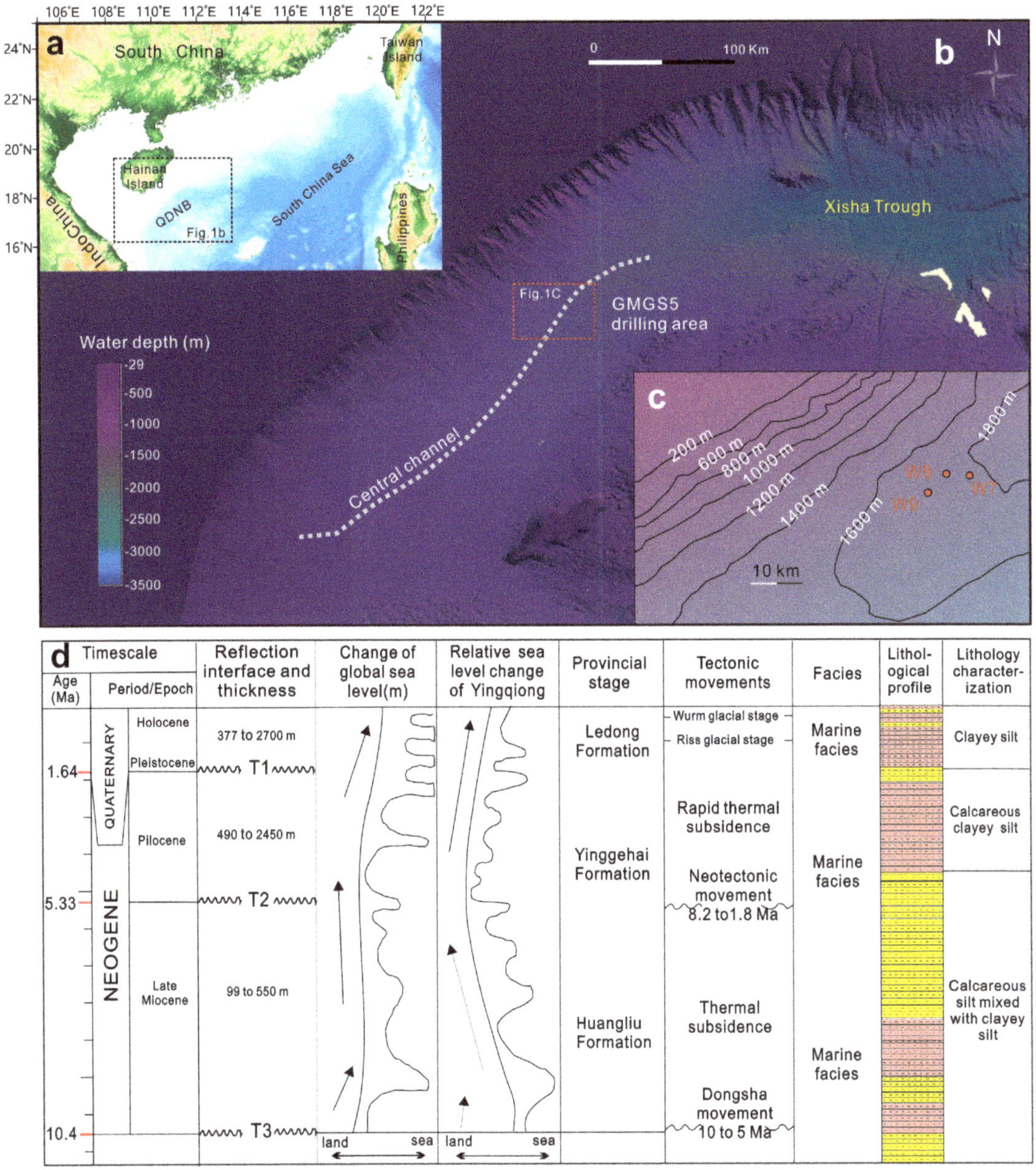

Figure 1. General geographic information and geological interpretation of the study area: (**a**) location of QDN area, (**b**) seafloor slope topography at GMSS5 drilling sites, (**c**) local seawater bathymetry contours, and (**d**) comprehensive stratigraphic histogram.

In recent years, remarkable progress has been made in understanding how local seismic chimneys and MTDs control fluid migration and determine the growth patterns of GH reservoirs from geologic, geochemical, and geophysical studies according to field investigation and the drilling core analysis results of GMGS5 [29,30]. In this study, we analyze GH reaction kinetics into the frame of a spatio-temporal dynamic evolution scale in a well constrained geological background. We considered that the evolution process of GH reservoirs is also controlled by environmental factors such as seafloor deposition, eustatic sea-level change, and organic gas supply. We use a flow-reaction model to explore how the interactions of sedimentary properties, fluid qualities, and reaction kinetics can lead to the current occurrence and distribution of GHs. We also focus on the growth and characteristics of fracture pathways in sediments within HSZ and how it propagates over GH content.

2. Geological Setting

The GMGS5 investigation area is located in the central channel of the upper slope of the QDN basin, adjacent to the Xisha Trough to the northeast (Figure 1b,c). Since the Neogene, the QDN basin has been in a post-rift thermal subsidence tectonic evolution stage. Wide-spreading Miocene (Huangliu formation), Pliocene (Yinggehai formation), and Quaternary (Ledong formation) marine sedimentary sequences formed in this period (Figure 1d). Since then, the local high deposition rate has led to the maximum sedimentary accumulation thickness along slope surface stretches about several kilometers [31–34]. The detrital sediments of the Holocene and Pleistocene age have a sand content of more than 50%. Elastoplastic unconsolidated argillaceous particles fill into skeleton matrixes. They serve as effective barriers to stop fluid from further smoothing upward seepage and provide sufficient space for GH storage. Previous exploration revealed that deep hydrocarbon source rocks developed below the Huangliu formation and have entered the mature hydrocarbon threshold. The thermal evolution degree of organic matter is in the range of a mature and high mature light oil and gas window, which has the geological conditions of forming thermogenic gas. Moreover, the strata above the Huangliu formation are within the formation condition of biogenic gas because most of the source rocks developed in the low mature–immature stage [35–38]. Intensive biogenic gas manifestation occur in petroleum drilling boreholes, especially those shallower than 2300 m below seafloor (mbsf). Both coal-measure mudstone (which generates thermogenic methane) and marine mud shale (which produces biogenic methane extensively) are potential source rocks here. The measured carbon isotope of methane is −43.65–39.92‰ the mixture origin of gas source [30].

A large number of gas chimneys are found in the study area, with a prevalent northeast direction. The acoustic blanking area is mushroom-shaped with the appearance status of a small lower part and a large upper part [24,39,40]. Most of these structures are firmly associated with the structures of faults and fractures [41]. Almost all gas chimneys here vertically extended from a T3 to T1 sequence, and their tops can be up to 200–150 mbsf gauging from two-way travel time [36]. Through characterizing the weak amplitude and chaotic seismic sections, three sets of MTD since the Late Quaternary have been confirmed above T1 interface, and laterally stretch across gas chimneys. With a high amplitude and reversed polarity, BSR here tends to sit above gas chimneys and just below the third MTD. Its wave impedance difference between the upper and lower layers is large and also shows a blank zone, indicating that probably abundant free gas is available within this study seabed [29,42].

3. Data and Methods

3.1. GHs in Site W08

The 2D seismic profile and instantaneous frequency profile collected in the GMGS5 expedition have been carefully inspected. From the characteristics of the BSRs, logging while drillings (LWDs), and lithologic coring results, the site W08 can be deemed as a representative well of this complex system because it is located exactly above the gas chimney section (Figure 2a,b). The deep seismic chimney is estimated to originate from 3 Ma BP. It is elliptic plane-shaped and has a diameter of several kilometers. The upward hydrocarbon fluids are reckoned to pass through this section and reach the T1 interface. The top of this seismic chimney shows the BSR occurrence and nearly coincides with the bottom of MTD3. A shallow fluid pipe shape in the seismic profile is observed, which has a diameter of 180 m. It is columnar and calculated to be merely 0.5 Ma years old. This pipe is located just above the gas chimney area. Fluids will likely escape to the seafloor through it because the dome and pockmarks on the local seafloor that surface from remotely operated vehicle (ROV) observation indicate signs of fluid activity. Local high sedimentation rates since the early Pliocene easily trigger disequilibrium pore compression and bring about the overpressure outcome. In many situations, the fluid flow is mainly driven by the pore overpressure [24,40]. The lithology of the MTD3 here has a 20–50% clay content which is inferred to be older than 0.5 Ma BP. GHs in drilling core images mainly appear messy and discontinuous, and they grow in the fractures within MTD3. With strata stress and thermodynamic condition change, the performance of gas-driven hydraulic fracturing is presumed to occur extensively in this GH system.

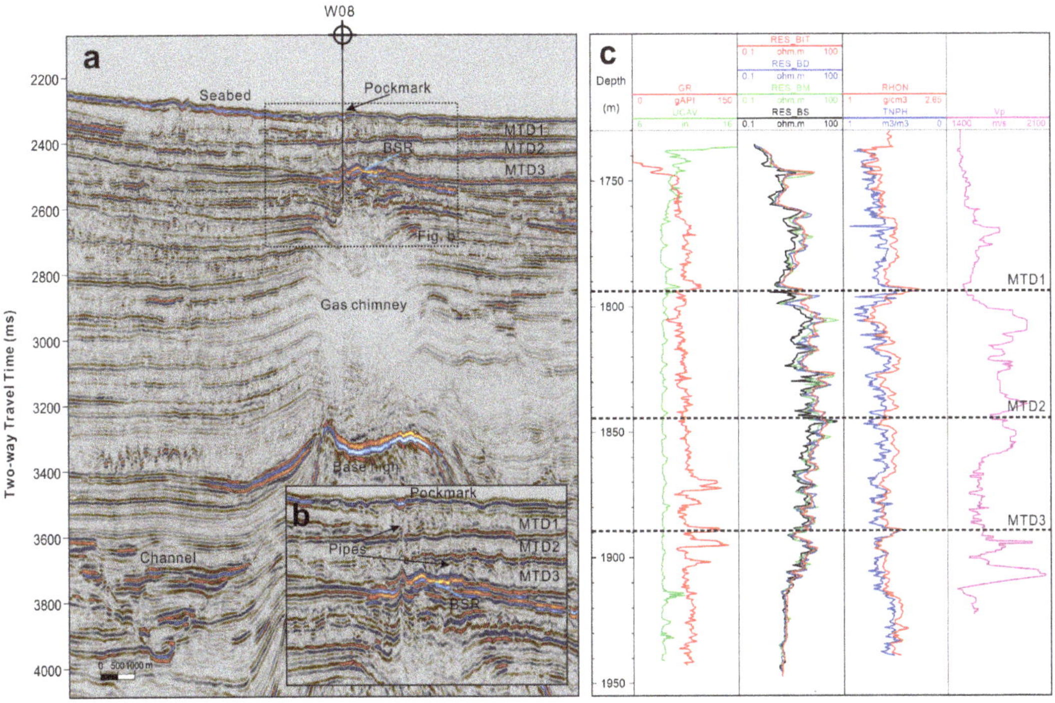

Figure 2. The structural configuration of site W08: (**a**) local seismic section, (**b**) zoomed seismic profile, and (**c**) three sets of MTD in LWD curves and their bottoms (MTD1-59, MTD2-112, and MTD3-157 mbsf).

3.2. Sea-Level Change and Deposition

The depth of seawater bottom in the GMGS5 investigation zone ranges from 1600 to 1800 m below sea level (mbsl). The deepest bottom boundary where GH core samples are drilled out is 237 mbsf. Nearly all GH layers occur above T1 interface (Figure 2b,c). The basin subsides since the Quaternary are roughly steady, and the sedimentary evolution is mainly controlled by the eustatic sea level change which is closely related to the Last Glacial Cycles that ended about 15,000 years ago. The detrital input is brought into the marginal sea by paleo-rivers [43]. Since seabed position changes can be calculated according to relative sea level and deposition rate v_{sed} (cm/kyr) (Figure 3a), the temperature and hydrostatic pressure at the bottom of the seawater are determined accordingly (Figure 3b). The recent rise in sea level in the area begins at about 30 kyr BP from historical inversion.

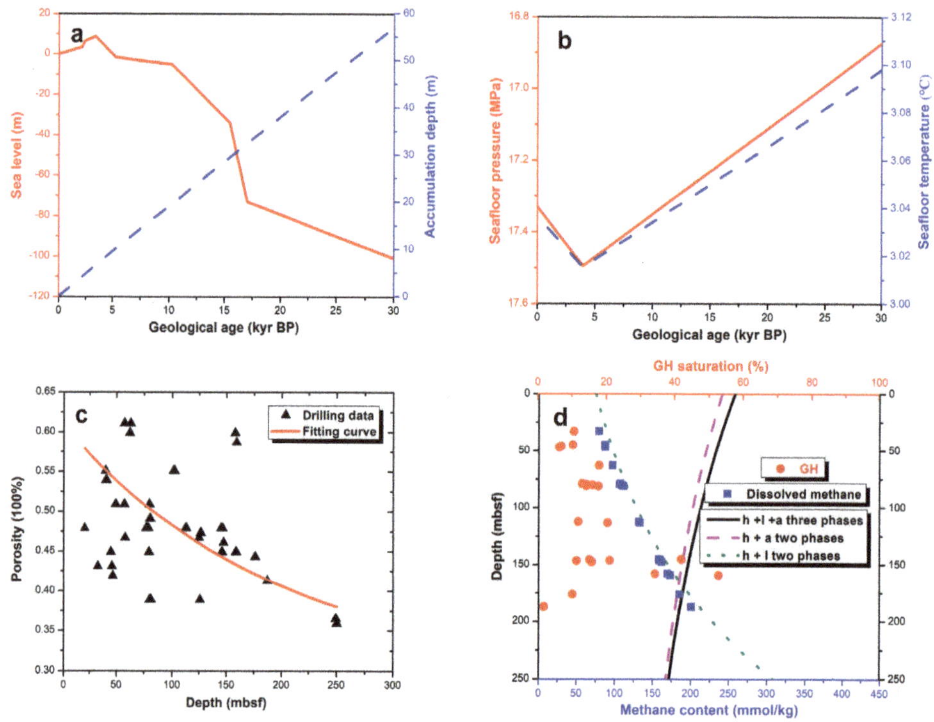

Figure 3. Local depositional conditions including: (**a**) sea level and sedimentation rate [44], (**b**) pressure and temperature change at seabed surface, (**c**) fitting on in situ porosity, and (**d**) methane concentration from drilling core test and phase lines in different situations (l-liquid, h-GH, and a-gas phase).

The distribution of deposit particle sizes and the skeleton in the seabed are often chaotic, and in particular, they suffer from the MTD influence. The measured local porosity distribution is plotted in Figure 3c. In order to simplify the treatment on the anomalistic sedimentary layers, initial porosity ϕ_0 in deposits without GH existence is gauged by field data and its expression with depth z (mbsf):

$$\phi_0(z) = 0.29 + (\phi_{top} - 0.29) \times \exp(-0.00505z) \ (R^2 = 0.2046), \tag{1}$$

where porosity at the sedimentary surface ϕ_{top} is 0.62. After GH forms in porous soil, we assume that the alteration of new porosity ϕ_h is suitable:

$$\phi_h(z) = \phi_0(z) \times (1 - S_h), \tag{2}$$

where S_h is GH saturation.

3.3. Methane in Pore Water

Through calculating the phase equilibrium curves of methane in liquid and gas states under local thermodynamic conditions [45–47], we compared how in situ GH is vertically distributed at GMGS5 site W08 (Figure 3d). It can be seen that methane supplies below HSZ bottom and even in the low part of HSZ are pretty abundant, so the current free gas is likely to be plentiful in these zones. We also inferred that maybe the GH reaction is still ongoing, while the methane gas in the middle and upper sections is completely converted into hydrates. The consequence is that merely GH and liquid water are preserved in local pores. According to the data from degassing cores [29], the initial methane concentration n_{diss} (mmol/kg) in pores is quantified as:

$$n_{diss} = \exp(4.17 + 0.007 \times z - 5.18 \times 10^{-6} \times z^2). \tag{3}$$

3.4. Sedimentary Soil-Water Properties

We continue to adopt traditional normalized relating content to gauge sedimentary soil-water characteristics curves (SWCC), which is [48,49] as follows:

$$S_e = \frac{S_l + S_g - S_{li} - S_{gi}}{1 - S_{li} - S_{gi}}, \tag{4}$$

where S_e is effective fluid saturation; S_l and S_g are liquid and gas saturation, respectively; S_{li} and S_{gi} are initial irreducible liquid and gas saturation, and here S_{gi} is always 0. Irreducible liquid water S_{li} is considered to be controlled by clay content x_c and ϕ_h as [50]:

$$S_{li} = 0.326{x_c}^{0.219} + 0.0262/\phi_h. \tag{5}$$

The saturation-dependent relative permeability for gas k_{rg} and liquid k_{rl} phases can be defined by Corey's model:

$$k_{rl} = (S_e)^{\frac{2+3\pi}{\pi}}, \tag{6}$$

and

$$k_{rg} = (1 - S_e)^2 \times [1 - (S_e)^{\frac{2+\pi}{\pi}}], \tag{7}$$

where the pore size profile exponent π is 7.3 in this simulation [51]. As the pressure difference between pore gas P_g (MPa) and liquid phases P_l (MPa), capillary pressure P_c (kPa) is defined by pore entry pressure P_{th} (kPa):

$$P_c = P_g - P_l = P_{th} \times {S_e}^{-1/\omega}, \tag{8}$$

where ω is another Brooks–Corey parameter. The demarcation on capillary drainage and the imbibition curves of shallow seabed sediments has been reasonably introduced by Daigle et al. (2020) [52] when tracing how gas-driven tensile fractures in marine hydrate-bearing layers. We use the same measurement to show the influence of clay fractions x_c, which are:

$$P_{th} = \exp[6.59(1 - \phi_h + S_{li})x_c - 2.76], \tag{9}$$

and

$$\omega = \exp(1.64x_c - 0.921). \tag{10}$$

The clay fraction scope in its seabed soil is 25–60%. When setting the layer porosity to be minimum 0.3 and maximum 0.7, the scope of the corresponding S_{li} ranges from 0.2781 to 0.3788.

3.5. Hydraulic Conductivity in Gas Chimney

Compared with the sedimentary transportation quality of fluids in other parts of northern SCS [53], we can use an exponential expression to depict the strata intrinsic permeability k_h (mD) after GH existence in the QDN area as:

$$k_h = k_0(1 - S_h)^3, \tag{11}$$

where k_0 (mD) is layer instinct permeability without GH existence. Based on the above description of the chimney–hydrate system, we set two different permeability values to reflect MTD influence. Within the 0–180 mbsf section, the permeability k_0 is reckoned as 2 mD. Below 180 mbsf, it is just within the gas chimney area and its value is 200 mD [30,36].

3.6. Tensile Strength Estimation

We also abide by the assumption that tensile failure can only occur when pore pressure steps across the resultant of minimum principle stress and the tensile strength of seabed soil together [52,54]. The final force difference criterion term S_v (MPa) in this work is:

$$S_v = P_c + P_l - T_{sp} - \sigma_h > 0. \tag{12}$$

The vertical stress σ_v (MPa) and horizontal stress σ_h (MPa) can be linked according to Poisson's ratio in the QDN area, and the expression is:

$$\frac{\sigma_h - P_l}{\sigma_v - P_l} = \frac{\nu}{1-\nu} > 0.8. \tag{13}$$

The sub-seafloor σ_v is simplified as $\sigma_v = \rho_b g z$, and here the bulk sedimentary density ρ_b is 1910 kg/m^3. The tensile strength T_{sp} (MPa) is determined as:

$$T_{sp} = -\frac{1}{2} \times 10^{[-6.36+2.45\times\log_{10}(0.86\times V_p - 1172)]} \times [m_i - \sqrt{m_i{}^2 + 4}], \tag{14}$$

where the Hoek–Brown constant m_i is 4, and compressional wave velocity V_p (m/s) at site W08 can be retrieved from the in situ logging curve as:

$$V_p = 2.47 \times z + 1470. \tag{15}$$

3.7. GH layer Formation Dynamics

From the comprehensive perspectives of gas source, kinetic reaction, fluid migration pathway, HSZ alteration, MTDs, and sedimentary soil properties, we established a one-dimensional dynamic evolution model, which is in line with the previous geological background, to demonstrate the accumulation process of hydrate-bearing sediments and put specific parsing steps and corresponding expressions in the Appendix A of this paper.

When the seawater depth is larger than 500 mbsl, the relationship of temperature at the seafloor surface T_{sf} (°C) with the depth L_w (mbsl) is [55]:

$$T_{sf} = \exp\{[2.0339 - \ln(L_w/1000)]/1.3361\} \tag{16}$$

The local temperature T_{se} (°C) in sediments is determined by geothermal gradient T_d (°C/m) and T_{sf} (°C), which is:

$$T_{se} = T_{sb} + z \times T_d. \tag{17}$$

All the parameter values in this basic simulation scene are listed in the following Table 1.

Table 1. Parameters used for deducing the basic scenario here.

Parameters	Values
Methane flux (kg·m^{-2}·yr^{-1}): q_m	0.25
Water flux (kg·m^{-2}·yr^{-1}): q_w	0
Geothermal heat (mW·yr^{-2}): q_e	80
Sedimentary rate (cm/kyr): v_s	19
Kinetic coefficient (mol·m^{-3}·MPa^{-1}·yr^{-1}): ζ_0	0.3
Total evolution time (kyr): t	30
Density (kg/m^3): ρ_s, ρ_w, ρ_{g0}, ρ_h	2650, 1025, 0.7 (standard temperature and pressure), 912
Molar mass (g/mol): η_w, η_m, η_h	18, 16, 124
Viscosity (Pa·s): μ_l, μ_g	1.3×10^{-3}, 1.5×10^{-5}
Diffusivity (m^2/s): $D_l{}^m$, $D_l{}^s$	1.3×10^{-9}, 1.0×10^{-9}
Initial salinity: $X_{l0}{}^s$	0.033
Fraction of clay particle volume: x_c	0.3
Geothermal gradient (°C/m): T_d	102
Heat conductivity (W·m^{-1}·K^{-1}): λ_r, λ_l, λ_h, λ_g	1.2, 0.58, 0.53, 0.05
Specific heat (J·kg^{-1}·K^{-1}): C_r, C_l, C_h, C_g	2.2×10^3, 4.2×10^3, 2.1×10^3, 3×10^3

4. Results

4.1. Evolution Process

The old seafloor depth at 30 kyr BP is calculated at1790 mbsl in the light of the current sea level position. The earliest HSZ bottom boundary is 147 mbsf at that time. After the sedimentary process piles up, porewater pressure in accumulated sediments increases slightly to 0.0101 MPa/m, related to hydrostatic pressure 0.01 MP/m [56–58]. To better exhibit how this complex system consisting of a gas chimney, GH, pipe, and MTDs evolves, the following six key controlling factors, including strata pressure, temperature, capillary pressure, phase saturations, permeability, and salinity were chosen to represent system characteristics, as they can provide enough insightful strata information. Table 1 tells the basic thermodynamic and hydraulic environment about this complex GH system. Both the initial gas and hydrate content are 0 in this basic scene.

Five moments including four divided times (30, 20, 10, and 5 kyr BP), and the present, are selected to show the system evolution at that time. In Figure 4a, the positive pressure change is set as the difference between sedimentary pore pressure and seawater hydrostatic pressure if there is no deposition activity. Similarly, in Figure 4b, positive temperature change is defined as the difference between sedimentary temperature and seawater temperature if there are no sediments. Both of them verify the previous judgment that the depth of the local seafloor position increases from 30 to 5 kyr BP and then decreases a bit until now. GH is calculated to accumulate considerably within the lower part of HSZ and the maximum saturation of GH from model simulation can reach 47% (Figure 4c). It can be seen that both the GH distribution close to the HSZ bottom area and its peak of saturation coincide with the drilling data very well. Simulation results reveal that free gas will gradually gather at the top zone of the gas chimney, which is also the bottom zone of HSZ and MTD3. The current gas saturation is 20.3%, to possibly provoke BSR signs. Our results also demonstrate that GH and free gas can hardly coexist in pores, according to the strata setting of the gas chimney quality. The alternation of salinity and capillary pressure has a strong positive correlation with GH saturation, while permeability displays a wholly negative correlation (Figure 4d–f). These phenomena are consistent with the general cognition of marine GH sediments. Overall, the flow-reaction process, strata features, and the combination of related key parameter values perfectly explain how gas chimneys and MTD determine the dynamic evolution and final occurrence of GH systems.

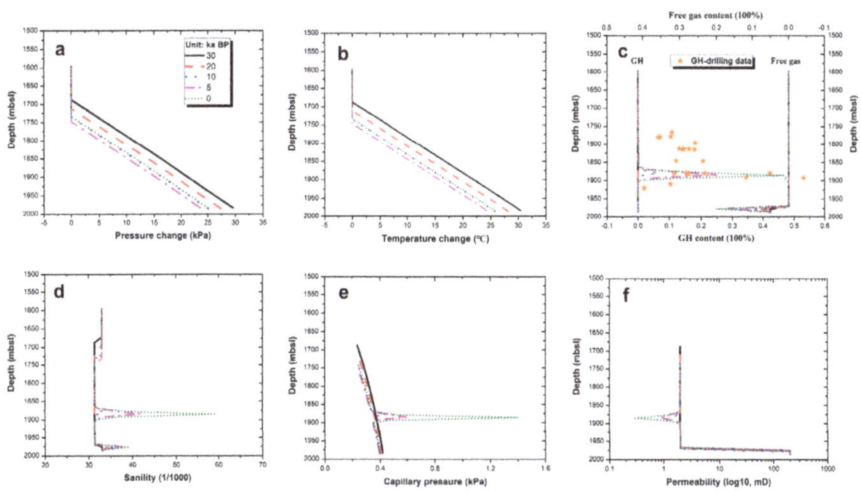

Figure 4. The evolution state of: (**a**) the change of pressure difference, (**b**) the change of temperature difference, (**c**) gas and hydrate content, (**d**) pore salinity, (**e**) capillary pressure, and (**f**) permeability, at five moments 30, 20, 10, 5 and 0 kyr BP.

4.2. Force Disequilibrium

The distribution of sedimentary stress and strength and corresponding force balance in this geological system attracted widespread attention in the academic community. Whether strata hydraulic fracturing occurs in clay-dominated fine-grained layers and how cracks are created by pore overpressure propagating in coarse-grained turbidities are essential to understanding GH existence and occurrence. Based on the stress and strength analysis in Equations (12)–(15), we calculated their values at these five moments in Figure 5a. The results show that the local tensile strength gradually decreases from the largest value at 30 kyr BP to the minimum at 5 kyr BP. After that, it increases rapidly. The current tensile strength is the same as that at 10 kyr BP. Meanwhile, the results of the criterion term, S_V, indicate that except for the bottom area of the studied strata at the beginning time of 30 kyr BP, the phenomenon of tensile fracture will always occur (Figure 5b). The strata are most likely to break at 5 kyr BP, and the existence of GH can greatly improve the possibility of fracturing behavior in the strata. GH saturation has an observably positive change with the term S_V.

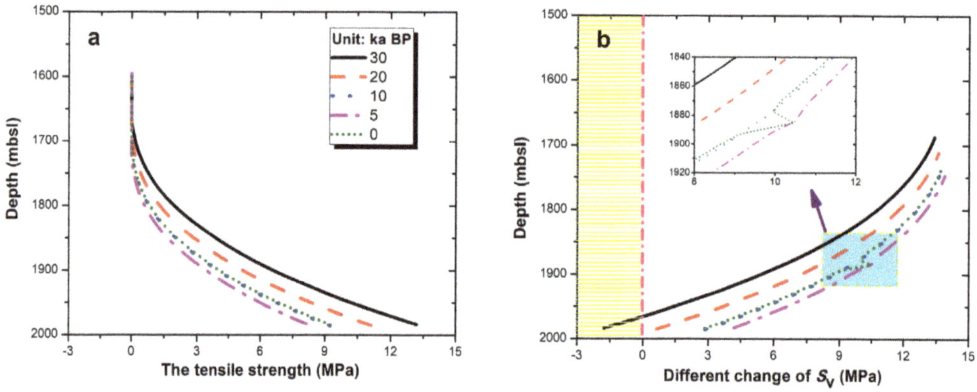

Figure 5. The evolution state of: (**a**) tensile strength, and (**b**) fracturing criterion term S_V at these five moments.

4.3. Different Sedimentary Conditions

Local gas supply, kinetic reaction, and permeability are three decisive variables in our model to simulate this GH system. We designed six other geological scenarios that can represent possible combinations of strata and fluid environments to cultivate diverse occurrences of hydrate-bearing sediments (Table 2). We presented the differences between the results of free gas, GH, and fracturing criterion term in these six scenarios and the previous basic condition (Figure 6a,b).

Table 2. Six possible geological scenarios for GH generation that may occur in the QDN area.

Scenarios	Gas Flux ($kg\cdot m^{-2}\cdot yr^{-1}$)	Kinetic Coefficient ($mol\cdot m^{-3}\cdot MPa^{-1}\cdot yr^{-1}$)	Permeability (in Chimney, in MTDs) (mD)
1	0.25	0.3	20,000, 2000
2	0.025	0.03	200, 2
3	2.5	0.03	200, 2
4	25	0.003	200, 20
5	250	0.03	200, 2
6	2500	0.03	2000, 2

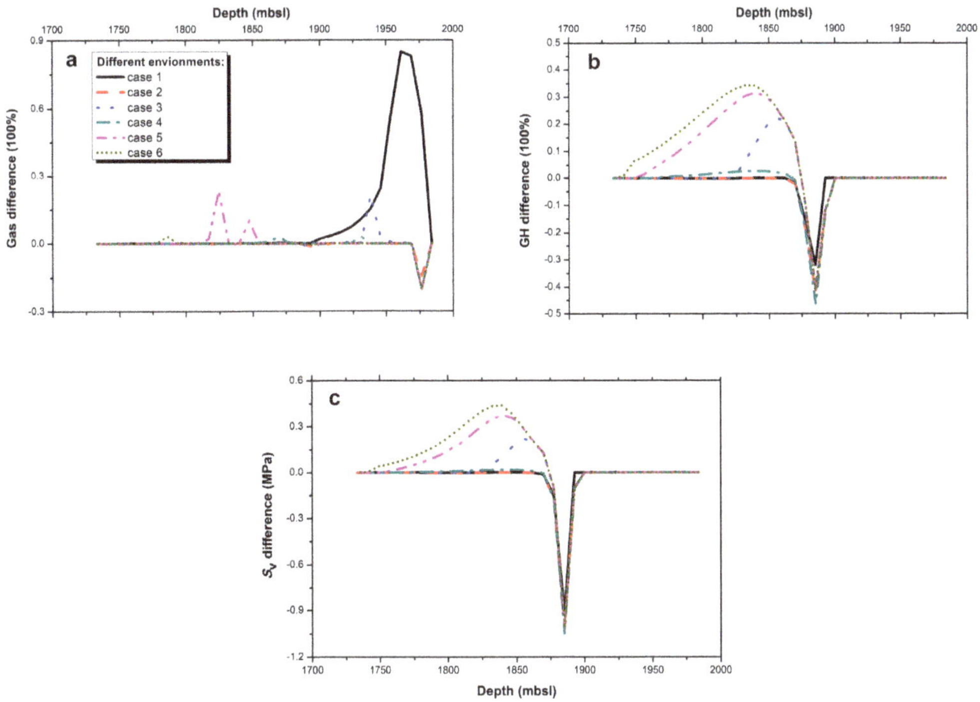

Figure 6. The differences of: (**a**) free gas, (**b**) GH, and (**c**) the criterion term S_V between the respective results of these six scenes and the previous basic condition.

Scenario 1 represents extremely high permeability in the gas chimney and the upper strata. The results show that the GH distribution range is nearly the same, but the saturation peak can only reach 15%. The gas peak can be up to 85%, and its distribution range is wider. GH and free gas can coexist in the HSZ. Scenario 2 shows less gas supply and kinetic reaction coefficient. The content of hydrates and free gas both decline dramatically. Scenario 3 shows a larger gas supply and a smaller reaction coefficient. The results indicate that the distribution of GH is wider. The highest position can reach about 95 mbsf, but the GH peak which shifts upward can only reach 23%. The free gas content is the same, but the peak position is also shifted higher. Scenario 4 represents a higher gas supply and the slowest reaction coefficient, while the upper sedimentary permeability increases. The results show that although the hydrate and free gas are more widely distributed, the contents of them are very low. Scenario 5 represents a higher gas supply and a slower kinetic reaction coefficient. The results show that although the hydrate distribution range is wide and the peak is high enough, there is also a wide range of free gas content in the HSZ with a peak value of 23%. Scenario 6 represents the largest gas supply, a smaller reaction coefficient, and a larger layer permeability in the gas chimney. GHs can fill the entire HSZ where sporadic free gas exists in the pores. By calculating these six fracturing criterion terms S_V, all the values are positive, indicating that fracturing will form or occur (Figure 6c). Meanwhile, the greater the hydrate content, the more likely that hydraulic fracturing will happen.

4.4. Palaeo-GH Existence

There is also a possibility that a certain amount of GH has already existed in the original HSZ before 30 kyr BP. With the rather big shift of the HSZ bottom boundary, the GH in the lower part decomposes until it vanishes, while the upper part remains there.

We set the three initial hydrate contents as 5% (case1), 10% (case2), and 15% (case3) to investigate the evolution process since 30 kyr BP. Similarly, we employ the difference of GH, free gas and fracturing criterion term S_V between these three cases and the corresponding value in the basic case. From the results, it can be seen that free gas contents also increase accordingly (Figure 7a). GH peaks at the HSZ bottom in the three cases can reach 53%, 57% and 59%, respectively. The possibility of a second peak is 9%, 19%, and 31% at about 105 mbsf where they nearly coincide with MTD2 bottom in Figure 7b. The criterion term curves are still positive, indicating that strata fracture will develop in these settings (Figure 7c). Furthermore, the higher the hydrate content, the more intensive the fracture behavior. On the whole, it is believed that the initial GH saturation of 10% is nearly perfect and more in line with the current observation about the GH existence close to MTD2 and MTD3 from drilling logging.

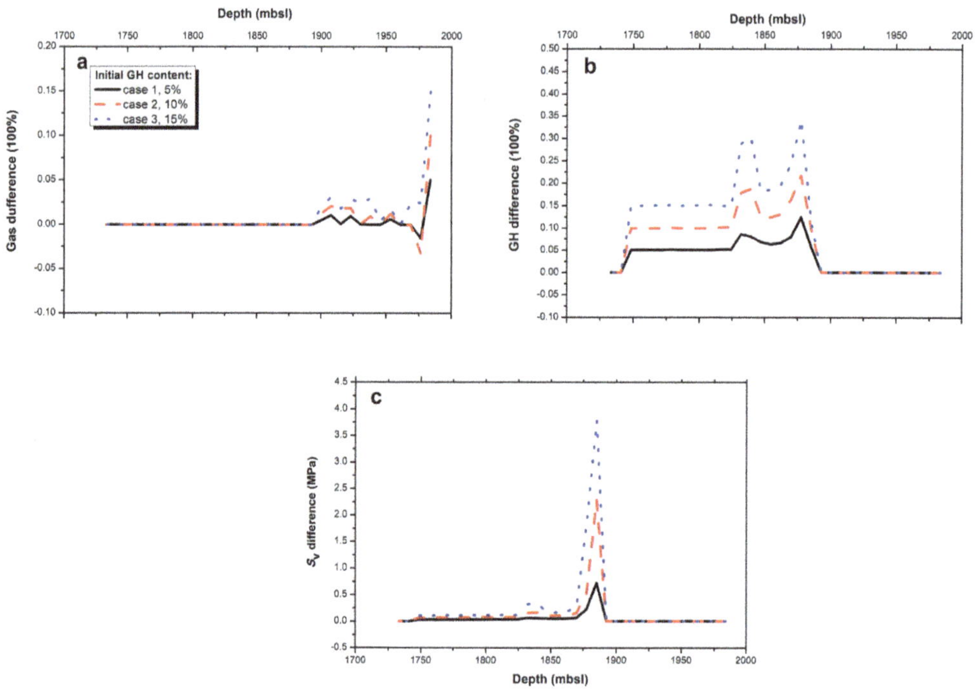

Figure 7. The difference of: (**a**) free gas, (**b**) GH, and (**c**) criterion term S_V with the corresponding value in the basic environment in three cases when initial GH saturation is 5%, 10%, and 15%, respectively.

5. Discussion

5.1. Petrophysical Variations

Except for the heterogeneity of sedimentary layers, other lithological variations of mixtures in the marine subsurface, including pore shape and connection, particle size distribution, and clay contents, exert strong impacts on fluid transport ability and the stiffness of skeleton endmembers. In our model, although the uniform lithology assumption is used to gauge this deposition-flow-reaction process, different values of the intrinsic permeability in separate gas chimneys and pipe zones have been set to roughly reflect the real lithological situation. One of the two uncertainties lies in the usage of the empirical parameters of SWCC properties in Equations (5)–(11). In our simulation, the general values of sedimentary factors are adopted, which define the drainage–imbibition performance of seabed soils. Another is the failure criterion to judge distinct Mohr–Coulomb hydraulic fracturing. We do not elaborate on the scopes of Poisson's ratio, clay content, and the Hoek–Brown constant here, because they are the mean values and can explain most ordinary cases in our research deposits.

5.2. Difference of MTDs

In our mode, we do not set any property discrepancies among the three MTDs. The simulation focus is on the lowest MTD3 because it almost marks both the bottom of HSZ and the top of BSR and dominates the unsaturated fluid flow qualities when passing into upper pipe space. Whether or not the overpressure created by upward fluids in gas chimneys gathering beneath MTD3 incites the pipe development via the hydraulic fracturing manner is still controversial. Undoubtedly, the latest emergence period of MTD2 and MTD3 is much earlier than 30 kyr BP because the calculated age of authigenic carbonate rocks at 52.1 mbsf is 0.53 Ma. The authigenic carbonate concretions lying along 3 mbsf level here are estimated to have formed at 3 kyr BP. The shallowest MTD1 is most likely to horizontally insert into the pipe structure. This activity impeded the upward fluid flowwhich just escaped from low MTD2 within the 30 kyr period. That is, multi-stage GH formation might have happened and led to the different occurrences of GHs among MTD1, MTD2, and MTD3 depth layers.

5.3. Multiple Accumulation Assumption

The analysis of paleo-GH existence suggests the great possibility that a multi-stage accumulation activity may have occurred. Rapid sedimentation together with sea level variations, landslides, and turbidity currents inevitably induces the simultaneous reaction of GH. The GH formation in the upper and middle layers of the sedimentary layer and decomposition in the lower part occurred [57,59]. Huge thick unconsolidated sedimentary seabed breeds and traps large amounts of methane-rich overpressure pore fluids. The mechanical disturbance in early MTD propagation easily triggered the internal rupture of the underlying muddy clay layer and prompted previously isolated pores to connect. The overpressure fluids subsequently penetrated the gas chimney channel, episodically migrated upward, and formed a hydrate-rich layer. After the MTD extension was completed, decreased permeability caused covered layers to serve as a sealing cap. These upward-migrating fluids were impeded. They had to gather largely at the bottom area of the MTDs and develop laterally. Local excellent GH resource sweet-spot layers with a certain thickness and depth were generated in this situation. This effect will be weakened with the temporal and spatial distribution sequence of the MTDs from bottom to top. From the difference in hydrate distribution at site W8, the oldest MTD3 has a GH abundance of up to 50%, and the second oldest MTD2 can also reach 20%. This phenomenon illustrates the rationality of our multiple accumulation speculations. Compared with other high-saturation hydrate reservoirs in oceanic hot reservoirs, such as the Alaminos Canyon in the Gulf of Mexico [60], Ulleung Basin, and the KG Basin offshore Kakinada, their drilling results are often firmly related to the MTD existence. Perhaps it provides a new point of view for hydrate resource exploration optimization.

6. Conclusions

The GH reservoirs firmly associated with gas chimney, pipe structure, and MTDs have been detected in GMGS5 investigation in the QDN area. How this gas chimney–hydrate geological system evolves and why GH-bearing sediments display the current distribution and occurrence attracts strong interest in probing its dynamic operation mechanism. Because site W08 shows representative system features, it is selected to elucidate the impact of environmental and geological factors, including sea-level change, sedimentation activity, soil-water qualities in and out of the gas chimney area, and MTDs on the evolution process. Through analyzing reasonable deposition-flow-reaction mode matching the local spatio-temporal geological background since 30 kyr BP, six key strata factors at five evolving moments are obtained. The sedimentary tensile fracture and occur at 5 kyr BP. Moreover, the evolution performance of this complex system in six other designed geological settings is compared, and the possibility of three-content palaeo-GH occurrences before 30 kyr ago is discussed. Finally, we found that about 10% pre-existing GH seems to have the exact distribution and occurrence in fine-grained clayey silt layers adjacent to MTD2 and MTD3. However, how the overpressure created by gas-bearing fluid accumulating at the top of gas chimneys splits the overlying MTD strata and triggers the development of a large number of internal fractures within these layers has not yet been resolved.

Author Contributions: Conceptualization, J.G., W.Z. and M.H.; Data curation, M.W. and L.W.; Formal analysis, M.W., L.W. and Q.W.; Funding acquisition, J.G.; Investigation, L.W.; Methodology, J.G., W.Z., M.H. and Q.W.; Resources, W.Z. and M.H.; Supervision, M.H.; Validation, J.G. and M.H.; Visualization, Q.W.; Writing—original draft, J.G. and M.H.; Writing—review and editing, J.G., W.Z. and Q.W. All authors have read and agreed to the published version of the manuscript.

Funding: This work was jointly sponsored by the National Science Foundation of China (42176215), the Natural Science Foundation of Guangdong Province (2021A1515011515), China Scholarship Council (202104910252), the Guangzhou Science and Technology Planning Project (202201011434, 202201011409), the First Batch of the "Nanhai New Star" Project (NHXXRCXM202357), the Guangdong Basic and Applied Basic Research Foundation (2019B030302004), Project of Sanya Yazhou Bay Science and Technology City (SCKJ-JYRC-2023-02), Sanya Science and Technology Innovation Project (2022KJCX14), and Key-Area Research and Development Program of Guangdong Province (2023B1111050014).

Institutional Review Board Statement: Not applicable.

Informed Consent Statement: Not applicable.

Data Availability Statement: The source code used in this manuscript is written by Matlab language and can be downloaded at an open-source online data repository (accessed on 12 March 2024, https://data.mendeley.com/datasets/vtkgd4zgsx; doi:10.17632/vtkgd4zgsx.1). Figures 4–6 were obtained from simulation data using this code and drawn with OriginPro 2016. The codes in this article are permitted for noncommercial or academic use, sharing, adaptation, distribution, and reproduction, and credit to the original authors is appreciated.

Acknowledgments: We would like to thank for the support and suggestions from Elke Kossel, and Christian Deusner throughout the whole research process.

Conflicts of Interest: The authors declare no conflicts of interest.

Appendix A. A Flow-Reaction Model

We conceived this gas chimney–hydrate compound geological system as a multi-phase transportation process in porous sediments, which includes three phases (h, l and a) and four components (h, water-w, methane-m, and salt-s). Firstly, GH kinetic reaction r_{h} (mol·m^{-3}·yr^{-1}) can be expressed as the gradient difference consequence of the fugacity under the local thermodynamic condition f (MPa) and equilibrium condition f_{eq} (MPa), respectively:

$$r_h = \varsigma_0 \exp[-\frac{\Delta E_{\mathrm{a}}}{R}(\frac{1}{T_{\mathrm{se}}} - \frac{1}{T_{\mathrm{eq}}})]\phi_h(f - f_{\mathrm{eq}}), \tag{A1}$$

where ζ_0 (mol·m^{-3}·MPa^{-1}·yr^{-1}) is the kinetic coefficient of GH reaction; $\Delta E_{\mathrm{a}}/R$ is 9752.73 K; T_{eq} (K) is the equilibrium temperature. The corresponding water r_{w} (mol·m^{-3}·yr^{-1}) and methane r_{m} (mol·m^{-3}·yr^{-1}) have the following relations:

$$r_w = 5.76 \times r_h, \tag{A2}$$

and

$$r_m = r_h. \tag{A3}$$

It is assumed that methane is immersed into pore gas and liquid, as follows:

$$\begin{aligned}&\frac{\partial}{\partial t}[\phi(\rho_g S_g + \rho_l S_l X_l^m)] + \nabla \cdot \begin{bmatrix} -\rho_g k_h \frac{k_{rg}}{\mu_g}(\nabla P_g - \rho_g g) \\ -\rho_l k_h \frac{k_{rl}}{\mu_l}(\nabla P_l - \rho_l g) X_l^m - \phi \rho_l D_l^m \nabla X_l^m \end{bmatrix} \\ &= -\eta_m r_m + \nabla \cdot q_m\end{aligned} \quad \text{(A4)}$$

Similar to the water component, we have:

$$\frac{\partial}{\partial t}[\phi \rho_l S_l X_l^w] + \nabla \cdot \left[-\rho_l k_h \frac{k_{rl}}{\mu_l}(\nabla P_l - \rho_l g) X_l^w \right] = -\eta_w r_w + \nabla \cdot q_w. \quad \text{(A5)}$$

The current salinity X_l^s is merely defined by initial liquid saturation S_{l0} and initial salinity X_{l0}^s:

$$X_l^s = \frac{S_{l\,0}}{S_l} X_{l\,0}^s. \quad \text{(A6)}$$

The expression of GH is as follows:

$$\frac{\partial}{\partial t}(\phi S_h \rho_h) = \eta_h r_h. \quad \text{(A7)}$$

In the above equations, ρ_κ (kg/m^3) (κ = a, l, h) is the phase density; X_l^m, X_l^w, and X_l^s are the mass fractions of methane, water, and salt in pore water; η_i is mole mass (i = m, w, h); μ_κ (Pa·s) (κ = a, l) is phase viscosity; g is 9.8 N/kg; t (yr) is time; D_l^m and D_l^s (m^2/s) are diffusion coefficients of methane and salt in liquids; q_i (kg·m^{-2}·yr^{-1}) is component flux (i = m, w).

The total heat energy balance is:

$$\begin{aligned}&[(1-\phi_h)\rho_r C_r + \phi_h(\rho_h C_h S_h + \rho_l C_l S_l + \rho_g C_g S_g)]\frac{\partial T_{se}}{\partial t} + [C_g \rho_g u_g + C_l \rho_l u_l]\frac{\partial T_{se}}{\partial z} \\ &= \frac{\partial}{\partial z}\left\{[(1-\phi_h)\lambda_r + \phi_h(S_h\lambda_h + S_g\lambda_g + S_l\lambda_l)]\frac{\partial T_{se}}{\partial z}\right\} + Q_h + \nabla \cdot q_e\end{aligned}, \quad \text{(A8)}$$

where C_κ (J·kg^{-1}·K^{-1}) and λ_κ (W·m^{-1}·K^{-1}) (κ = r, h, l, and a) are phase-specific heat and heat conductivity, respectively; q_e is local heat flow (mW/m^2). Darcy velocity of phases u_κ (κ = l, a) are:

$$u_g = -k_h \frac{k_{rg}}{\mu_g}(\nabla P_g - \rho_g g), \quad \text{(A9)}$$

and

$$u_l = -k_h \frac{k_{rl}}{\mu_l}(\nabla P_l - \rho_l g). \quad \text{(A10)}$$

Finally, the latent heat of GH in reaction Q_h (J·m^{-3}·s^{-1}) is:

$$Q_h = \eta_h r_h \times (446.12 \times 10^3 - 132.638 \times T_{se}). \quad \text{(A11)}$$

References

1. Collett, T.; Bahk, J.; Baker, R.; Boswell, R.; Divins, D.; Frye, M.; Goldberg, D.; Husebø, J.; Koh, C.; Malone, M.; et al. Methane hydrates in nature—Current knowledge and challenges. *J. Chem. Eng. Data* **2015**, *60*, 319–329. [CrossRef]
2. Boswell, R.; Shipp, C.; Reichel, T.; Shelander, D.; Saeki, T.; Frye, M.; Shedd, W.; Collett, T.S.; McConnell, D.R. Prospecting for marine gas hydrate resources. *Interpretation* **2016**, *4*, SA13–SA24. [CrossRef]
3. Cook, A.E.; Portnov, A.; Heber, R.C.; Vadakkepuliyambatta, S.; Bunz, S. Widespread subseafloor gas hydrate in the Barents Sea and Norwegian Margin. *Earth Planet. Sci. Lett.* **2023**, *604*, 117993. [CrossRef]
4. Liu, J.; Haeckel, M.; Rutqvist, J.; Wang, S.; Yan, W. The mechanism of methane gas migration through the gas hydrate stability zone: Insights from numerical simulations. *JGR Solid Earth* **2019**, *124*, 4399–4427. [CrossRef]
5. Bello, A.; Heggland, R.; Peacock, D.C.P. Pressure significance of gas chimneys. *Mar. Pet. Geol.* **2017**, *86*, 402–407. [CrossRef]
6. Elger, J.; Berndt, C.; Rupke, L.; Krastel, S.; Gross, F.; Geissler, W.H. Submarine slope failures due to pipe structure formation. *Nat. Commun.* **2018**, *9*, 715. [CrossRef] [PubMed]

7. Robinson, A.H.; Callow, B.; Bottner, C.; Yilo, N.; Provenzano, G.; Falcon-Suarez, I.H.; Marin-Moreno, H.; Lichtschlag, A.; Bayrakci, G.; Gehrmann, R.; et al. Multiscale characterization of chimneys/pipes: Fluid escape structures within sedimentary basins. *International J. Greenh. Gas Control* **2021**, *106*, 103245. [CrossRef]
8. Dentzer, J.; Bruel, D.; Delescluse, M.; Chamot-Rooke, N.; Beccaletto, L.; Lopez, S.; Courrioux, G.; Violette, S. Thermal and seismic hints for chimney type cross-stratal fluid flow in onshore basins. *Sci. Rep.* **2018**, *8*, 15330. [CrossRef]
9. Han, W.; Chen, L.; Liu, C. Distribution and characteristics of gas chimneys in the passive margin offshore SW Taiwan. *Mar. Geophys. Res.* **2021**, *42*, 25. [CrossRef]
10. Hovland, M.; Sommerville, J.H. Characteristics of two natural gas seepages in the North Sea. *Mar. Pet. Geol.* **1985**, *2*, 319–326. [CrossRef]
11. Hovland, M. The formation of pockmarks and their potential influence on offshore construction. *Q. J. Eng. Geol. Hydrogeol.* **1989**, *22*, 131–138. [CrossRef]
12. Hovland, M.; Svensen, H. Submarine pingoes: Indicators of shallow gas hydrates in a pockmark at Nyegga, Norwegian Sea. *Mar. Geol.* **2006**, *228*, 15–23. [CrossRef]
13. Cheng, C.; Jiang, T.; Kuang, Z.; Ren, J.; Liang, J.; Lai, H.; Xiong, P. Seismic characteristics and distributions of Quaternary mass transport deposits in the Qiongdongnan Basin, northern South China Sea. *Mar. Pet. Geol.* **2021**, *129*, 105118. [CrossRef]
14. Yang, J.; Davies, R.J.; Huuse, M. Gas migration below gas hydrates controlled by mass transport complexes, offshore Mauritania. *Mar. Pet. Geol.* **2013**, *48*, 366–378. [CrossRef]
15. Chatterjee, S.; Bhatnagar, G.; Dugan, B.; Dickens, G.R.; Chapman, W.G.; Hirasaki, G.J. The impact of lithologic heterogeneity and focused fluid flow upon gas hydrate distribution in marine sediment. *J. Geophys. Res. Solid Earth* **2014**, *119*, 6705–6732. [CrossRef]
16. Guan, J.; Wan, L.; Liang, D. Gauging formation dynamics of structural-seepage methane hydrate reservoirs in Shenhu area of northern South China Sea: Impact of seafloor sedimentation and assessment of controlling factors. *Mar. Pet. Geol.* **2019**, *107*, 185–197. [CrossRef]
17. Yang, S.; Liang, J.; Lu, J.; Qu, C.; Liu, B. New understandings on characteristic and controlling factors of gas hydrate reservoirs in Shenhu area on northern slope of South China Sea. *Earth Sci. Front.* **2017**, *24*, 1–14, (In Chinese with English abstract).
18. Cathles, L.M.; Su, Z.; Chen, D. The physics of gas chimney and pockmark formation, with implications for assessment of seafloor hazards and gas sequestration. *Mar. Pet. Geol.* **2010**, *27*, 82–91. [CrossRef]
19. Sun, Q.; Cartwright, J.; Wu, S.; Chen, D. 3D seismic interpretation of dissolution pipes in the South China Sea: Genesis by subsurface, fluid induced collapse. *Mar. Geol.* **2013**, *337*, 171–181. [CrossRef]
20. Wan, Z.; Zhang, W.; Ma, C.; Liang, J.; Li, A.; Meng, D.; Huang, W.; Yang, C.; Zhang, J.; Sun, Y. Dissociation of gas hydrates by hydrocarbon migration and accumulation-derived slope failures: An example from the South China Sea. *Geosci. Front.* **2021**, *12*, 101345. [CrossRef]
21. Argentino, C.; Conti, S.; Fioroni, C.; Fontana, D. Evidences for Paleo-Gas Hydrate Occurrence: What We Can Infer for the Miocene of the Northern Apennines (Italy). *Geosciences* **2019**, *9*, 134. [CrossRef]
22. Chun, J.H.; Bahk, J.J.; Um, I.K. Ulleung basin gas hydrate drilling expeditions, Korea: Lithologic characteristics of gas hydrate-bearing sediments. In *World Atlas of Submarine Gas Hydrates in Continental Margins*; Mienert, J., Berndt, C., Tréhu, A.M., Camerlenghi, A., Liu, C.S., Eds.; Springer: Cham, Switzerland, 2022; pp. 155–161.
23. Collett, T.S.; Chopra, K.; Bhardwaj, A.; Boswell, R.; Waite, W.F.; Misra, A.K.; Kumar, P. A review of the exploration, discovery and characterization of highly concentrated gas hydrate accumulations in coarse-grained reservoir systems along the eastern continental margin of India. In *World Atlas of Submarine Gas Hydrates in Continental Margins*; Mienert, J., Berndt, C., Tréhu, A.M., Camerlenghi, A., Liu, C.S., Eds.; Springer: Cham, Switzerland, 2022; pp. 139–154.
24. Liang, C.; Liu, C.; Xie, X.; Yu, X.; He, Y.; Su, M.; Chen, H.; Zhou, Z.; Tian, D.; Mi, H.; et al. Basal shear zones of recurrent mass transport deposits serve as potential reservoirs for gas hydrates in the Central Canyon area, South China Sea. *Mar. Geol.* **2021**, *441*, 106631. [CrossRef]
25. Zhao, Z.; Sun, Z.; Wan, Z.; Sun, Z.; Liu, J.; Zhang, C. The high resolution sedimentary filling in Qiongdongnan Basin, northern South China Sea. *Mar. Geol.* **2015**, *361*, 11–24. [CrossRef]
26. Zhao, Z.; Sun, Z.; Sun, L.; Wang, Z.; Sun, Z. Cenozoic tectonic subsidence in the Qiongdongnan Basin, northern South China Sea. *Basin Res.* **2018**, *30* (Suppl. S1), 269–288. [CrossRef]
27. Deng, W.; Liang, J.; Zhang, W.; Kuang, Z.; Zhong, T.; He, Y. Typical characteristics of fracture-filling hydrate-charged reservoirs caused by heterogeneous fluid flow in the Qiongdongnan Basin, northern South China Sea. *Mar. Pet. Geol.* **2021**, *124*, 104810.
28. Lai, H.; Fang, Y.; Kuang, Z.; Ren, J.; Liang, J.; Lu, J.; Wang, G.; Xing, C. Geochemistry, origin and accumulation of natural gas hydrates in the Qiongdongnan Basin, South China Sea: Implications from site GMGS5-W08. *Mar. Pet. Geol.* **2021**, *123*, 104774. [CrossRef]
29. Wei, J.; Liang, J.; Lu, J.; Zhang, W.; He, Y. Characteristics and dynamics of gas hydrate systems in the northwestern South China Sea-Results of the fifth gas hydrate drilling expedition. *Mar. Pet. Geol.* **2019**, *110*, 287–298. [CrossRef]
30. Ye, J.; Wei, J.; Liang, J.; Lu, J.; Lu, H.; Zhang, W.; all the participants of GMGS5. Complex gas hydrate system in a gas chimney, South China Sea. *Mar. Pet. Geol.* **2019**, *104*, 29–39. [CrossRef]
31. Gong, C.; Wang, Y.; Zhu, W.; Li, W.; Xu, Q.; Zhang, J. The central submarine canyon in the Qiongdongnan Basin, northwestern South China Sea: Architecture, sequence stratigraphy, and depositional processes. *Mar. Pet. Geol.* **2011**, *28*, 1690–1702. [CrossRef]

32. Li, W.; Alves, T.M.; Wu, S.; Volker, D.; Zhao, F.; Mi, L.; Kopf, A. Recurrent slope failure and submarine channel incision as key factors controlling reservoir potential in the South China Sea (Qiongdongnan Basin, South Hainan Island). *Mar. Pet. Geol.* **2015**, *64*, 17–30. [CrossRef]
33. Ren, J.; Cheng, C.; Xiong, P.; Kuang, Z.; Liang, J.; Lai, H.; Chen, Z.; Chen, Y.; Li, T.; Jiang, T. Sand-rich gas hydrate and shallow gas systems in the Qiongdongnan Basin, northern South China Sea. *J. Pet. Sci. Eng.* **2022**, *215 Pt B*, 110630. [CrossRef]
34. Wang, Z.; Jiang, T.; Zhang, D.; Wang, Y.; Zuo, Q.; He, W. Evolution of deepwater sedimentary environments and its implication for hydrocarbon exploration in Qiongdongnan Basin, northwestern South China Sea. *Acta Oceanol. Sin.* **2015**, *34*, 1–10. [CrossRef]
35. Huang, B.; Tian, H.; Li, X.; Wang, Z.; Xiao, X. Geochemistry, origin and accumulation of natural gases in the Deepwater area of the Qiongdongnan Basin, South China Sea. Mar. *Pet. Geol.* **2016**, *72*, 254–267. [CrossRef]
36. Liang, J.; Zhang, W.; Lu, J.; Wei, J.; Kuang, Z.; He, Y.; all participants of GMGS5 expedition. Geological occurrence and accumulation mechanism of natural gas hydrates in the eastern Qiongdongnan Basin of the South China Sea: Insights from site GMGS5-W9-2018. *Mar. Pet. Geol.* **2019**, *418*, 106042. [CrossRef]
37. Meng, M.; Liang, J.; Lu, J.; Zhang, W.; Kuang, Z.; Fang, Y.; He, Y.; Deng, W.; Huang, W. Quaternary deep-water sedimentary characteristics and their relationship with the gas hydrate accumulations in the Qiongdongnan Basin, Northwest South China Sea. *Deep-Sea Res.* **2021**, *177*, 103628. [CrossRef]
38. Zhang, W.; Liang, J.; Yang, X.; Su, P.; Wan, Z. The formation mechanism of mud diapirs and gas chimneys and their relationship with natural gas hydrates: Insights from the deep-water area of Qiongdongnan Basin, northern South China Sea. *Int. Geol. Rev.* **2018**, *62*, 789–810. [CrossRef]
39. Zhang, W.; Liang, J.; Zhang, R.; Deng, W.; Gu, Y.; He, Y.; Gong, Y.; Meng, M.; Feng, J.; Liang, J. Gas hydrate accumulation in shelf break setting: Example from the Qiongdongnan Basin in the northern slope of the South China Sea. *Geol. J.* **2022**, *57*, 1153–1171. [CrossRef]
40. Yang, W.; Kuang, Z.; Ren, J.; Liang, J.; Lu, H.; Ning, Z.; Xu, C.; Lai, H.; Chen, R.; Zhao, B.; et al. The controlling factors of the natural gas hydrate accumulation in the Songnan Low Uplift, Qiongdongnan Basin, China. *Front. Earth Sci.* **2022**, *10*, 882080.
41. Conti, S.; Fontana, D.; Lucente, C.C.; Pini, G.A. Relationships between seep-carbonates, mud volcanism and basin geometry in the Late Miocene of the northern Apennines of Italy: The Montardone mélange. *Int. J. Earth Sci. (Geol. Rundsch.)* **2014**, *103*, 281–295. [CrossRef]
42. Liu, S.; Feng, X.; Feng, Z.; Xiao, X.; Feng, L. Geochemical evidence of methane seepage in the sediments of the Qiongdongnan Basin, South China Sea. *Chem. Geol.* **2020**, *543*, 119588. [CrossRef]
43. He, Y.; Liang, J.; Kuang, Z.; Deng, W.; Ren, J.; Lai, H.; Meng, M.; Zhang, W. Migration and accumulation characteristics of natural gas hydrates in the uplifts and their slope zones in the Qiongdongnan Basin, China. *China Geol.* **2022**, *5*, 234–250. [CrossRef]
44. Deng, Y.; Chen, F.; Guo, Q.; Hu, Y.; Chen, D.; Yang, S.; Cao, J.; Chen, H.; Wei, R.; Cheng, S.; et al. Possible links between methane seepages and glacial-interglacial transitions in the South China Sea. *Geophys. Res. Lett.* **2021**, *48*, e2020GL091429. [CrossRef]
45. Davie, M.K.; Zatsepina, O.Y.; Buffett, B.A. Methane solubility in marine hydrate environments. *Mar. Geol.* **2004**, *203*, 177–184. [CrossRef]
46. Duan, Z.; Moller, N.; Greenberg, J.; Weare, J. The prediction of methane solubility in natural waters to high ionic strength from 0 to 250 °C and from 0 to 1600 bar. *Geochim. Cosmochim. Acta* **1992**, *56*, 1451–1460. [CrossRef]
47. Tishchenko, P.; Hensen, C.; Wallmann, K.; Wong, C.S. Calculation of the stability and solubility of methane hydrate in seawater. *Chem. Geol.* **2005**, *219*, 37–52. [CrossRef]
48. Fredlund, D.G.; Xing, A.; Huang, S. Predicting the permeability function for unsaturated soils using the soil-water characteristic curve. *Can. Geotech. J.* **1994**, *31*, 533–546. [CrossRef]
49. Murphy, Z.W.; DiCarlo, D.A.; Flemings, P.B.; Daigle, H. Hydrate is a nonwetting phase in porous media. *Geophys. Res. Lett.* **2020**, *47*, e2020GL089289. [CrossRef]
50. Daigle, H.; Ghanbarian, B.; Henry, P.; Conin, M. Universal scaling of the formation factor in clays: Example from the Nankai Trough. *J. Geophys. Res. Solid Earth* **2015**, *120*, 7361–7375. [CrossRef]
51. Guan, J.; Liang, D.; Wu, N.; Fan, S. The methane hydrate formation and the resource estimate resulting from free gas migration in seeping seafloor hydrate stability zone. *J. Asian Earth Sci.* **2009**, *36*, 277–288. [CrossRef]
52. Daigle, H.; Cook, A.; Fang, Y.; Bihani, A.; Song, W.; Flemings, P.B. Gas-driven tensile fracturing in shallow marine sediments. *JGR Solid Earth* **2020**, *125*, e2020JB020835. [CrossRef]
53. Wei, J.; Yang, L.; Liang, Q.; Liang, J.; Lu, J.; Zhang, W.; Zhang, X.; Lu, X. Geomechanical properties of gas hydrate-bearing sediments in Shenhu Area of the South China Sea. *Energy Rep.* **2021**, *7*, 8013–8020. [CrossRef]
54. Crutchley, G.J.; Mountjoy, J.J.; Hillman, J.I.T.; Turco, F.; Watson, S.; Flemings, P.B.; Davy, B.; Woelz, S.; Gorman, A.R.; Bialas, J. Upward-Doming Zones of Gas Hydrate and Free Gas at the Bases of Gas Chimneys, New Zealand's Hikurangi Margin. *JGR Solid Earth* **2021**, *126*, e2020JB021489. [CrossRef]
55. Su, P.; Liang, J.; Peng, J.; Zhang, W.; Xu, Z. Petroleum systems modeling on gas hydrate of the first experimental exploitation region in the Shenhu area, northern South China Sea. *J. Asian Earth Sci.* **2018**, *16*, 57–76. [CrossRef]
56. Bi, G.; Lyu, C.; Li, C.; Chen, G.; Zhang, G.; Zhou, Q.; Li, C.; Zhao, Y. Impact of early hydrocarbon charge on the diagenetic history and reservoir quality of the Central Canyon sandstones in the Qiongdongnan Basin, South China Sea. *J. Asian Earth Sci.* **2019**, *185*, 104022. [CrossRef]

57. Guan, J.; Liang, Y.; Wang, S.; Wan, L.; Fan, S.; Su, P.; Zhang, W.; Liang, D. New insight on the stratigraphic-diffusive gas hydrate system since the Pleistocene in the Dongsha area of the Northeastern South China Sea. *J. Mar. Sci. Eng.* **2022**, *10*, 434. [CrossRef]
58. Riboulot, R.; Cattaneo, A.; Sultan, N.; Ker, S.G.; Imbert, P.; Voisset, M. Sea-level change and free gas occurrence influencing a submarine landslide and pockmark formation and distribution in deepwater Nigeria. *Earth Planet. Sci. Lett.* **2013**, *375*, 78–91. [CrossRef]
59. Wang, B.; Lei, H.; Huang, F.; Kong, Y.; Pan, F.; Cheng, W.; Chen, L.; Guo, L. Effect of Sea-Level Change on Deep-Sea Sedimentary Records in the Northeastern South China Sea over the past 42 kyr. *Geofluids* **2020**, *2020*, 8814545. [CrossRef]
60. Boswell, R.; Collett, T.S.; Frye, M.; Shedd, W.; McConnell, D.R.; Shelander, D. Subsurface gas hydrates in the northern Gulf of Mexico. *Mar. Pet. Geol.* **2012**, *34*, 4–30. [CrossRef]

Journal of
Marine Science and Engineering

MDPI

Article

Mechanical Characteristics of Gas Hydrate-Bearing Sediments: An Experimental Study from the South China Sea

Qingmeng Yuan [1,2,3], Liang Kong [3,*], Qianyong Liang [1,2,*], Jinqiang Liang [1,2], Lin Yang [1,2], Yifei Dong [1,2], Zhigang Wang [1,2] and Xuemin Wu [1,2]

1 Guangzhou Marine Geological Survey, Guangzhou 511400, China; yqm905@126.com (Q.Y.); lljinqiang@petalmail.com (J.L.); yl_gmgs@163.com (L.Y.); yfdong1987@163.com (Y.D.); wangzg0106@163.com (Z.W.); wxm9211@163.com (X.W.)
2 National Engineering Research Center for Gas Hydrate Exploration and Development, Guangzhou 511400, China
3 School of Science, Qingdao University of Technology, Qingdao 266520, China
* Correspondence: qdkongliang@163.com (L.K.); tomlqy@163.com (Q.L.)

Abstract: Clarifying the mechanical characteristics of gas hydrate-bearing sediments (GHBS) from a mechanical perspective is crucial for ensuring the long-term, safe, and efficient extraction of natural gas hydrates. In this study, seabed soft clay from the northern South China Sea was utilized to prepare clayey silt samples, aligning with gradation curves related to hydrate extraction projects in the Shenhu area of the South China Sea. Utilizing the high-pressure low-temperature hydrate triaxial testing system (ETAS), twelve sets of triaxial shear tests were conducted. The results highlight that increases in hydrate saturation and confining pressure significantly enhance GHBS' strength and stiffness, with more pronounced volume expansion observed during shearing. These tests have elucidated the mechanical responses of GHBS. Subsequently, empirical formulas were developed to characterize their properties under varying conditions. Additionally, based on the experimental data, the micro-mechanisms of GHBS were analyzed, suggesting that hydrates notably contribute to the filling and cementing effects in GHBS, with these effects varying with changes in hydrate saturation and confining pressure. This study contributes to a deeper understanding of the fundamental mechanical properties of GHBS.

Keywords: gas hydrate-bearing sediments; triaxial shear test; mechanical properties; empirical formulas; seabed soft clay; South China Sea

1. Introduction

Gas hydrate-bearing sediments (GHBS) are deep-sea sedimentary bodies that contain natural gas hydrates. These hydrates are cage-like crystalline compounds formed from natural gas and water under high-pressure and low-temperature conditions, predominantly comprising methane [1]. They are widely described in geological records [2], and are extensively found in terrestrial permafrost regions, along continental margins, and in the deep-sea abyssal plains [3]. The discovery of natural gas hydrate deposits in numerous countries [4,5] signifies their potential as a new energy source, especially given the declining availability of traditional hydrocarbons. However, GHBS are subject to stringent stability requirements. Environmental changes or human activities that induce minor shifts in soil pressure or temperature can destabilize these hydrates, leading to gas release. Such events can drastically alter the soil's mechanical behavior, affecting characteristics like compressibility [6], failure features [7], and the strength and stiffness of the soil [8]. These changes can precipitate extensive marine geological disasters, exacerbate the greenhouse effect [9], and harm the marine ecosystem [10], thereby posing significant risks to hydrate exploitation and environmental safety. Consequently, investigating the mechanical properties of GHBS is vital for understanding their deformation responses to ensure the sustainable and safe extraction of natural gas hydrates.

In situ mechanical testing is deemed the most reliable approach to ensure accuracy. Winters et al. [11,12] examined the physical properties of GHBS from various locations, including the North Slope of Alaska, Mackenzie Delta, and regions off the east coast of India and the Andaman Islands, using techniques such as infrared imaging, acoustic testing, thermal conductivity, resistivity testing, and shear strength assessments. Japan's "Tokai-oki to Kumano-nada Project" in 2004 [13] involved drilling up to 250–400 m below the seabed and employed a pressure–temperature corer (PTCS) for sampling. Bondarenko et al. [14] identified the optimal conditions for natural gas hydrate deposits in the Black Sea at depths of 500–750 m (8–9 °C, 20 MPa) and summarized their characteristics. However, due to the unique conditions required for hydrate storage, deep-sea in situ mechanical testing is costly. Consequently, most current research on GHBS involves synthesizing hydrates in labs and analyzing their mechanical properties. Methods like triaxial shear tests [15,16], one-dimensional consolidation tests [17], and analogous material tests [18] simulate various hydrate saturations, particle types, and temperature and pressure conditions in studying the behavior of GHBS. These experiments yield crucial insights but also have limitations, such as difficulty in replicating natural high-pressure, low-temperature conditions and differences in microstructure and properties between synthetic and natural GHBS samples. Therefore, while lab tests offer valuable perspectives on the mechanical properties of GHBS, their results should be interpreted with caution and complemented by other research approaches for a fuller understanding.

In this study, surface soil from the northeastern South China Sea was gathered to investigate the mechanical properties of silty fine sandy GHBS using configured sandy soil. We focused on the quantitative relationships between the mechanical properties, strength, and stiffness of GHBS, leading to the development of empirical equations. Finally, we analyzed the unique mechanical characteristics of GHBS and discussed the hydrates' mechanisms of action.

2. Experimental Study of GHBS

2.1. Material and Test Equipment

In August 2017, we procured surface soil samples from a natural gas hydrate deposit in the northern region of the South China Sea. This effort was supported by the third leg of the shared voyage initiative sponsored by the National Science Foundation Committee in the northeastern South China Sea and Luzon Strait. The sampling occurred at the YD2-04 station (115°16.407′ E, 19°57.056′ N), located at a depth of 1077 m. The sample comprised soft clay, as depicted in Figure 1.

Figure 1. Preparation of argillaceous silt.

Figure 1 illustrates the natural gas hydrate deposit in the South China Sea of China [19,20]. Previously, two pilot mining projects [21] demonstrated that the primary reservoir soil in the South China Sea's hydrate deposit area is silty mud. Due to sample limitations in quantity and volume, our study primarily utilizes the particle size and mineral composition analysis data from Liu et al. [22] on sediments post-hydrate decomposition in the Shenhu area of the South China Sea. We synthesized silty mud, combining high-purity quartz sand with the obtained soft clay. Initially, we cleaned and dried the quartz sand, then separated particles into different sizes using a vibrating sieve, as shown in Figure 2a. Based on Liu et al. [22], we determined an 8% clay content and incorporated it into the residual soil sample from prior tests. The final composition of the prepared silty mud after thorough mixing is presented in Figure 2b.

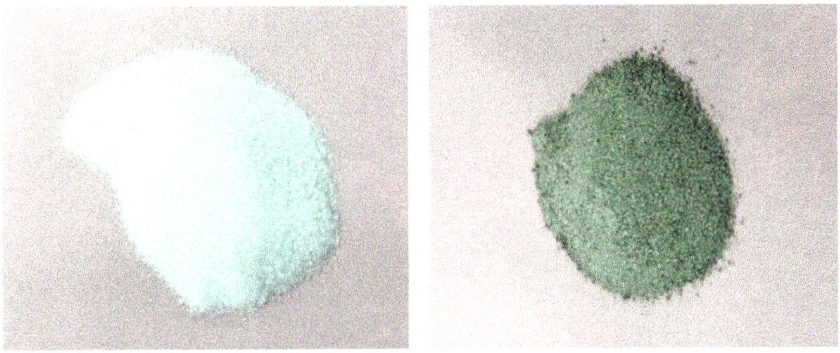

(**a**) Screened quartz sand (**b**) Prepared silty mud

Figure 2. Details the preparation process of silty mud.

Figure 3 displays the grading curve, while Table 1 compares the grading indices of the experimental materials from Liu et al. [22] and our study. This comparison indicates a close resemblance between the prepared silty mud and the particle grading of the mining area's reservoir soil, aligning well with the grading characteristics of the South China Sea mining area's reservoir soil. Beyond the metrics listed in Table 1, the silty mud prepared in our study exhibited a specific gravity of 2.6590, with maximum and minimum porosity ratios of 1.2117 and 0.7693, respectively.

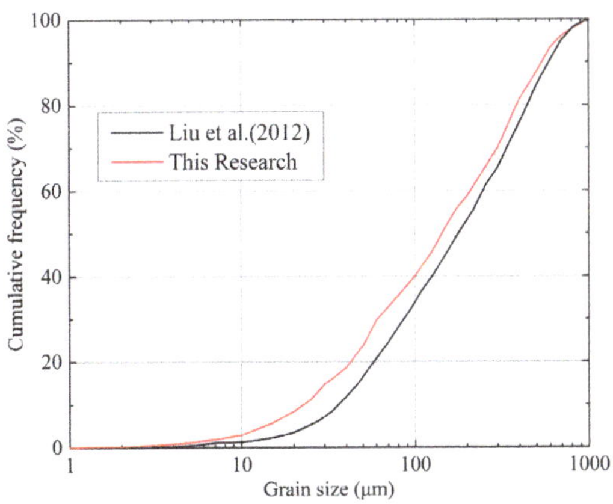

Figure 3. Experimental material grading curve [22].

Table 1. Grading index.

Material	d_{10}/μm	d_{30}/μm	d_{50}/μm	d_{60}/μm	$C_u = d_{60}/d_{10}$	$C_c = d_{30}^2/(d_{10} \cdot d_{60})$
[22]	0.0430	0.1012	0.21479	0.2932	6.8198	0.8123
This study	0.0299	0.0768	0.1737	0.2360	7.8824	0.8349

Our experimental setup utilized the high-pressure low-temperature hydrate triaxial testing system (ETAS system) produced by Europa and America Geotechnical Company. Figure 4a illustrates the physical assembly of the system, while its schematic is provided in Figure 4b. For experimental safety, the CH_4 gas tank was housed in an explosion-proof cabinet and complemented with flammable gas concentration detection and communication devices, including a handheld flammable gas concentration detector.

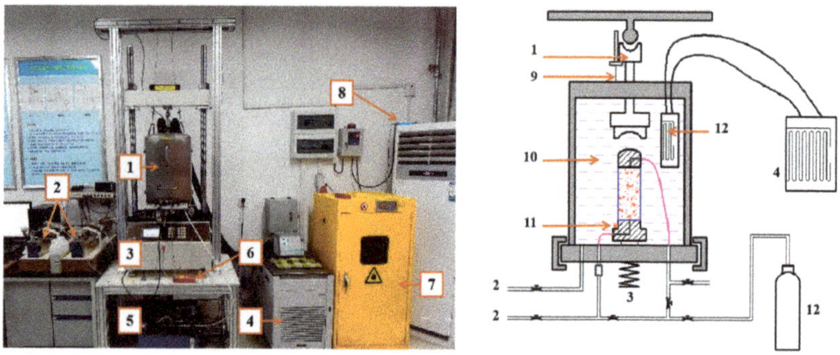

(**a**) ETAS system physical diagram (**b**) ETAS system schematic

Figure 4. Physical and schematic diagrams of the ETAS system. 1—pressure chamber; 2—standard controller (32 MPa/200 mL); 3—axial loading (100 kN); 4—water bath (−20 °C~65 °C); 5—oil pump; 6—gas detector; 7—explosion-proof cabinet; 8—gas alarm communication instrument; 9—displacement sensor; 10—silicon oil; 11—temperature sensor; 12—methane gas.

The ETAS system comprises an axial loading system, a stainless steel high-pressure triaxial pressure chamber, a water bath, two standard controllers, and other essential components and spare parts. The axial loading system, motor servo-controlled with a 100 kN range and a precision of 0.1% of the full scale, facilitates pressure control either manually or via GDSLAB (v2) software. Manual operation offers speed selection, enabling easy specimen-load sensor contact. The axial load is exerted externally; the pressure chamber, crafted from high-strength stainless steel, can withstand up to 32 MPa. Silicone oil is pumped into the chamber to control the confining pressure. The water bath, maintaining the chamber's liquid medium temperature, ranges from −20 °C to 65 °C, with a temperature sensor at the sample base's interface providing real-time feedback. The two standard pressure–volume controllers operate within a 32 MPa/200 mL range with a 0.1% full range/volume accuracy. The pressure transmission mediums can be gas, liquid, or a mixture, with the confining pressure controller utilizing silicone oil and the back pressure controller using methane gas. Qingdao Ludong Gas Co., Ltd. (Qingdao, China) supplied the methane, with a factory pressure of 10 ± 0.5 MPa and a purity of 99.999%.

2.2. Specimen Preparation

The formation of hydrates necessitates specific thermobaric conditions, requiring both the temperature and pressure to be situated within the confines of the phase equilibrium curve. This curve is mathematically represented as:

$$\ln(1000p_{\text{eq}}) = 38.98 - 8533.8/T_{\text{eq}} \tag{1}$$

where p_{eq} denotes the critical equilibrium pressure for the stability of natural hydrates (in Pascals), and T_{eq} represents the critical equilibrium temperature for the stability of natural hydrates (in Kelvins).

Converting p_{eq} and T_{eq} from Equation (1) into MPa and °C, respectively, yields the phase equilibrium curve for natural gas hydrates, depicted in Figure 5. The point (T_{eq}, p_{eq}) must reside within the shaded region of this curve. The lines and arrows in the figure trace the hydrate formation pathway utilized in this study. Figure 6 illustrates a schematic of the gas channel structure, facilitating the preparation of geotechnical samples infused with energy sources through controlled valve and pipeline manipulation.

For sample preparation using the excess gas method, a matrix sample with predetermined water content was initially prepared, followed by the introduction of an excess of methane. The required mass of water was calculated based on the chemical equation governing the formation of natural gas hydrates, represented as:

$$CH_4 + nH_2O \Leftrightarrow CH_4 \cdot nH_2O \tag{2}$$

Here, n is the reaction stoichiometry, typically ranging between 5 and 6 [23], and can be approximated as 5.75 [24,25].

Assuming a sample volume of V, with a matrix sample porosity of n, hydrate saturation denoted as S_h, and a hydrate density (ρ_h) of 0.91 g/cm^3, the mass of water required can be derived as:

$$m_{H_2O} = \frac{207}{239} n V S_h \rho_h \tag{3}$$

To prepare GHBS samples, dried clayey silt was placed in a tray. The water mass, exclusive of gas, was computed using Equation (3). This water was then added to a sprayer and carefully sprayed into the material while simultaneously mixing with a spoon. The mixture was then wrapped in cling film and left to rest for 24 h, ensuring uniform water distribution, before proceeding with GHBS sample preparation.

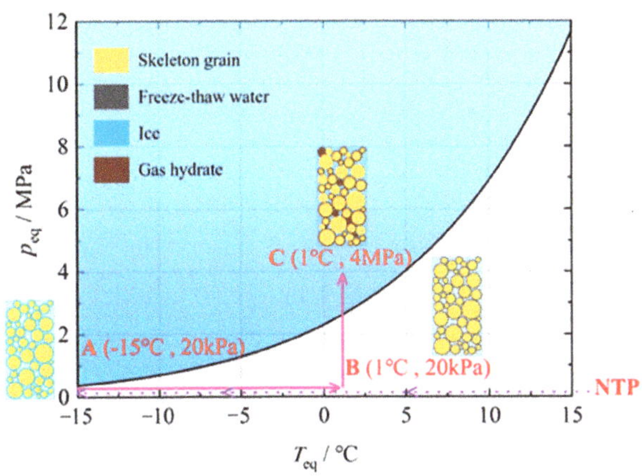

Figure 5. Phase equilibrium curve and hydrate formation path.

Utilizing ice meltwater as the water source for hydrate sediments, a method proven to expedite hydrate formation, has been validated by multiple studies. For instance, Takeya et al. [26] and Chen et al. [27] observed that freeze–thaw cycles considerably reduce the hydrate formation induction period. Hyodo et al. [28] and Khlebnikov et al. [29] employed meltwater from ice as the water source in energy layers, significantly enhancing the rate of natural gas hydrate formation. This study adopts similar methodologies, employing controlled temperature and valve adjustments to generate meltwater from ice and manage the direction of methane gas flow, thereby facilitating rapid and stable hydrate synthesis. The specific experimental procedures are as follows:

(1) Initiate the oil pump to fill the pressure chamber with silicone oil, applying a confining pressure of 200 kPa. Connect the water bath to the pressure chamber and activate it, achieving the conditions at point A in Figure 5 (−15 °C, 20 kPa). Commence sample freezing, maintaining this state for 6 h.

(2) Upon completion of freezing, open the gas tank and channels L_A and L_0 (Figure 6), where L_0 is attached to a 2 L gasbag. Monitor the pore pressure readings and adjust the gas tank's pressure relief valve to maintain pore pressure around 20 kPa, ensuring the gas flow rate is controlled to prevent sample disturbance. This process allows methane to thoroughly flush the sample, displacing the internal air. When the volume of the gasbag is maximized, indicating complete displacement, close the L_0 channel.

(3) Proceed with thawing by heating, aligning with point B in Figure 5 (1 °C, 20 kPa). Keep channels L_A and L_B open to ensure the even distribution of meltwater and gas pressure within the sample.

(4) Induce hydrate formation by adjusting the temperature and pressure to match point C in Figure 5 (1 °C, 4 MPa), while concurrently raising the confining pressure to 5 MPa. To maintain steady back pressure, open the L_C channel connected to the back pressure controller, keeping channels L_A and L_B open to maintain a consistent 4 MPa pressure at both the top and bottom of the sample. Continue this process for a minimum of 48 h, ensuring adequate hydrate formation time and sufficient gas volume during the formation phase. After 48 h, close the gas tank, keep only the L_C channel open, and use the back pressure controller to continue stabilizing the pressure for over 10 h. Once the back pressure controller volume remains largely unchanged, it can be concluded that no further hydrate formation will occur, signifying the completion of GHBS sample preparation.

(5) Follow the experimental plan for subsequent procedures, including consolidation, shearing, and other test stages.

(6) Post experiment, close all channels, attach a flowmeter to the L_0 channel, and gradually open the L_0 valve. The gas released by the decomposition of hydrates within the GHBS is collected in the gasbag through the flowmeter.

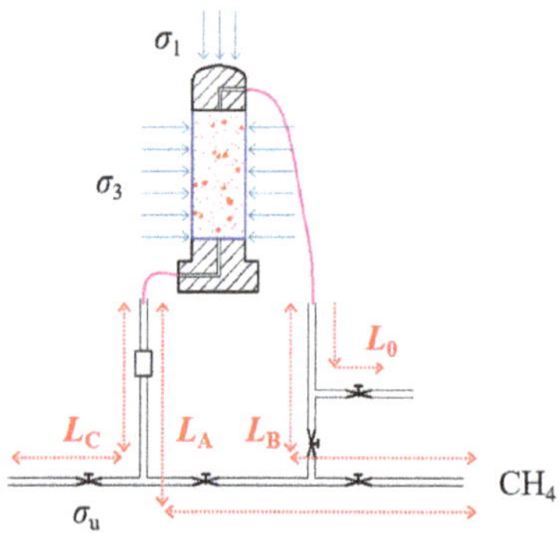

Figure 6. Schematic of pipeline channels during GHBS sample preparation.

2.3. Test Scheme

This study examines two variables across 12 sets of consolidation and drained tests: effective confining stress and hydrate saturation, detailed in Table 2. The effective confining stresses (σ_3') were set at three levels: 1 MPa, 3 MPa, and 5 MPa. For each level of effective confining stress, three different hydrate saturations were established: 10%, 20%, and 30%.

In the course of the experiment, the pore pressure was consistently maintained at 4 MPa, as indicated in Figure 5. Consequently, the total confining pressures were established at 5 MPa, 7 MPa, and 9 MPa, respectively.

Table 2. Test Scheme.

Number	Hydrate Saturation (S_h, %)	Effective Confining Stress (σ_3', MPa)
GHBS01	0, 10, 20, 30	1
GHBS03	0, 10, 20, 30	3
GHBS05	0, 10, 20, 30	5

3. Results and Interpretation

3.1. Stress–Strain Relationship

Figure 7 presents the stress–strain curve at $S_h = 0$, where samples devoid of hydrates demonstrate characteristics akin to loose or medium dense sand, uniformly exhibiting strain hardening without noticeable peak strength. This suggests that GHBS, in the absence of hydrates, share similarities with typical soil types. However, hydrate saturation markedly influences the mechanical properties of GHBS. Figure 8, representing S_h = 10%, continues to display strain hardening. Under identical axial strains, the deviatoric stress is notably higher than that in Figure 7, especially under the confining pressures of 3 MPa and 5 MPa. It was observed that at a confining pressure of 1 MPa and S_h = 10%, the GHBS maintained stable deviatoric stress at an axial strain of 8%, attributable to hydrates enhancing the compactness of the GHBS, which in turn strengthens particle frictional interactions and facilitates the earlier attainment of peak strength.

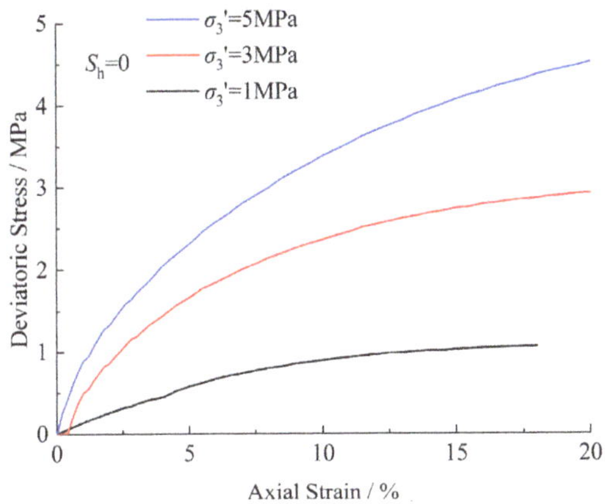

Figure 7. Stress–strain curve (S_h = 0).

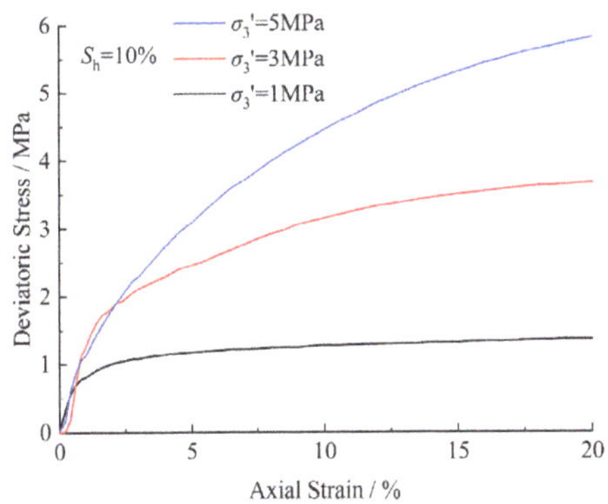

Figure 8. Stress–strain curve (S_h = 10%).

In Figure 9, at S_h = 20%, under a 1 MPa confining pressure, it can be seen that peak strength is achieved at an axial strain of 3%, and at a 3 MPa confining pressure, the curve nears peak strength at an axial strain of 15%. This behavior is attributed to hydrates enhancing the frictional properties within GHBS particles, while increased confining pressure additionally boosts GHBS compactness. This indicates the dual contribution of hydrate saturation and confining pressure to GHBS strength.

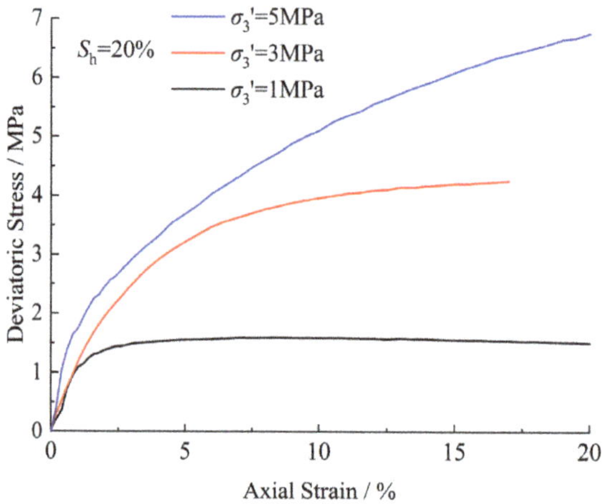

Figure 9. Stress–strain curve (S_h = 20%).

Figure 10, corresponding to S_h = 30%, aligns with Figures 7–9 in that increased confining pressure leads to higher deviatoric stress. Notably, the curve shape differs: the initial stiffness of the sample sets is more pronounced, and at 1 MPa confining pressure, the curve exhibits strain softening. Moreover, hydrates contribute additional strength to GHBS, but this strength diminishes more swiftly under lower confining pressures (1 MPa), resulting in strain softening. This elucidates two impacts of hydrates: the filling effect, where greater S_h enhances initial stiffness and shear strength, and the cementing effect, where hydrates alter a GHBS's internal structure, boosting its additional strength.

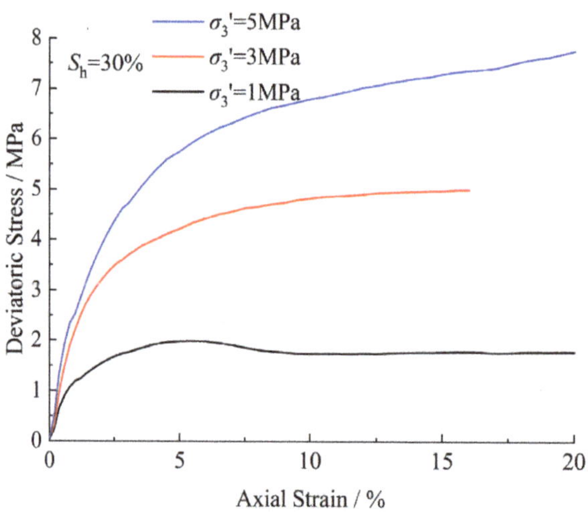

Figure 10. Stress–strain curve (S_h = 30%).

3.2. Volumetric Strain

Figure 11 illustrates the volumetric strain curve of silty fine sand without hydrates, where under various σ_3' conditions, a consistent pattern of shear shrinkage is observed. As σ_3' intensifies, the restraint imposed on the sample strengthens. This enhanced confining pressure tightens particle contacts, demanding greater frictional resistance to facilitate particle rotation and sliding, thus leading to increased volume shrinkage during shearing.

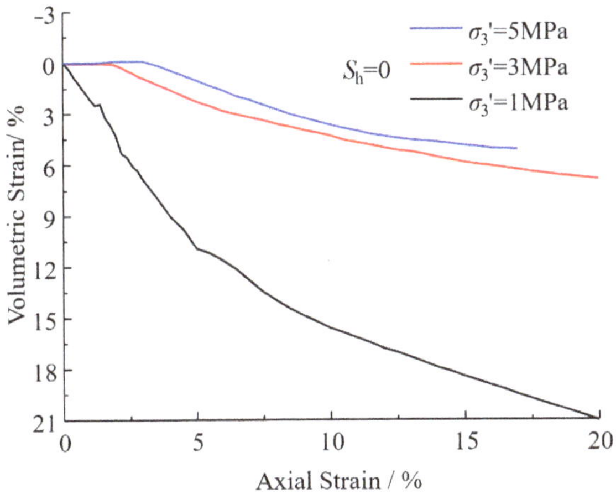

Figure 11. Volumetric Strain Curve (S_h = 0).

For GHBS at S_h = 10%, the volumetric strain curve shown in Figure 9 markedly contrasts with Figure 12. At σ_3' = 5 MPa, the shear shrinkage in GHBS is considerably lower than in non-hydrate-bearing silty fine sand. At σ_3' values of 1 MPa and 3 MPa, GHBS exhibits behavior akin to dense sand, where shear dilation escalates with decreasing σ_3'.

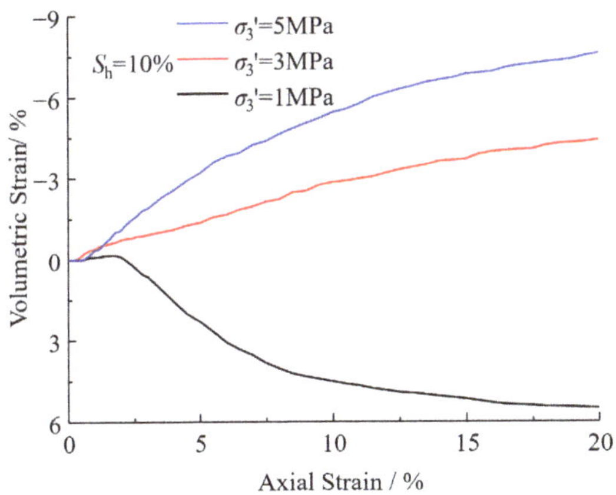

Figure 12. Volumetric strain curve (S_h = 10%).

Figure 13 depicts the volumetric strain curve for GHBS at S_h = 20%. A comparison with Figure 8 (S_h = 0) and Figure 9 (S_h = 10%) reveals that with increasing S_h, GHBS exhibits reduced shear shrinkage and augmented shear dilation, indicating that hydrates enhance the shear dilation characteristics of GHBS. The presence of hydrates in GHBS increases soil compactness, effectively altering GHBS grading. This grading shift impacts the internal structure of the particles and their spatial arrangement, leading to varied volumetric responses under identical stress paths.

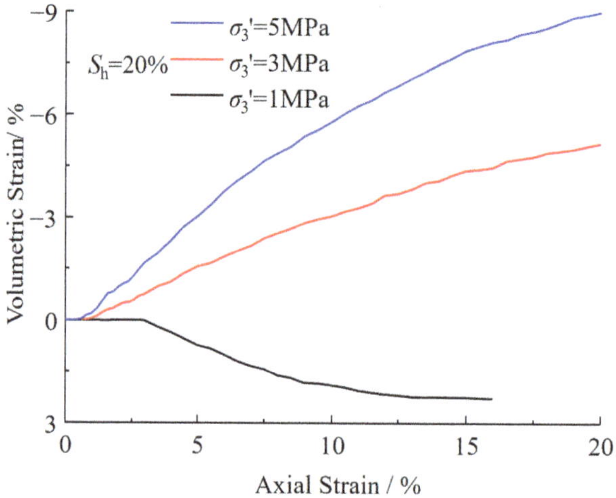

Figure 13. Volumetric strain curve (S_h = 20%).

Additionally, the shear dilation observed in Figures 12 and 13 predominantly occurred under the lower confining pressures of σ_3' = 1 MPa and 3 MPa. This is attributed to the reduced constraining effect of confining pressure on GHBS. Under shear stress, internal soil particles within GHBS undergo rotation and flipping, resulting in shear dilation deformation. This phenomenon is more pronounced under conditions of lower confining pressure and higher hydrate particle content. Concurrently, the relative motion among particles weakens the additional strength of GHBS, leading to a softening effect. This is corroborated by the experimental data shown in Figure 8 (σ_3' = 1 MPa, S_h = 30%), which indicate that GHBS shear dilation becomes more pronounced with increasing σ_3' and S_h.

Figure 14 presents the volumetric strain at S_h = 30%, where the GHBS consistently demonstrate shear dilation. Higher hydrate content increases the likelihood and extent of particle inter-rotation and movement, thereby accentuating the shear dilation properties.

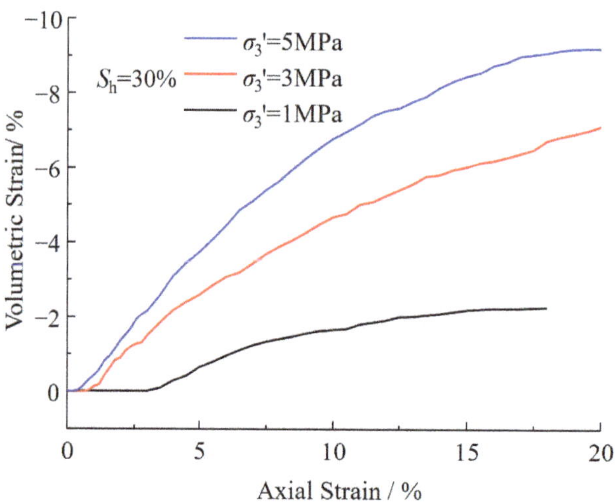

Figure 14. Volumetric strain curve (S_h = 30%).

4. Discussion

4.1. Shear Strength and Peak Strength

We analyzed the strength of clayey silt GHBS based on the previous test data. The so-called strength is the shear strength of materials. The function that characterizes the failure conditions of materials is called the strength criterion. Among many strength criteria, the Mohr–Coulomb criterion (M-C criterion) is the most widely applied in engineering practice because of its simplicity, practicality, and simple parameter acquisition [30]. The M-C criterion states that when the ratio of shear stress to normal stress on the shear plane of a material reaches the maximum, the material will yield and collapse. The M-C criterion can be expressed as:

$$\tau = c + \sigma' \tan(\varphi) \tag{4}$$

where τ is the shear strength, that is, the shear stress on the failure surface; σ' is the normal stress on the failure surface; c is the cohesion; and φ is the friction angle.

Taking the value corresponding to the axial strain ε_a = 15% as the failure strength value, according to the data shown in Figures 7–10, we drew the corresponding Mohr circle to obtain the common tangent; that is, the strength envelope, the slope, and intercept of the envelope. The shear strength index of GHBS under the M-C criterion can be obtained by applying Equation (4), as shown in Figure 15.

According to the data in Figure 15, the quantitative relationship between c and φ can be obtained as shown in Figure 16. The expression can be recorded as:

$$\begin{cases} c = \left(-3.72S_h^2 + 167.10S_h\right)p_a \\ \varphi = (17.20 + 0.22S_h)^\circ \end{cases} \tag{5}$$

where p_a is atmospheric pressure, which can be simply taken as 0.1 MPa.

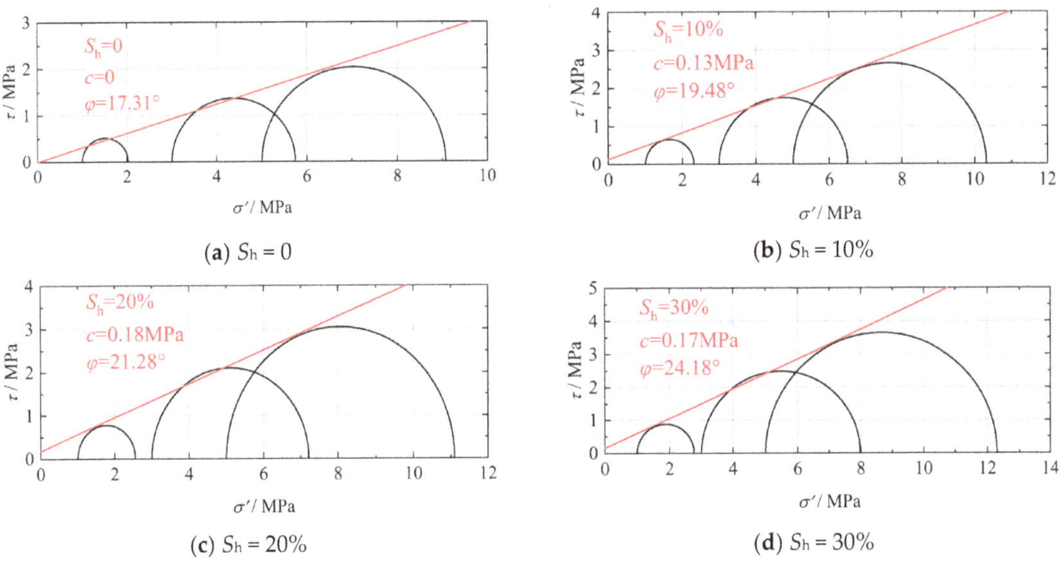

Figure 15. Mohr circle of GHBS.

When substituting Equation (5) into Equation (4) to obtain the strength of GHBS described by the M-C criterion, it can be seen that the friction angle φ rises continuously with the increase in S_h, but the increase is small. The cohesion c has a parabolic relationship with S_h, and when S_h is small, c rises gradually. However, when S_h increases to a certain extent, c begins to decrease.

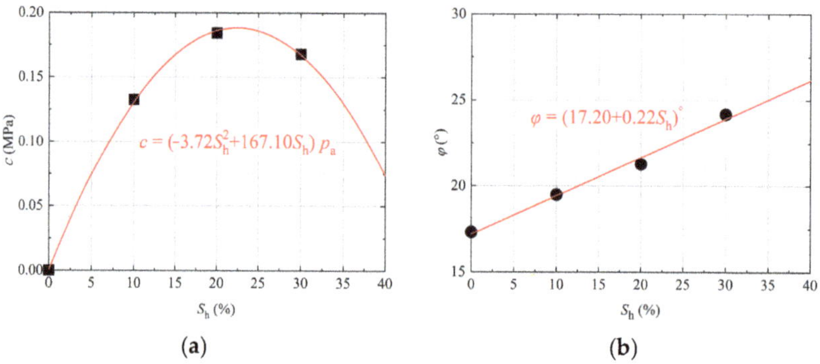

Figure 16. Relationship curve between shear strength and hydrate saturation. (**a**) Cohesion and S_h; (**b**) friction angle and S_h.

Figure 17 shows the peak strength of the 12 sets of trials, plotted in three dimensions. It is apparent that the peak intensity σ_p exhibits a monotonically rising relationship with both S_h and $\sigma_3{}'$, and the slope rises with increasing S_h and $\sigma_3{}'$. This means that both S_h and $\sigma_3{}'$ contribute to the peak strength.

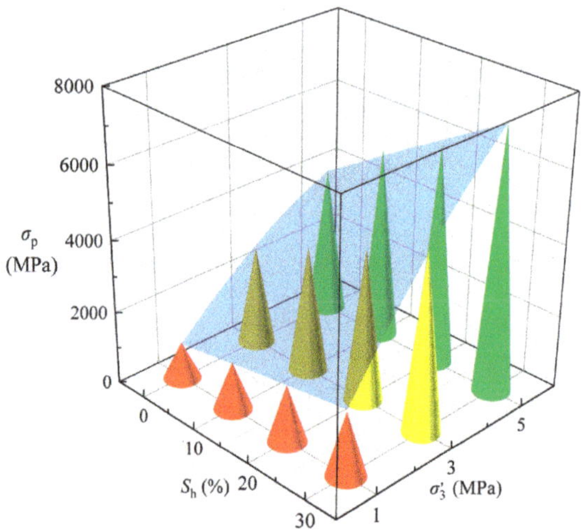

Figure 17. Initial tangent modulus of GHBS and effective confining stress.

The empirical equation used for fitting to obtain the peak strength is as follows:

$$\sigma_p = S_h p_a\left(78.78 + 19.13\frac{\sigma_3'}{p_a}\right) + 28.56\sigma_3' + \sigma_{p0} \tag{6}$$

where σ_{p0} is the peak strength of the specimen without hydrate.

4.2. Initial Stiffness

In order to describe the mechanical properties of GHBS quantitatively, it was observed that GHBS exhibited strain hardening, and the Duncan–Chang model (D-C model) was used to describe this type of soil. The D-C model notates the stress–strain curve as:

$$q = \frac{\varepsilon_a}{a + b\varepsilon_a} \tag{7}$$

in which

$$\begin{cases} a = 1/E_i \\ b = R_f/q_f \end{cases} \tag{8}$$

where q is the deviator stress, MPa; a and b are the fitting parameters; E_i is the initial tangent modulus, MPa; R_f is the damage ratio; and q_f is the shear stress at failure, MPa.

Using Equations (7) and (8) to fit the experimental data, the damage ratio R_f of 0.85 was obtained. The initial tangential stiffness E_i of GHBS under different initial conditions is shown in Table 3 and Figure 18. From Figure 18, it can be seen that E_i has a good linear relationship with S_h and $\sigma_3{}'$ has basically no effect on the slope, but $\sigma_3{}'$ determines the intercept of the curve, i.e., E_{i0}. This can be uniformly notated as:

$$E_i = 851.62 S_h p_a + E_{i0} \tag{9}$$

where 851.62 is the proportional coefficient and E_{i0} is the initial stiffness of the sample without hydrate, MPa, as shown in Table 3.

Table 3. Initial tangent modulus E_i (MPa) of GHBS.

S_h	$\sigma_3{}'$ = 1 MPa	$\sigma_3{}'$ = 3 MPa	$\sigma_3{}'$ = 5 MPa
0	262	773	1100
10%	1241	1442	1483
20%	2087	1771	1878
30%	2599	3181	3497

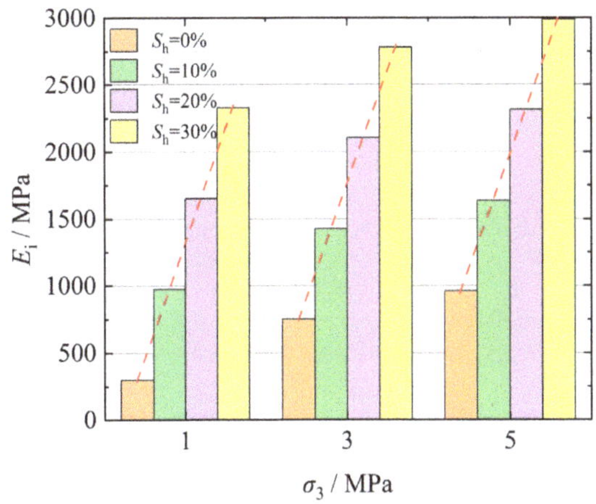

Figure 18. Initial tangent modulus of GHBS and effective confining stress.

4.3. Discussion of Mechanism

Using the interactions between the microscopic particles of GHBS, the patterns of variation in their macroscopic mechanical properties can be better described. Many scholars have investigated the microstructure of GHBS using techniques such as CT and SEM. CT can observe the process of hydrate formation and decomposition in GHBS and identify the pore structure of the sediment [31,32]; the current observations show that hydrates are not uniformly distributed in GHBS and are usually concentrated in the center of the particle pores. The mode of assignment changes during formation, and hydrates may appear in various states, such as filling and connection, at the same moment. The filling mode dominates in the early phase of formation, where hydrates are suspended in the pore space, multiple modes coexist in the middle phase of generation; and the connection and cementation modes dominate in the late phase of generation [33]. Observations made using a Scanning Electron Microscope (SEM) revealed that at temperatures as low as −180 °C, hydrates initially begin to adhere to soil particles. As the concentration of hydrates increases, they progressively envelop the soil particles, a process illustrated in Figure 19 [34]. During the generation of GHBS in this study, the temperature was meticulously maintained at 1 °C. This ensured the absence of ice grains, thereby isolating hydrates as the sole influencer on the structural integrity of the GHBS.

(a) (b) (c)

Figure 19. SEM images of GHBS [34]. (**a**) $S_h = 0$; (**b**) $S_h = 30\%$; (**c**) $S_h = 50\%$.

Based on these observation results and the empirical formula given in this paper, our analysis shows that:

(1) The internal friction angle reflects the frictional resistance between microscopic particles, while cohesion indicates their cementation capability. An increase in S_h leads to an initial rise and subsequent fall in cohesion, with a continuous increase in the internal friction angle, suggesting an enhancement in particle cementation and friction capabilities during early hydrate formation.

(2) The formation of hydrates densifies the soil skeleton of GHBS, leading to a tighter particle arrangement and increased resistance to particle rotation and dislocation, as shown in Figure 20a. This results in an overall increase in stiffness and strength.

(3) As the hydrate content increases, as seen in Figure 20b, the formation of a cementation structure envelops soil particles, as depicted in Figure 20c, restricting their deformation and sliding, and providing additional strength. This process enhances the rigidity and strength of GHBS. At higher S_h levels, the formation and accumulation of hydrates can expel soil particles, leading to an uneven distribution of the cementation structure, which may initially be damaged during shearing, causing a decrease in cohesion as illustrated in Figure 20d.

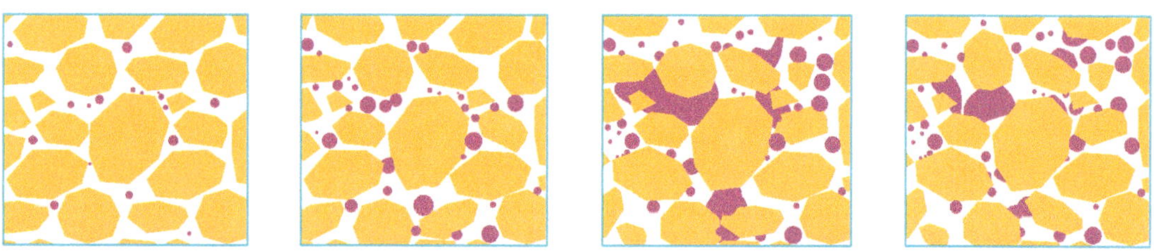

(a) (b) (c) (d)

Figure 20. Microscopic mechanism of the mechanical properties of GHBS. (**a**) The initial phase of hydrate formation; (**b**) the filling of hydrate in pores; (**c**) the formation of the cemented structure; (**d**) the shearing of GHBS.

To sum up, hydrate has two effects on GHBS: filling and cementing. These two effects together lead to changes in the density, stiffness, and strength of GHBS.

5. Conclusions

This study conducted detailed experimental tests and analyses on clayey silt-type gas hydrate-bearing sediments (GHBS), thoroughly investigating the role of hydrates in GHBS. We discovered that the filling and cementing effects of hydrates not only alter the density and structure of GHBS but also significantly enhance its stiffness and strength. With increasing hydrate saturation, GHBS exhibits a marked trend towards densification and mechanical property improvement, especially in the early stages of formation, where these effects substantially boost inter-particle friction and cementation capabilities. Our findings offer new insights into the mechanisms by which hydrates influence GHBS' mechanical properties, which is crucial for future research in related fields. Additionally, the developed experimental methods and empirical formulas provide essential quantitative tools for analyzing GHBS' mechanical characteristics, laying a solid foundation for engineering applications and further studies.

Author Contributions: Conceptualization, Q.Y., Q.L., L.K. and J.L.; methodology, Q.L. and L.K.; writing—original draft preparation, Q.Y., L.K., L.Y., Y.D., Z.W. and X.W.; writing—review and editing, Q.L., L.K., J.L., L.Y., Y.D., Z.W. and X.W.; project administration, L.K., Q.L. and J.L.; funding acquisition, L.K., Q.L. and J.L. All authors have read and agreed to the published version of the manuscript.

Funding: This research was funded by the National Natural Science Foundation of China (No. 12172187 and 11572165), the Marine Geological Survey Program of China Geological Survey (No. DD20221706), the Science and Technology Program of Guangzhou (No. 202103040003), the Guangdong Basic and Applied Basic Research Foundation (No. 2019B030302004), and the Director's research fund of the Guangzhou Marine Geological Survey (No. 2023GMGSJZJJ00002).

Institutional Review Board Statement: Not applicable.

Informed Consent Statement: Not applicable.

Data Availability Statement: The data that support the findings of this study are available from the corresponding author upon reasonable request.

Acknowledgments: The authors wish to thank Xuhui Zhang from the Chinese Academy of Sciences for fruitful discussions and for helping to revise the manuscript, and Rui Xu, Yapeng Zhao, and Jiaqi Liu from Qingdao University of Technology for helping with the laboratory operation.

Conflicts of Interest: The authors declare no conflicts of interest.

References

1. Makogon, Y.F. Natural Gas Hydrates—A Promising Source of Energy. *J. Nat. Gas Sci. Eng.* **2010**, *2*, 49–59. [CrossRef]
2. Argentino, C.; Conti, S.; Fioroni, C.; Fontana, D. Evidences for Paleo-Gas Hydrate Occurrence: What We Can Infer for the Miocene of the Northern Apennines (Italy). *Geosciences* **2019**, *9*, 134. [CrossRef]
3. Max, M. *Natural Gas Hydrate in Oceanic and Permafrost Environments*; Coastal Systems Cont. Margins; Springer: Dordrecht, The Netherlands, 2000.
4. Farahani, M.V.; Hassanpouryouzband, A.; Yang, J.; Tohidi, B. Development of a Coupled Geophysical–Geothermal Scheme for Quantification of Hydrates in Gas Hydrate-Bearing Permafrost Sediments. *Phys. Chem. Chem. Phys.* **2021**, *23*, 24249–24264. [CrossRef] [PubMed]
5. Merey, S.; Sinayuc, C. Investigation of Gas Hydrate Potential of the Black Sea and Modelling of Gas Production from a Hypothetical Class 1 Methane Hydrate Reservoir in the Black Sea Conditions. *J. Nat. Gas Sci. Eng.* **2016**, *29*, 66–79. [CrossRef]
6. Miyazaki, K.; Masui, A.; Sakamoto, Y.; Aoki, K.; Tenma, N.; Yamaguchi, T. Triaxial Compressive Properties of Artificial Methane-Hydrate-Bearing Sediment. *J. Geophys. Res. Solid Earth* **2011**, *116*, B06102. [CrossRef]
7. Du, H.; Chen, H.; Kong, F.; Luo, Y. Failure Mode and the Mechanism of Methane Hydrate-Bearing Clayey Sand Sediments under Depressurization. *Energy* **2023**, *279*, 128110. [CrossRef]
8. Liu, Z.; Dai, S.; Ning, F.; Peng, L.; Wei, H.; Wei, C. Strength Estimation for Hydrate-Bearing Sediments from Direct Shear Tests of Hydrate-Bearing Sand and Silt. *Geophys. Res. Lett.* **2018**, *45*, 715–723. [CrossRef]
9. Archer, D. Methane Hydrate Stability and Anthropogenic Climate Change. *Biogeosciences* **2007**, *4*, 521–544. [CrossRef]
10. Bondarenko, V.I.; Sai, K.S. *Process Pattern of Heterogeneous Gas Hydrate Deposits Dissociation*; Naukovyi Visnyk Natsionalnoho Hirnychoho Universytetu: Dnipro, Ukraine, 2018.
11. Winters, W.J.; Waite, W.F.; Mason, D.H.; Gilbert, L.Y.; Pecher, I.A. Methane Gas Hydrate Effect on Sediment Acoustic and Strength Properties. *J. Pet. Sci. Eng.* **2007**, *56*, 127–135. [CrossRef]

12. Winters, W.J.; Wilcox-Cline, R.W.; Long, P.; Dewri, S.K.; Kumar, P.; Stern, L.; Kerr, L. Comparison of the Physical and Geotechnical Properties of Gas-Hydrate-Bearing Sediments from Offshore India and Other Gas-Hydrate-Reservoir Systems. *Mar. Pet. Geol.* **2014**, *58*, 139–167. [CrossRef]
13. Nagakubo, S. Pictorial 1: Meti Exploratory Test Wells "Tokai-Oki to Kumano-Nada" Conducted in Fy2003. *J. Geogr.* **2009**, *118*, xxi. [CrossRef]
14. Bondarenko, V.; Sai, K.; Petlovanyi, M. Peculiarities of Geological and Thermobaric Conditions for the Gas Hydrate Deposits Occurence in the Black Sea and the Prospects for Their Development. *J. Geol. Geogr. Geoecol.* **2019**, *28*, 395–408. [CrossRef]
15. Dong, L.; Li, Y.; Zhang, Y.; Hu, G.; Liao, H.; Chen, Q.; Wu, N. Deformation Characteristics of Hydrate-Bearing Sediments. *J. Ocean. Univ. China* **2024**, *23*, 149–156. [CrossRef]
16. Wu, P.; Li, Y.; Liu, W.; Sun, X.; Kong, X.; Song, Y. Cementation Failure Behavior of Consolidated Gas Hydrate-Bearing Sand. *J. Geophys. Res. Solid Earth* **2020**, *125*, e2019JB018623. [CrossRef]
17. Kim, H.-S.; Cho, G.-C. Experimental Study on the Compressibility of Gas Hydrate-Bearing Sediments. In Proceedings of the 2014 World Congress on Advances in Civil Environmental and Materials Research, Busan, Republic of Korea, 24–28 August 2014.
18. Hyodo, M.; Li, Y.; Yoneda, J.; Nakata, Y.; Yoshimoto, N.; Nishimura, A. Effects of Dissociation on the Shear Strength and Deformation Behavior of Methane Hydrate-Bearing Sediments. *Mar. Pet. Geol.* **2014**, *51*, 52–62. [CrossRef]
19. Zhao, Y.; Kong, L.; Xu, R.; Liu, J.; Sang, S. Strength Behaviors of Hydrate-Bearing Clayey-Silty Sediments with Multiple Factors. *J. Pet. Sci. Eng.* **2022**, *219*, 111035. [CrossRef]
20. Qin, X.; Liang, Q.; Ye, J.; Yang, L.; Qiu, H.; Xie, W.; Liang, J.; Lu, J.; Lu, C.; Lu, H.; et al. The Response of Temperature and Pressure of Hydrate Reservoirs in the First Gas Hydrate Production Test in South China Sea. *Appl. Energy* **2020**, *278*, 115649. [CrossRef]
21. Ye, J.; Qin, X.; Xie, W.; Lu, H.; Ma, B.; Qiu, H.; Liang, J.; Lu, J.; Kuang, Z.; Lu, C. Main Progress of the Second Gas Hydrate Trial Production in the South China Sea. *Geol. China* **2020**, *47*, 557–568.
22. Liu, C.L.; Ye, Y.G.; Meng, Q.G.; He, X.L.; Lu, H.L.; Zhang, J.; Liu, J.; Yang, S.X. The Characteristics of Gas Hydrates Recovered from Shenhu Area in the South China Sea. *Mar. Geol.* **2012**, *307*, 22–27. [CrossRef]
23. Ren, S.; Liu, J.; Liu, Y.; Fan, Z.; Liu, Y.; Zuo, J.; Yu, H. Experimental Study on Formation and Dissociation of Methane Hydrate in Porous Media. *Acta Pet. Sin.* **2009**, *30*, 583.
24. Yan, C.; Cheng, Y.; Li, M.; Han, Z.; Zhang, H.; Li, Q.; Teng, F.; Ding, J. Mechanical Experiments and Constitutive Model of Natural Gas Hydrate Reservoirs. *Int. J. Hydrog. Energy* **2017**, *42*, 19810–19818. [CrossRef]
25. Zhao, Y.; Liu, J.; Sang, S.; Hua, L.; Kong, L.; Zeng, Z.; Yuan, Q. Experimental Investigation on the Permeability Characteristics of Methane Hydrate-Bearing Clayey-Silty Sediments Considering Various Factors. *Energy* **2023**, *269*, 126811. [CrossRef]
26. Takeya, S.; Hori, A.; Hondoh, T.; Uchida, T. Freezing-Memory Effect of Water on Nucleation of Co2 Hydrate Crystals. *J. Phys. Chem. B* **2000**, *104*, 4164–4168. [CrossRef]
27. Chen, J.; Yan, K.-L.; Jia, M.-L.; Sun, C.-Y.; Zhang, Y.-Q.; Si, S.; Ma, Q.-L.; Yang, L.-Y.; Wang, X.-Q.; Chen, G.-J. Memory Effect Test of Methane Hydrate in Water+ Diesel Oil+ Sorbitan Monolaurate Dispersed Systems. *Energy Fuels* **2013**, *27*, 7259–7266. [CrossRef]
28. Hyodo, M.; Nakata, Y.; Yoshimoto, N. Challenge for Methane Hydrate Production by Geotechnical Engineering. *Jpn. Geotech. Soc. Spec. Publ.* **2016**, *2*, 62–75. [CrossRef]
29. Khlebnikov, V.N.; Antonov, S.V.; Mishin, A.S.; Liang, M.; Khamidullina, I.V.; Zobov, P.M.; Likhacheva, N.V.; Gushchin, P.A. Major Factors Influencing the Formation of Natural Gas Hydrates in Porous Media. *Tianranqi Gongye* **2017**, *37*, 38–45.
30. Fjaer, E.; Holt, R.M.; Horsrud, P.; Raaen, A.M. Failure Mechanics. In *Petroleum Related Rock Mechanics;* Elsevier: Amsterdam, The Netherlands, 2021; pp. 89–155.
31. Mikami, J.; Masuda, Y.; Uchida, T.; Satoh, T.; Takeda, H. Dissociation of Natural Gas Hydrates Observed by X-ray Ct Scanner. *Ann. N. Y. Acad. Sci.* **2006**, *912*, 1011–1020. [CrossRef]
32. Qin, J.; Kuhs, W.F. Quantitative Analysis of Gas Hydrates Using Raman Spectroscopy. *AIChE J.* **2000**, *59*, 2155–2167. [CrossRef]
33. Yoneda, J.; Jin, Y.; Katagiri, J.; Tenma, N. Strengthening Mechanism of Cemented Hydrate-Bearing Sand at Microscales. *Geophys. Res. Lett.* **2016**, *43*, 7442–7450. [CrossRef]
34. Hyodo, M.; Yoneda, J.; Yoshimoto, N.; Nakata, Y. Mechanical and Dissociation Properties of Methane Hydrate-Bearing Sand in Deep Seabed. *Soils Found.* **2013**, *53*, 299–314. [CrossRef]

Journal of
Marine Science and Engineering

MDPI

Article

Numerical Simulation of Gas Production Behavior Using Stepwise Depressurization with a Vertical Well in the Shenhu Sea Area Hydrate Reservoir of the South China Sea

Tinghui Wan [1,2], Zhanzhao Li [1,2], Hongfeng Lu [1,2], Mingming Wen [1,2], Zongheng Chen [1,2], Lieyu Tian [1,2], Qi Li [1,2], Jia Qu [1,2] and Jingli Wang [1,2,*]

1 Guangzhou Marine Geology Survey, China Geological Survey, Ministry of Natural Resources, Guangzhou 511458, China; atomion@126.com (T.W.); 13650780173@163.com (Z.L.); gmgslhf@126.com (H.L.); wmingming@mail.cgs.gov.cn (M.W.); czhgs@126.com (Z.C.); tianlieyu23@163.com (L.T.); liqi3412@163.com (Q.L.); qujia2261520@163.com (J.Q.)

2 National Engineering Research Center for Gas Hydrate Exploration and Development, Guangzhou 511458, China

* Correspondence: wjl06012527@126.com

Abstract: Stepwise depressurization is an important depressurization strategy in the development of natural gas hydrates. This work numerically analyzes the effects of different depressurization gradients and constant pressure durations on gas and water production during stepwise depressurization extraction with a vertical well in the Shenhu Sea area hydrate reservoir of the South China Sea. The results indicate that stepwise depressurization can reduce water production and raise the gas-to-water ratio in the early stages of production while ensuring cumulative gas output. When the vertical well is deployed at the model's center with a completion length of 70 m and a constant pressure duration of 10 days, a depressurization gradient of 0.5 MPa, stepwise depressurization by 6 MPa, and continuous production for one year is achieved. Compared with direct depressurization, its cumulative gas production is 2.966×10^6 ST m^3, which only decreases by 2.94%. However, it maintains a higher gas-to-water ratio in the early stages of production. Considering factors such as engineering operability, cumulative gas output, and gas-to-water ratio, it is recommended to use a small pressure gradient and a medium constant pressure stabilization time for stepwise depressurization Stepwise depressurization can maintain a high gas-to-water ratio while ensuring gas production and reducing water production can alleviate sand production problems and improve economic efficiency. The understanding gained from this work has reference value for the development of similar hydrate reservoirs worldwide.

Keywords: natural gas hydrate; vertical well; stepwise depressurization; numerical simulation; TOUGH+HYDRATE

1. Introduction

Sustainable development of the global economy and technology relies on a stable energy supply. Currently, going green and low-carbon is the main direction for global energy structure adjustment, and there is an urgent need to develop clean energy. Natural gas hydrates (NGHs) are widely distributed in the permafrost and marine continental shelf. Its industrial exploitation may effectively promote global energy structure adjustment [1–7]. Japan and China's offshore NGH testing production projects confirm the technological superiority of the depressurization method for NGH extraction [8–11]. However, the average daily gas output remains far below the commercial development standard of 50×10^4 m^3/d [2]. Therefore, many scholars focus on how to significantly increase production capacity and achieve effective recovery of NGHs, such as Too et al., 2018, who confirmed the possibility of forming artificial fractures in synthetic methane hydrate-bearing sand, which provides an opportunity for the efficient development of NGH [12]. Nair et al., 2018 conducted experimental studies on the dissociation behavior of hydrates in clay hydrate reservoirs using depressurization, thermal stimulation, and combination methods, and the results showed that the combination method had the best effect [13]. Aghajari et al., 2019 found that methane production during NGH exploitation is highly correlated with reservoir porosity and permeability, while the dissociation rate is positively correlated with depressurization and inversely correlated with reservoir temperature [14]. Vedachalam et al., 2020 believe that synchronous depressurization of 40 wells can achieve commercial exploitation of NGH in the Krishna Godavari Basin of India [15]. Terzariol et al., 2021 found that there is a synergistic depressurization effect between wells when multiple wells are synchronously depressurized, and its efficiency is higher than that of an equal number of independent working wells [16]. Wu et al., 2021 reviewed the relevant work and concluded that adopting different production well designs—such as horizontal wells, complex structure wells, etc., combined with varying strategies of depressurization and using different stimulation methods such as in situ heating or hot water injection, reservoir hydraulic fracturing, or near wellbore reservoir reconstruction—can significantly improve production capacity [17]. Optimizing the depressurization method is a noteworthy approach [18].

In recent years, efforts have been put into research on NGH development with the stepwise depressurization strategy. Phillips et al., 2019 experimentally studied the dissociation characteristics of hydrates with stepwise depressurization. In reservoirs with high salinity and saturation, salt diffusion and heat transfer slow down the dissociation of hydrates [19]. Yin et al., 2020 conducted a depressurization experiment in hydrate sediments and found that the water production rate grew linearly with the depressurization rate [20]. Zhao et al., 2020 found that dividing direct depressurization into several stages can accelerate hydrate decomposition [21]. Guo et al., 2020 conducted a set of experiments and found that stepwise depressurization helps alleviate initial water production; the fine pressure gradient during the high-yield water stage can increase the gas production rate to 31%, and the cumulative water production depends on the degree of depressurization [22]. Ravesh et al., 2019 studied the dissociation behavior of methane hydrates in porous media using single-step and multi-stage depressurization in a 25 L reactor and found that multi-stage depressurization can improve the gas recovery rate [23]. Yoon et al., 2021 conducted a numerical study on the geomechanical response of gas hydrate deposits in the Ulleung Basin of the Korea East Sea with various depressurization schemes and found that the productivity is similar among different depressurization schemes, with periodic depressurization schemes having relatively small subsidence [24]. Li et al., 2021 found that stepwise depressurization helps to increase the integrated gas-to-water ratio and avoid a significant decrease in reservoir temperature [25]. Yin et al., 2022 established a multi-field coupling model and investigated horizontal well depressurization's gas production behavior. The results show that stepwise and cyclic depressurization can improve production capacity to varying degrees [26]. Wang et al., 2022 proposed an optimized simulation model based on the reservoir characteristics of the Shenhu Sea area and analyzed the impact of seepage characteristics caused by stepwise depressurization on hydrate extraction. The results indicate that relative gas permeability and porosity are key parameters affecting hydrate production [27]. Yin et al., 2022 found that the optimal duration of constant pressure for single-stage depressurization is 5–10 days. The optimal depressurization gradient for long-term stable production is 0.2 MPa/d [28]. Xue et al., 2023 conducted a numerical analysis of the gas production behavior of vertical wells with stepwise depressurization. The results show that the gas production increased by more than 10% [29]. Ge et al., 2023 proposed combining the hydrate dissociation model with a genetic algorithm to optimize the stepwise depressurization strategy. Research has found that increasing depressurization steps is detrimental to exploitation efficiency [30]. Wang et al., 2023 numerically analyzed the production behavior of stepwise depressurization extraction of NGH with a horizontal well. The results showed a significant decrease in water production [31]. Various studies in recent years likewise contribute relevant information and context on gas hydrate stability, destabilization, and changes to temperature and/or pressure influencing this hydrate stability (e.g., Biastoch et al., 2011; Burton et al., 2020; Kim and Zhang, 2022; Burton & Dafov, 2023) [32–35].

The insights gained from the above research on the stepwise depressurization strategy provide valuable theoretical references for practical application. Multi-well or group well exploitation is an important development direction for the commercial development of natural gas hydrates, and the vertical well is one of the important basic well types [17]. Currently, there is limited research on the stepwise depressurization of NGH reservoirs in the Shenhu Sea area with vertical wells. In this work, we established an ideal model of NGH reservoirs based on China's first offshore NGH testing data, numerically studied the gas production behavior and reservoir physical characteristics' evolution of vertical well stepwise depressurization, and focused on the effects of depressurization gradients and constant pressure durations on gas production behavior.

2. Methodology

2.1. Simulation Code

TOUGH+HYDRATE V1.0 can simulate various exploitation methods such as depressurization, thermal stimulation, and inhibitors, focusing on well-scale or reservoir-scale simulation (instead of basin-scale simulation, e.g., Piñero et al., 2016; Burton, 2022) [36–38]. The accuracy of the code has been verified through laboratory and field testing [39–43]. Specifically, the equilibrium model was adopted to simulate the stepwise depressurization by a vertical well [44]. This work assumes that sand production is controllable, Darcy's law is effective, and the geological media does not deform. The main control equations for the simulation code are as follows [36]:

$$\frac{d}{dt}\int_{V_n} M^{\kappa} dV = \int_{\Gamma_n} F^{\kappa} \cdot \mathrm{n} d\Gamma + \int_{V_n} q^{\kappa} dV \tag{1}$$

Here, M^{κ} is the mass accumulation term, F^{κ} is the Darcy flux vector, and q^{κ} is the source/sink term. V_n, Γ_n, κ, and t represent the grid volume (m^3), grid surface area (m^2), components, and time (s), respectively, whereas V and Γ are volume (m^3) and surface area (m^2), respectively.

2.2. Model Construction

The SHSC4 well in the Shenhu Sea area was selected for the simulations (Figure 1) [45]. The NGH reservoir at this site belongs to Class 1-type, which consists of three parts: the gas hydrate-bearing layer (GHBL, contains water and hydrates, with thicknesses of 35 m); the three-phase layer (TPL, contains water, free gas and hydrates, with thicknesses of 15 m); and the free gas layer (FGL, contains water and free gas, with thicknesses of 27 m) [10,46,47]. A multi-layer reservoir model with a size of 500 × 500 × 137 m was constructed under the *x-y-z* coordinate system, as shown in Figure 2a. Ten simulation cases were simulated to assess the stepwise depressurization with different depressurization parameters, including a 70 m long vertical wellbore deployed at the center of the model (model's −21 and −91 m). Table 1 gives the detailed parameters for these cases.

The model is discretized into 23 × 23 × 81 grids in (x, y, z), with a total of 42,849 grids (Figure 2b). Specifically, corresponding to the radius of 0.1 m of the wellbore, the grid size of the wellbore in the x-y plane is 0.2 m × 0.2 m. To properly capture heat conduction, hydrate dissociation, and fluid flow, the vertical grid sizes of the GHBL, TPL, and FGL in the model were set to be 1.0 m.

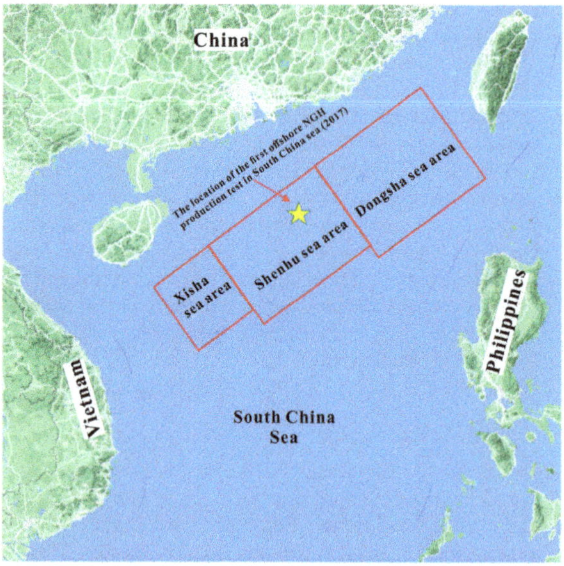

Figure 1. SHSC4 well location map (adapted from Reference [45]).

Table 1. The detailed settings of simulation cases.

Scenarios	Production Pressure Difference Regulation Process (MPa)	Depressurization Gradient (MPa)	Constant Pressure Duration (Day)
Case 01-DD-Ref	6.0	None	None
Case 02-SD1-20d	3.0→3.5→4.0→4.5→5.0→5.5→6.0	0.5	20
Case 03-SD1-10d	3.0→3.5→4.0→4.5→5.0→5.5→6.0	0.5	10
Case 04-SD1-05d	3.0→3.5→4.0→4.5→5.0→5.5→6.0	0.5	5
Case 05-SD2-20d	3.0→4.0→5.0→6.0	1.0	20
Case 06-SD2-10d	3.0→4.0→5.0→6.0	1.0	10
Case 07-SD2-05d	3.0→4.0→5.0→6.0	1.0	5
Case 08-SD3-20d	3.0→4.5→6.0	1.5	20
Case 09-SD3-10d	3.0→4.5→6.0	1.5	10
Case 10-SD3-05d	3.0→4.5→6.0	1.5	5

Note: DD is direct depressurization; SD is stepwise depressurization.

2.3. Model Initialization and Boundary Conditions

The model is divided into three subdomains: the GHBL, TPL, and FGL. The top and bottom of the TPL are considered as the boundaries of the hydrate interface (i.e., the three-phase conditions for the coexistence of hydrates, gas, and water). Each subdomain is simulated separately to determine the initial pressure, temperature, and saturation steady-state conditions; during the process, the heat flux between each subdomain is adjusted by fine-tuning the geothermal gradient. When the heat flux of the contact surface between the three subdomains remains consistent, the combination is performed to complete model initialization, as shown in Figure 3 [48–53].

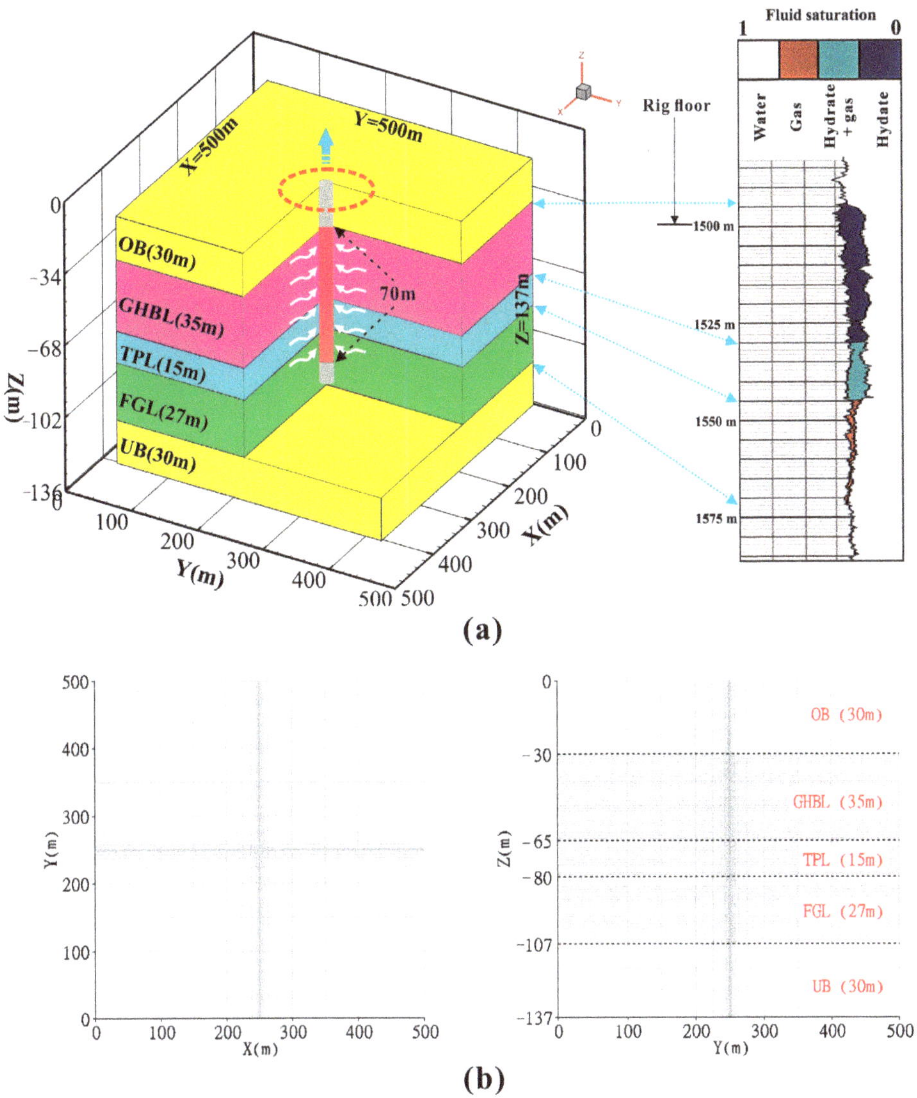

Figure 2. Model schematic diagram: (**a**) Well design. (**b**) Mesh discretization.

During the simulation process, the wellbore is considered the internal boundary, and fixed or stepwise production pressure differences are given to the wellbore grids according to different simulation cases. The wellbore grid is treated as a pseudo-porous medium with a porosity and permeability of 1 and 10,000 D, respectively [54]. The parameters for the reservoir and model are shown in Tables 2 and 3, respectively.

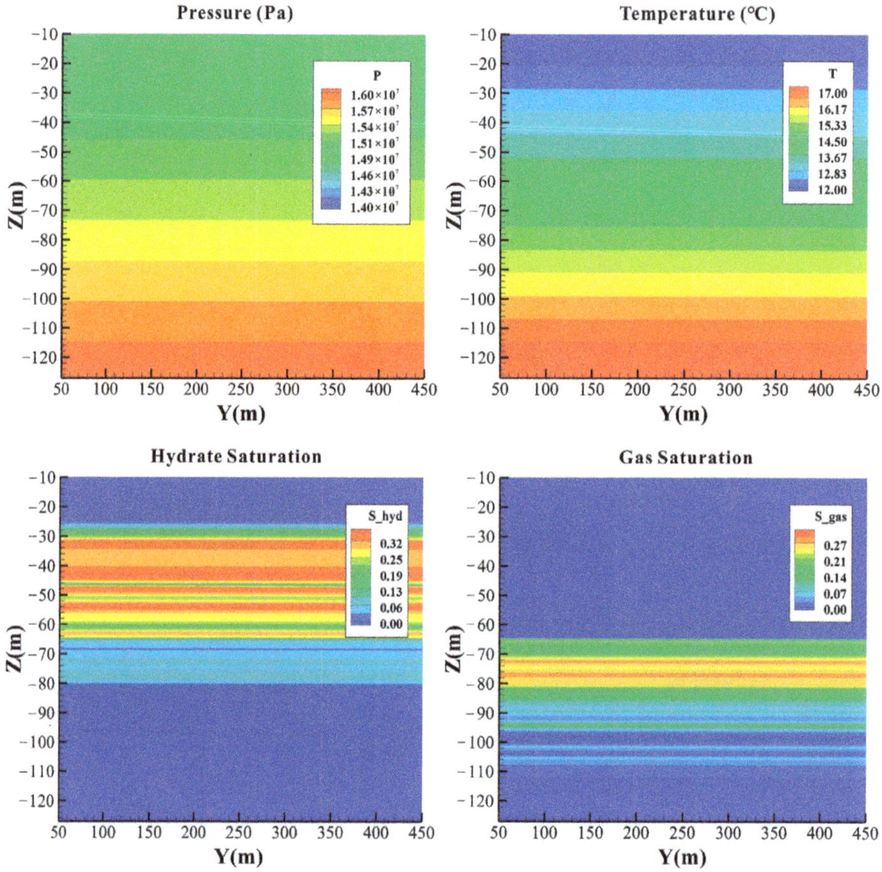

Figure 3. Initial conditions of the model.

Table 2. Parameters of the reservoir.

Layer	Parameter	Value and Unit
OB [55–59]	Thickness Porosity Initial permeability	30 m 0.30 2.0 mD
GHBL [10,56–59]	Thickness Porosity Initial permeability Initial hydrate saturation	35 m 0.35 2.9 mD Extracted from logging curve
TPL [10,56–59]	Thickness Porosity Initial permeability Initial hydrate saturation	15 m 0.33 1.5 mD Extracted from logging curve
FGL [10,56–59]	Thickness Porosity Initial permeability Initial free gas saturation	27 m 0.32 7.4 mD Extracted from logging curve
UB [55–59]	Thickness Porosity Initial permeability	30 m 0.30 2.0 mD

Table 3. Parameters of the model.

Parameter	Value and Unit
Wellbore radius [56–59]	0.1 m
Salinity [56–59]	3.5%
Grain density [56–59]	2600 kg/m^3
Geothermal gradient [56–59]	43.653 °C/km
Gas composition [56–59]	100% CH_4
Grain specific heat [56–59]	1000 $J{\cdot}kg^{-1}{\cdot}K^{-1}$
Dry thermal conductivity [56–59]	1.0 $W{\cdot}m^{-1}{\cdot}K^{-1}$
Wet thermal conductivity [56–59]	3.1 $W{\cdot}m^{-1}{\cdot}K^{-1}$
Capillary pressure model [56–59]	$P_{cap} = -P_0\left[(S^*)^{-1/\lambda} - 1\right]^{1-\lambda}$, $S^* = \frac{(S_A - S_{irA})}{(S_{mxA} - S_{irA})}$
S_{mxA} (maximum water saturation)	1
λ (porosity distribution index)	0.45
P_0 (initial capillary pressure)	104 Pa
Relative permeability model [56–59]	$K_{rA} = [(S_A - S_{irA})/(1 - S_{irA})]^{nA}$, $K_{rG} = [(S_G - S_{irG})/(1 - S_{irA})]^{nG}$
nA (aqueous phase permeability reduction index)	3.5
nG (gas phase permeability reduction index)	2.5
S_{irG} (residual gas saturation)	0.03

2.4. Model Validation

Based on China's first offshore NGH testing data, a vertical wellbore was deployed at the model's center with a completion length of 70 m [10,56–59]. Under the production pressure difference of 3 MPa, a 60-day gas production fitting was conducted to verify the model's accuracy [10,56–59]. As shown in Figure 4, the fitting results are within the acceptable range.

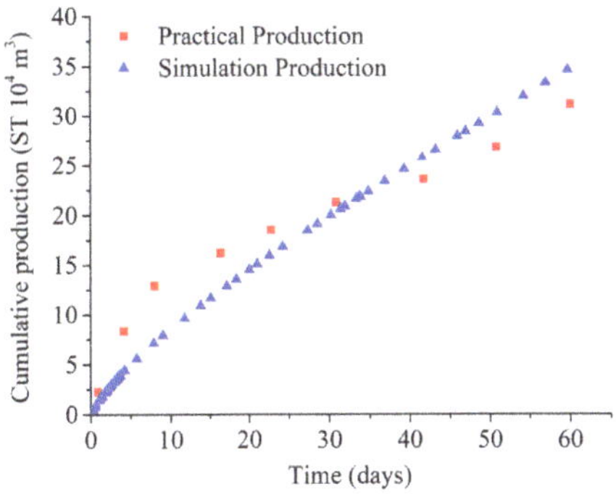

Figure 4. On-site gas production fitting.

3. Results and Analysis

3.1. Gas and Water Production

Two simulation cases were set up to compare the evolution characteristics of gas and water production behavior and reservoir physical properties between direct and stepwise depressurization: Case 01, which has direct depressurization by 6 MPa and continuous production for one year; Case 03, which has a constant pressure duration of 10 days, a depressurization gradient of 0.5 MPa, stepwise depressurization by 6 MPa, and continuous production for one year. From the gas production rate (Q_g) curves and cumulative gas production (V_g) curves in Figure 5a,b, it can be seen that Q_g of Case 01 reached its maximum value at the beginning, then slowly decreased to about 0.7 × 10^4 ST m^3/d, and the V_g reached 3.056 × 10^6 ST m^3. Direct and significant depressurization can quickly reduce reservoir pore pressure, promote rapid dissociation of hydrates, and rapidly decrease reservoir temperature. The huge pressure difference can cause the fluid to flow too quickly to the wellbore, which can easily lead to sand production problems. Usually, the seepage and heat transfer processes in the reservoir are slower than pressure propagation; so, significant depressurization cannot effectively exert the depressurization effect. The Q_g curve of Case 03 showed a stepwise increase followed by a gradual decrease within the initial 70 days of production and tended to be consistent with Case 01 around 135 days. The V_g reached 2.966×10^6 ST m^3, only a decrease of 2.94% compared to Case 01. Stepwise depressurization divides the significant depressurization into several small cycles with a small depressurization gradient, and the peak value of Q_g is much smaller than the direct depressurization, and a small depressurization gradient is beneficial for sand production control. Different from the above results, Xue et al., 2023 numerically studied natural gas hydrate production in the Nankai Trough of Japan with a vertical well and found that the stepwise depressurization resulted in a 10% increase in gas production compared to direct depressurization, which is due to the differences in hydrate reservoir types [29]. From Figure 5c,d, it can be seen that the water production rate (Q_w) curves of Case 03 in the early stages of production also show a stepwise increase followed by a gradual decrease trend and tends to be consistent with Case 01 around 135 days. Compared to Case 01, Case 03 can reduce the Q_w, this is similar to the research results of Wang et al., 2023 where stepwise depressurization can significantly reduce water production [31]. While lower water production is beneficial for sand production control, throughout the entire production cycle, Case 03 has better gas-to-water ratio (R_{gw}) performance compared to Case 01, especially during the initial 120 days of production, and a higher R_{gw} means better exploitation efficiency. The analysis results indicate that stepwise depressurization has a more significant impact on water production. This phenomenon may be related to the "gas slippage effect" [60–62].

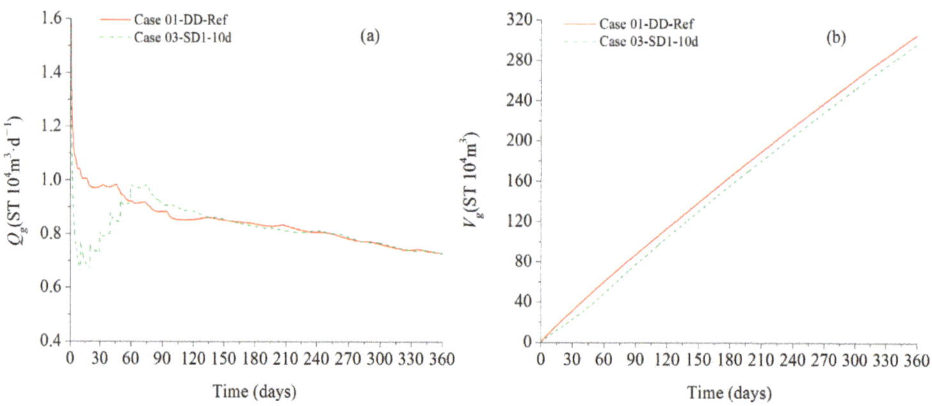

Figure 5. *Cont.*

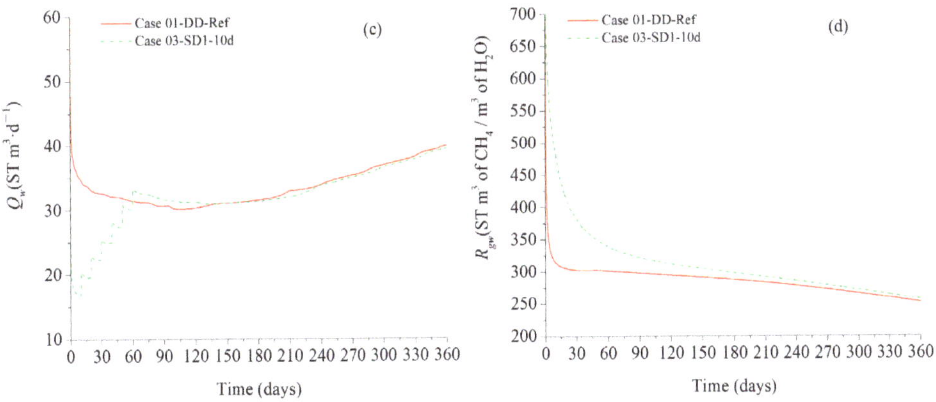

Figure 5. Gas and water production: (**a**) Q_g curves; (**b**) V_g curves; (**c**) Q_w curves; and (**d**) R_{gw} curves.

3.2. Physical Properties

3.2.1. Pressure

From Figure 6, it can be observed that Case 01 and Case 03 have limited pressure propagation in the reservoir around the TPL. This is because the permeability of the reservoir around the TPL decreases with the formation of secondary hydrates, and the pressure propagation range of Case 03 is smaller in the early production stage. The same phenomenon can be observed in Case 01 and Case 03. Compared to the reservoir located in the GHBL, the pressure propagation range of the reservoir located in the bottom free gas layer is significantly smaller, which is caused by the gas expansion effect.

3.2.2. Temperature

From Figure 7, it can be observed that both Case 01 and Case 03 have formed low-temperature zones in the reservoir around the wellbore. This is due to the dual effects of hydrate dissociation in the GHBL and the Joule–Thomson effect. The Joule–Thomson effect caused by the significant direct depressurization in Case 01 during the initial production stage is stronger, resulting in a larger range of temperature drops in the reservoir around the wellbore. However, stepwise depressurization avoids a rapid drop in reservoir temperature, providing favorable conditions for later production. This phenomenon is similar to the research results of Li et al., 2021 where stepwise depressurization helps to avoid a significant decrease in reservoir temperature [25].

3.2.3. Hydrate and Gas Saturation

From Figure 8, it can be observed that during the initial stage of production within 60 days, Case 01 has a larger range of hydrate dissociation in the GHBL. After one year of production, Case 01 and Case 03 both have a hydrate dissociation radius of about 2 m at the GHBL. Affected by the strong Joule–Thomson effect, secondary hydrates were formed in the reservoir around the wellbore at the TPL in Case 01 and Case 03. As production progressed, the amount of the secondary hydrates gradually increased. At the end of production, the saturation and amount of the secondary hydrates in Case 01 and Case 03 tended to be consistent. It is worth noting that the amount of secondary hydrates formed in the TPL of Case 03 is relatively small, which also means that the stepwise depressurization method can help alleviate the formation of secondary hydrates in the early stages of production.

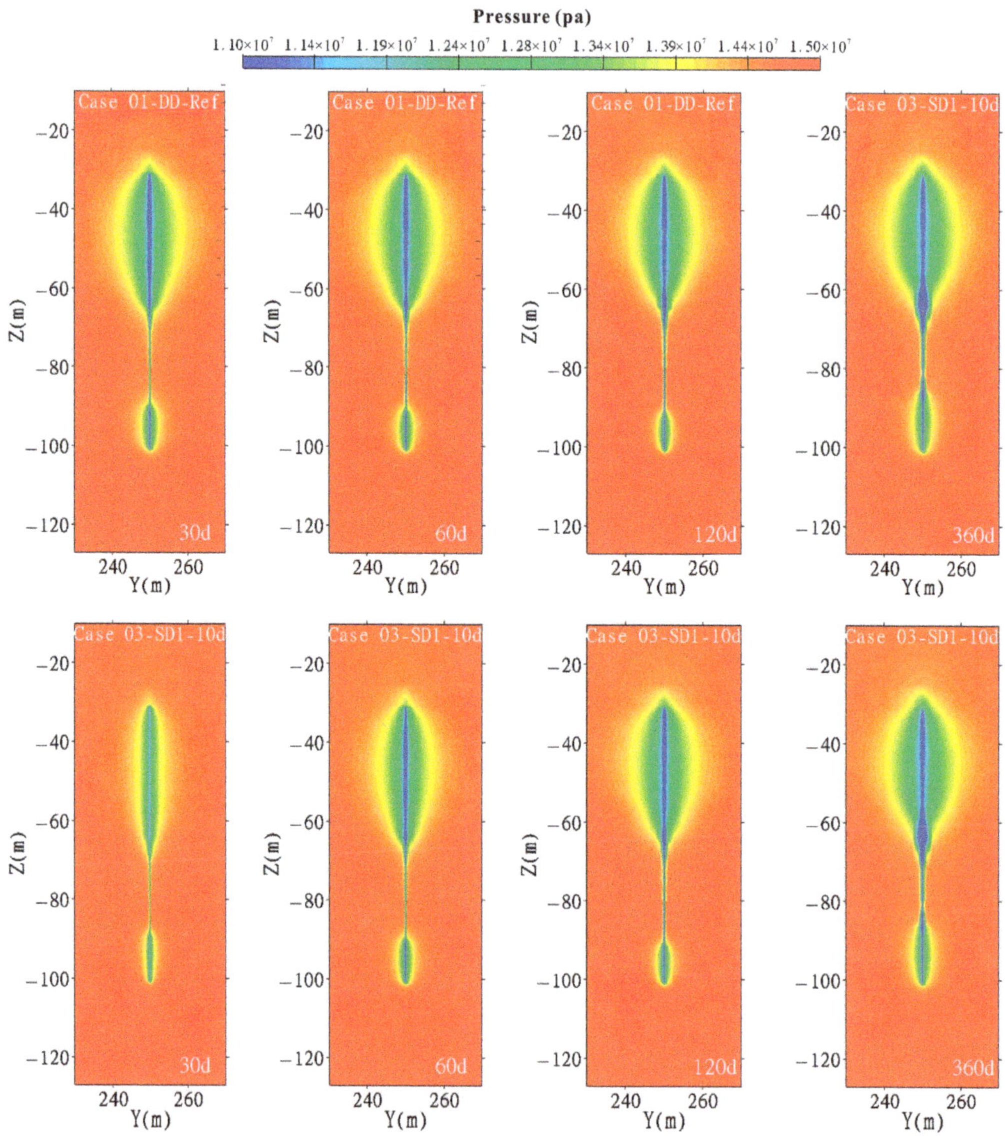

Figure 6. Pressure evolution diagram of direct and stepwise depressurization.

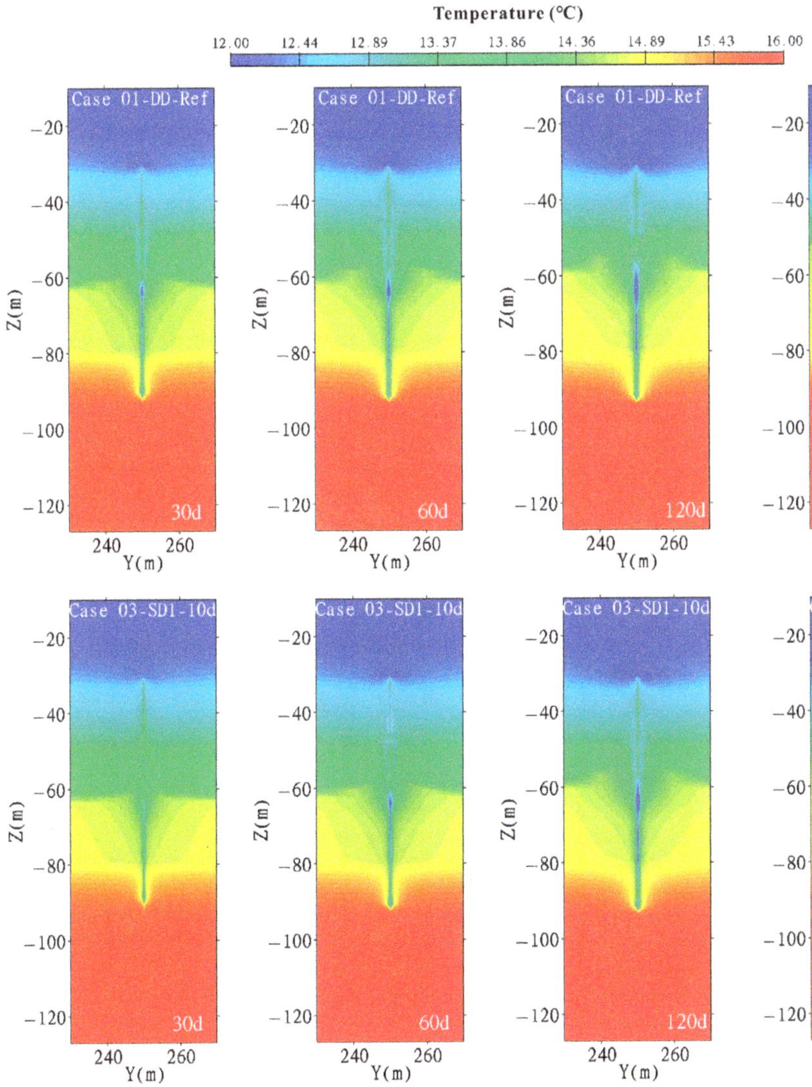

Figure 7. Temperature evolution diagram of direct and stepwise depressurization.

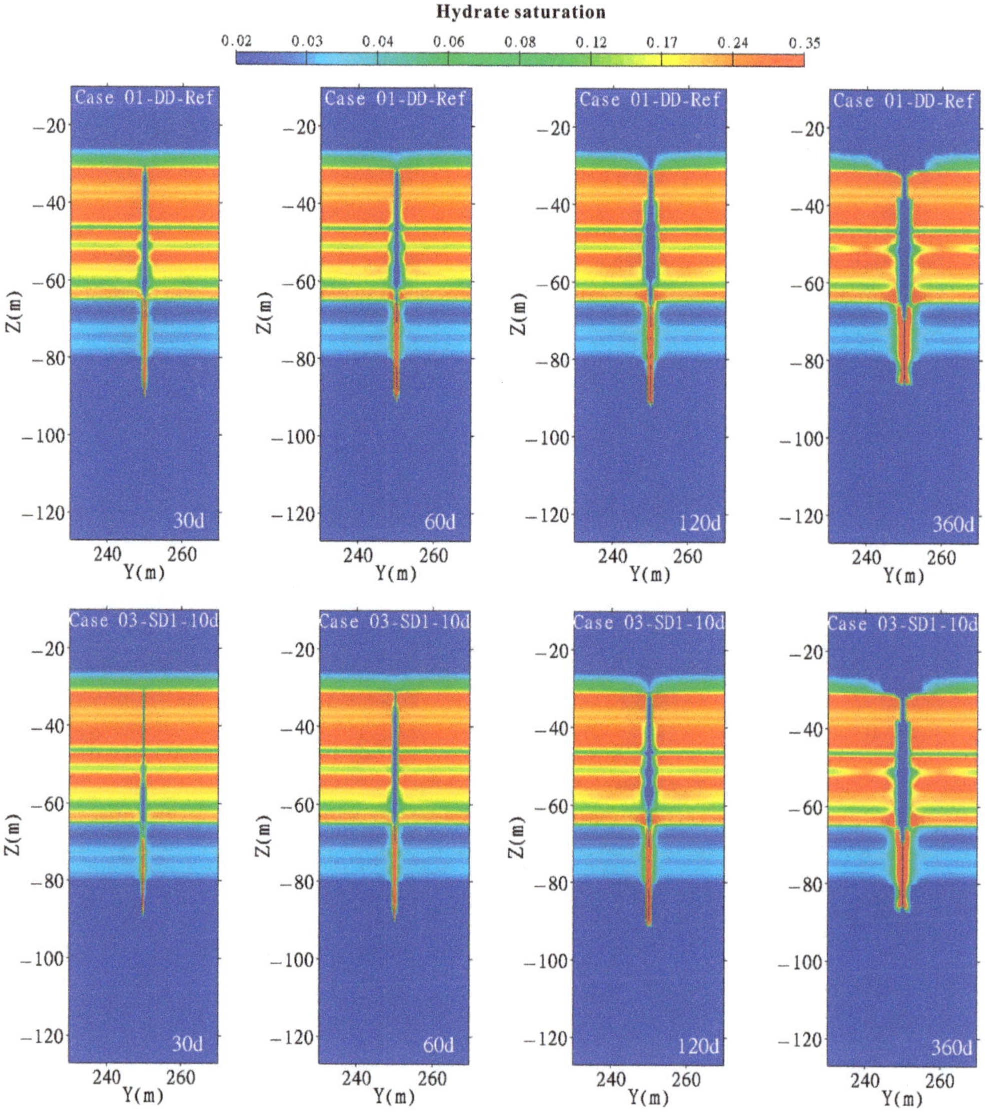

Figure 8. Hydrate saturation evolution diagram of direct and stepwise depressurization.

From Figure 9, it can be observed that as the hydrate dissociation in the GHBL and the free gas entered the wellbore in the TPL, both Case 01 and Case 03 formed gas at a relatively high saturation zone and gas at a relatively low saturation zone in the GHBL and TPL, respectively. In both cases, it can be observed that free gas moves upwards and downwards along the wellbore. The closer the free gas is to the wellbore, the faster it migrates, forming conical and inverted conical gas-bearing zones, respectively. After 120 days of production, the dissociation of hydrates located in the upper part of the TPL was observed in both Case 01 and Case 03, resulting in local high saturation gas accumulation around the wellbore.

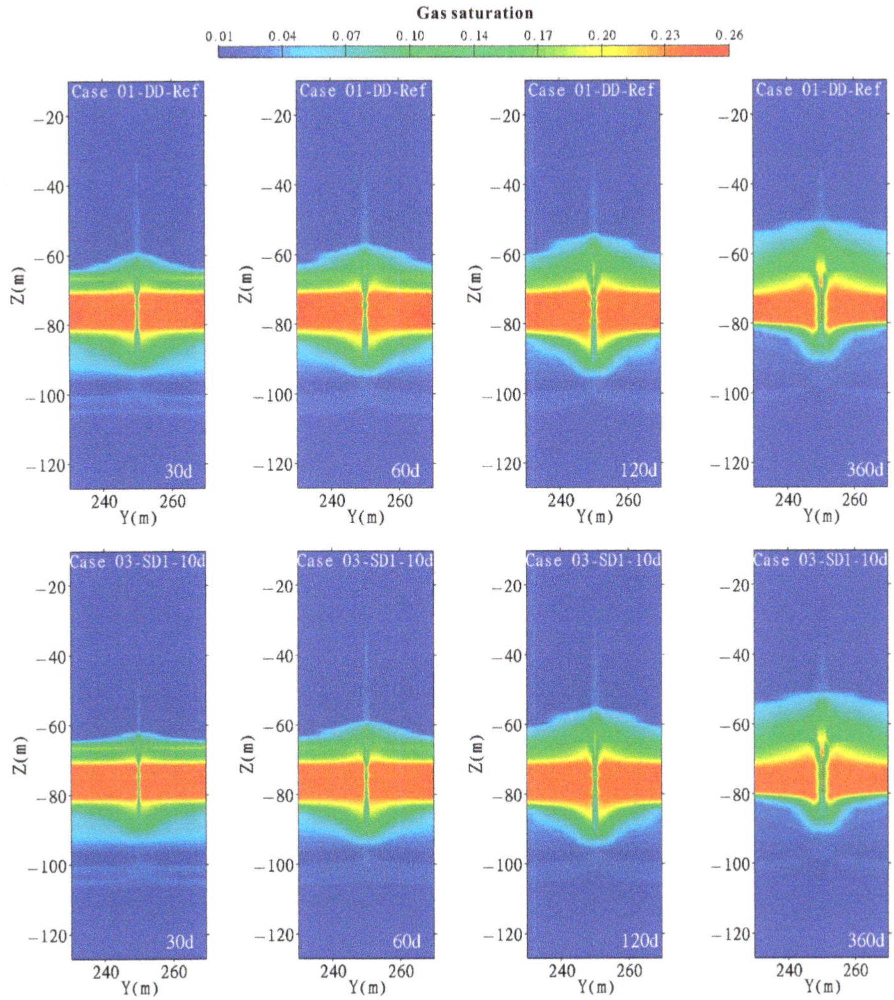

Figure 9. Gas saturation evolution diagram of direct and stepwise depressurization.

4. Discussion

4.1. Effects of Depressurization Gradient

As shown in Figure 10, Case 04, Case 07, and Case 10 have the same constant pressure duration of 5 days, with depressurization gradients of 0.5 MPa, 1.0 MPa, and 1.5 MPa, respectively. From Figure 10b, it can be seen that under the premise of a fixed constant pressure duration, the V_g increases with the increase in the depressurization gradient. The V_g after one year of production is 3.019, 3.039, and 3.045 $\times$ 10^6 ST m^3, respectively. Compared with direct depressurization, it decreases by 1.21%, 0.55%, and 0.35%, respectively, and there is no significant difference between them. As shown in Figure 10d, the R_{gw} decreases with the increase in the depressurization gradient. Throughout the entire production cycle, Case 04 has the best R_{gw} performance. Under the premise of fixed constant pressure duration and considering engineering operability, V_g, and R_{gw}, Case 04 is the optimal choice among them.

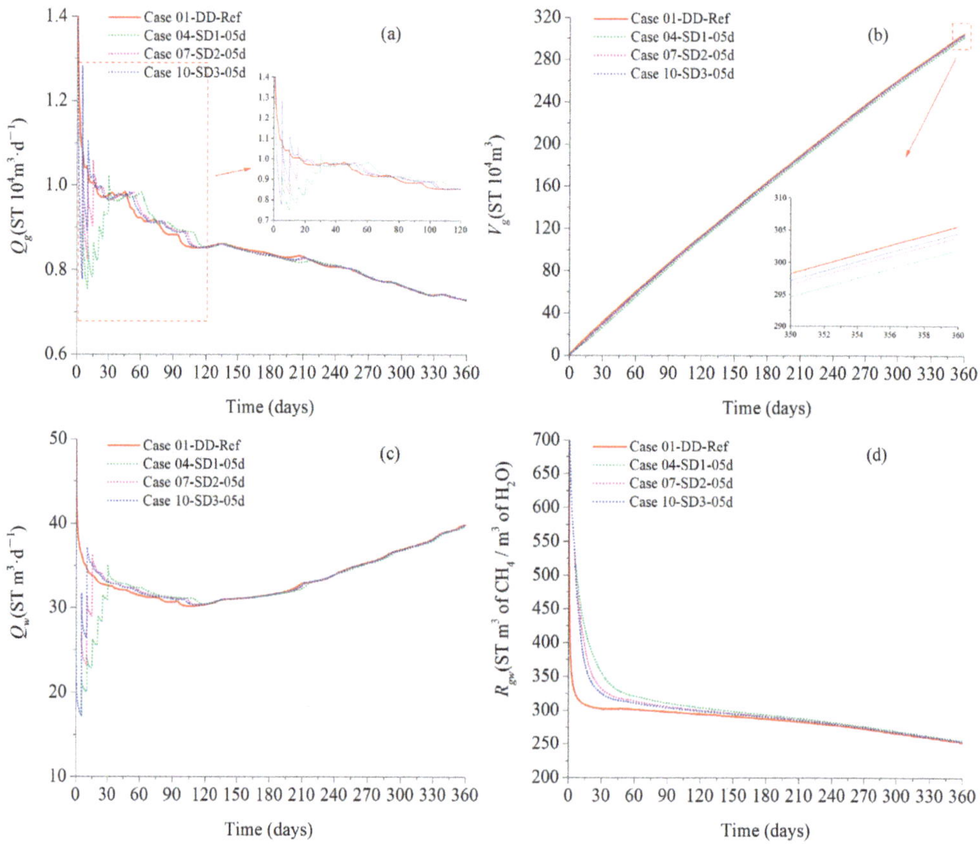

Figure 10. Gas and water production: (**a**) Q_g curves; (**b**) V_g curves; (**c**) Q_w curves; and (**d**) R_{gw} curves.

4.2. Effects of Constant Pressure Duration

As shown in Figure 11, the depressurization gradients for Case 02, Case 03, and Case 04 are all set to 0.5 MPa, and the constant pressure durations are set to 20 days, 10 days, and 5 days, respectively. From Figure 11b, it can be seen that under the premise of a fixed depressurization gradient, the V_g decreases with the increase in constant pressure duration. The V_g after one year of production is 2.864, 2.966, and 3.019 $\times$ 10^6 ST m^3, respectively. Compared with direct depressurization, it decreased by 6.28%, 2.94%, and 1.21%, respectively. As shown in Figure 11d, the R_{gw} increases with the increase in constant pressure duration. Throughout the entire production cycle, Case 02 has the best R_{gw} performance. Under the premise of fixed depressurization gradient and considering the V_g and R_{gw}, Case 03 is the optimal choice among them.

4.3. Comparison of Production Performance

Figures 12 and 13 show the R_{gw} and V_g for all simulated cases, respectively. Case 02 has the best R_{gw} performance, but its V_g of 2.864 $\times$ 10^6 ST m^3 is the lowest among all stepwise depressurization cases, a decrease of 6.28% compared to Case 01. The V_g of 3.045 $\times$ 10^6 ST m^3 in Case 10 is the highest among all stepwise depressurization cases, with only a decrease of 0.35% compared to Case 01, but its R_{gw} is the worst among all stepwise depressurization cases. The V_g of Case 03 is 2.966 $\times$ 10^6 ST m^3, which is only 2.94% lower than Case 01. Its R_{gw} ranks among the top in all stepwise depressurization cases, and it is also easy to achieve stepwise depressurization with small depressurization gradients in engineering, making it the best choice in all cases.

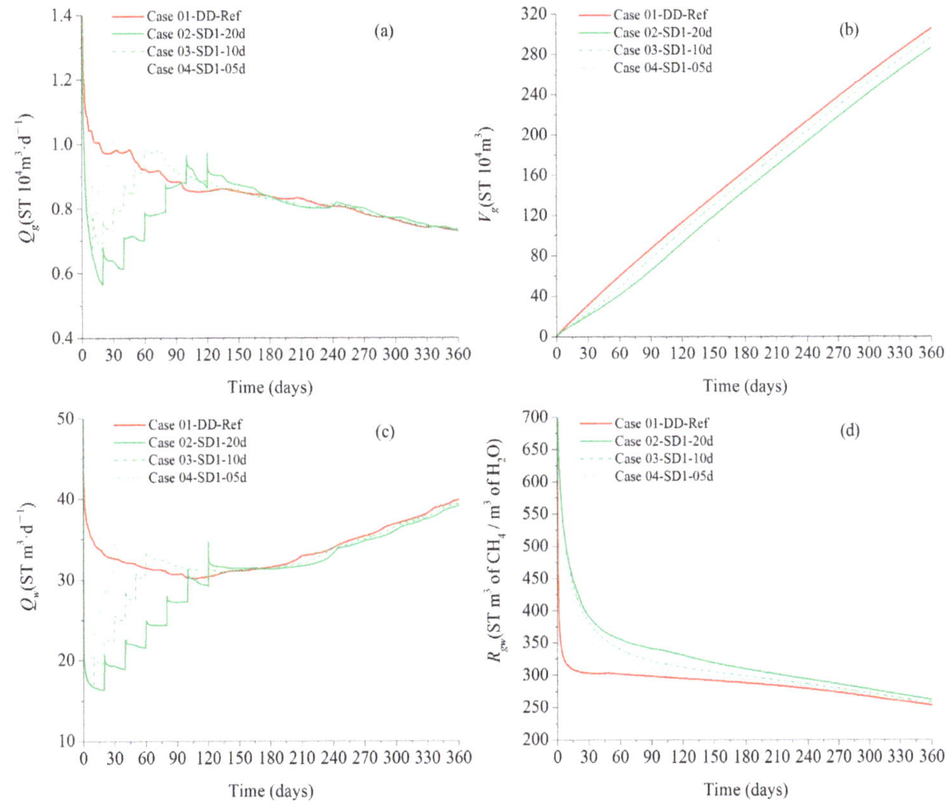

Figure 11. Gas and water production: (**a**) Q_g curves; (**b**) V_g curves; (**c**) Q_w curves; and (**d**) R_{gw} curves.

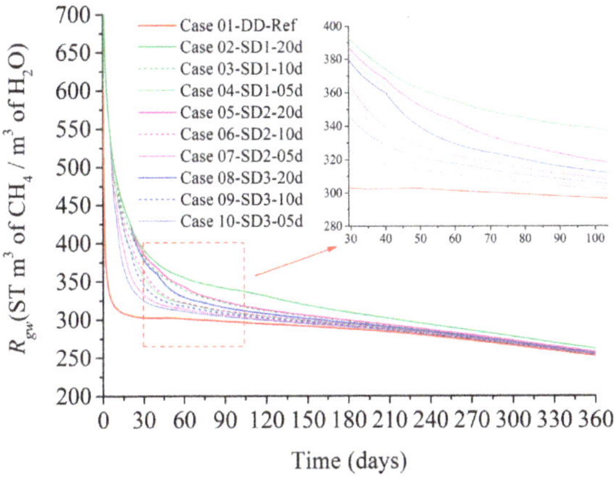

Figure 12. R_{gw} curves with different depressurization strategies.

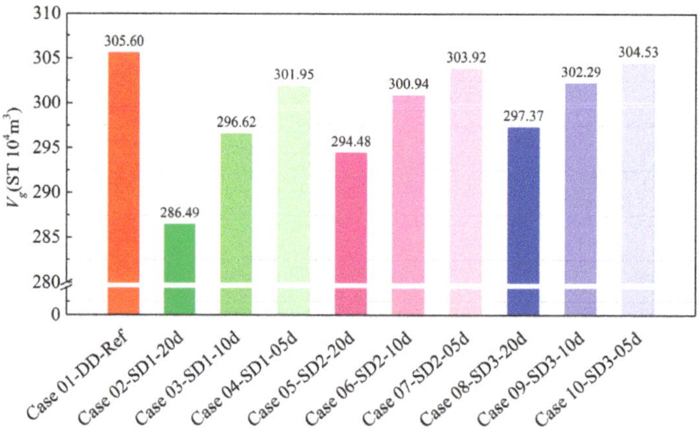

Figure 13. V_g with different depressurization strategies.

4.4. Implications and Future Recommendations

Multi-well or group well exploitation is an important development direction for the commercial development of NGHs, and the vertical well is one of the important basic well types. However, there is limited research on the stepwise depressurization of NGH reservoirs in the Shenhu Sea area with vertical wells. This work is based on SHSC4 well-logging data to establish an ideal numerical model and study the impact of different depressurization methods on gas and water production in Shenhu Sea area's NGH reservoirs with a vertical well. The results show that compared to direct depressurization, stepwise depressurization can increase the gas-to-water ratio while ensuring gas production. Maintaining a high gas-to-water ratio during natural gas hydrate exploitation has the following positive effects: 1. A decrease in water production can reduce the energy consumed for lifting or pumping water in engineering, which is beneficial for improving economic efficiency; 2. Related studies have shown that water production is an important factor leading to sand production in reservoirs [63–66]. A decrease in water production can alleviate sand production in reservoirs and avoid a series of engineering problems caused by sand production, such as wear of wellbore pipelines due to sand production and failure of sand control due to excessive sand production. This is similar to the numerical analysis results of Yin et al., 2022 using horizontal wells to stepwise-depressurize natural gas hydrates in the Shenhu Sea area [28]. It is recommended to use a smaller pressure gradient and a medium constant pressure duration for stepwise depressurization in Class 1-type hydrate reservoirs with a vertical well. This work helps us better understand the Class 1-type hydrate reservoirs and has reference value for the development of similar hydrate reservoirs worldwide. Furthermore, this work contributes to gas hydrate, natural gas, and petroleum system studies broadly speaking (e.g., Magoon & Dow, 1994; Burton et al., 2018, 2019; Jang et al., 2020; Almashwali et al., 2022) [67–71].

5. Conclusions

Based on the NGH testing data in the Shenhu Sea area, an ideal NGH reservoir model of the SHSC4 well was established, and the vertical well depressurization production simulation was conducted. A comprehensive analysis was made between direct and stepwise depressurization, including gas and water production behavior, as well as the evolution characteristics of reservoir physical properties. The sensitivity of the depressurization gradient and constant pressure duration were analyzed, and the following conclusions can be drawn:

(1) Stepwise depressurization can divide direct depressurization into several small cycles with small amplitude depressurization. While ensuring V_g, it can reduce water

production and increase R_{gw} in the early stages of production, which is beneficial for sand production control; in the early stage of production, stepwise depressurization can avoid a rapid decrease in reservoir temperature and alleviate the formation of secondary hydrates. Compared to the depressurization gradient, the effect of constant pressure duration on V_g and R_{gw} is more significant.

(2) When the vertical well is deployed at the model's center with a completion length of 70 m and a constant pressure duration of 10 days, a depressurization gradient of 0.5 MPa is used for depressurization by 6 MPa for continuous production for one year. Compared to direct depressurization, its V_g is 2.966×10^6 ST m^3, a decrease of only 2.94%, but it maintains a higher R_{gw} in the early stages of production. It is recommended to use a small pressure gradient and a medium constant pressure stabilization time for stepwise depressurization exploitation, which can effectively improve production efficiency.

(3) Stepwise depressurization can maintain a high gas-to-water ratio while ensuring gas production, improving economic efficiency, and reducing water production can alleviate sand production problems. These understandings have reference value for the development of similar hydrate reservoirs worldwide. In the future, we will combine auxiliary heating, near-wellbore reservoir reconstruction, hydraulic fracturing, and other methods to further study the impact of stepwise depressurization on gas and water production in Class 1-type hydrate reservoirs.

Author Contributions: T.W.: Conceptualization, Methodology, Software, and Writing—Original Draft. Z.L.: Formal analysis and Investigation. H.L.: Resources and Funding acquisition. M.W.: Formal analysis and Investigation. Z.C.: Formal analysis and Investigation. L.T.: Resources. Q.L.: Data curation and Visualization. J.Q.: Data curation and Visualization. J.W.: Writing—review and editing, Supervision, and Project administration. All authors have read and agreed to the published version of the manuscript.

Funding: National Key Research and Development Program of China (No. 2021YFB3401405 and No. SQ2023YFC2800361); Guangzhou Science and Technology Program (No. 202206050002); Guangdong Basic and Applied Basic Research Foundation (No. 2022A1515011902); and the Director General's Scientific Research Fund of Guangzhou Marine Geological Survey, China (No. 2023GMGSJZJJ00027).

Data Availability Statement: Data will be made available on request.

Conflicts of Interest: The authors declare that they do not have any commercial or associative interest that represents conflicts of interest in connection with the submitted work.

References

1. Collett, T. Energy resource potential of natural gas hydrates. *AAPG Bull.* **2002**, *86*, 1971–1992.
2. Sloan, E. Fundamental principles and applications of natural gas hydrates. *Nature* **2003**, *426*, 353–359. [CrossRef]
3. Boswell, R. Is gas hydrate energy within reach? *Science* **2009**, *325*, 957–958. [CrossRef]
4. Boswell, R.; Collett, T.S. Current perspectives on gas hydrate resources. *Energy Environ. Sci.* **2011**, *4*, 1206–1215. [CrossRef]
5. Chong, Z.; Yang, S.; Babu, P.; Linga, P.; Li, X. Review of natural gas hydrates as an energy resource: Prospects and challenges. *Appl. Energy* **2016**, *162*, 1633–1652. [CrossRef]
6. Makogon, Y. Natural gas hydrates–A promising source of energy. *J. Nat. Gas Sci. Eng.* **2010**, *2*, 49–59. [CrossRef]
7. Burton, Z.; Dafov, L. Testing the sediment organic contents required for biogenic gas hydrate formation: Insights from synthetic 3-D basin and hydrocarbon system modelling. *Fuels* **2022**, *3*, 555–562. [CrossRef]
8. Yamamoto, K.; Terao, Y.; Fujii, T.; Ikawa, T.; Seki, M.; Matsuzawa, M.; Kanno, T. Operational Overview of the First Offshore Production Test of Methane Hydrates in the Eastern Nankai Trough. In Proceedings of the Offshore Technology Conference, Houston, TX, USA, 5–8 May 2014; OnePetro: Richardson, TX, USA, 2014.
9. Yamamoto, K.; Wang, X.; Tamaki, M.; Suzuki, K. The second offshore production of methane hydrate in the Nankai Trough and gas production behavior from a heterogeneous methane hydrate reservoir. *RSC Adv.* **2019**, *9*, 25987–26013. [CrossRef]
10. Li, J.; Ye, J.; Qin, X.; Qiu, H.; Wu, N.; Lu, H.; Xie, W.; Lu, J.; Peng, F.; Xu, Z.; et al. The first offshore natural gas hydrate production test in South China Sea. *China Geol.* **2018**, *1*, 5–16. [CrossRef]
11. Ye, J.; Qin, X.; Xie, W.; Lu, H.; Ma, B.; Qiu, H.; Liang, J.; Lu, J.; Kuang, Z.; Lu, C.; et al. The second natural gas hydrate production test in the South China Sea. *China Geol.* **2020**, *3*, 197–209. [CrossRef]
12. Too, J.; Cheng, A.; Khoo, B.; Palmer, A.; Linga, P. Hydraulic fracturing in a penny-shaped crack. Part II: Testing the frackability of methane hydrate-bearing sand. *J. Nat. Gas Sci. Eng.* **2018**, *52*, 619–628. [CrossRef]

13. Nair, V.; Prasad, S.; Kumar, R.; Sangwai, J. Energy recovery from simulated clayey gas hydrate reservoir using depressurization by constant rate gas release, thermal stimulation and their combinations. *Appl. Energy* **2018**, *225*, 755–768. [CrossRef]
14. Aghajari, H.; Moghaddam, M.; Zallaghi, M. Study of effective parameters for enhancement of methane gas production from natural gas hydrate reservoirs. *Green Energy Environ.* **2019**, *4*, 453–469. [CrossRef]
15. Vedachalam, N.; Ramesh, S.; Jyothi, V.; Ramadass, G.; Atmanand, M.; Manivannan, P. Techno-economic viability studies on methane gas production from gas hydrates reservoir in the Krishna-Godavari basin, east coast of India. *J. Nat. Gas Sci. Eng.* **2020**, *77*, 103253. [CrossRef]
16. Terzariol, M.; Santamarina, J. Multi-well strategy for gas production by depressurization from methane hydrate-bearing sediments. *Energy* **2021**, *220*, 119710. [CrossRef]
17. Wu, N.; Li, Y.; Wan, Y.; Sun, J.; Huang, L.; Mao, P. Prospect of marine natural gas hydrate stimulation theory and technology system. *Nat. Gas Ind. B.* **2021**, *40*, 173–187. [CrossRef]
18. Moridis, G.; Reagan, M. Estimating the Upper Limit of Gas Production from Class 2 Hydrate Accumulations in the Permafrost: 1. Concepts, System Description, and the Production Base Case. *J. Pet. Sci. Eng.* **2011**, *76*, 194–204. [CrossRef]
19. Phillips, S.; Flemings, P.; You, K.; Meyer, D.; Dong, T. Investigation of in situ salinity and methane hydrate dissociation in coarse-grained sediments by slow, stepwise depressurization. *Mar. Petrol. Geol.* **2019**, *109*, 128–144. [CrossRef]
20. Yin, Z.; Wan, Q.; Qiang, G.; Linga, P. Effect of pressure drawdown rate on the fluid production behaviour from methane hydrate-bearing sediments. *Appl. Energy* **2020**, *271*, 115195. [CrossRef]
21. Zhao, J.; Liu, Y.; Guo, X.; Wei, R.; Yu, T.; Xu, L.; Sun, L.; Yang, L. Gas production behavior from hydrate-bearing fine natural sediments through optimized step-wise depressurization. *Appl. Energy* **2020**, *260*, 114275. [CrossRef]
22. Guo, X.; Xu, L.; Wang, B.; Sun, L.; Liu, Y.; Wei, R.; Yang, L.; Zhao, J. Optimized gas and water production from water-saturated hydrate-bearing sediment through step-wise depressurization combined with thermal stimulation. *Appl. Energy* **2020**, *276*, 115438. [CrossRef]
23. Ravesh, R.; Ansari, A.; Mohapatra, S.; Panigrahi, P.; Das, M. Methane Hydrate Dissociation in Porous Media using Multistep Depressurization: An Experimental Study. In Proceedings of the 25th National and 3rd International ISHMT-ASTFE Heat and Mass Transfer Conference (IHMTC-2019), Roorkee, India, 28–31 December 2019; Begel House Inc.: Danbury, CT, USA, 2019.
24. Yoon, H.; Yoon, S.; Lee, J.; Kim, J. Multiple porosity model of a heterogeneous layered gas hydrate deposit in Ulleung Basin, East Sea, Korea: A study on depressurization strategies, reservoir geomechanical response, and wellbore stability. *J. Nat. Gas Sci. Eng.* **2021**, *96*, 104321. [CrossRef]
25. Li, Y.; He, C.; Wu, N.; Chen, Q.; Liu, C.; Sun, Z.; Jin, Y.; Meng, Q. Laboratory study on hydrate production using a slow, multistage depressurization strategy. *Geofluids* **2021**, *2021*, 4352910. [CrossRef]
26. Yin, F.; Gao, Y.; Zhang, H.; Sun, B.; Chen, Y.; Gao, D.; Zhao, X. Comprehensive evaluation of gas production efficiency and reservoir stability of horizontal well with different depressurization methods in low permeability hydrate reservoir. *Energy* **2022**, *239*, 122422. [CrossRef]
27. Wang, J.; He, J.; Dong, H.; Ge, K. Association between multiphase seepage and exploitation of natural gas hydrate based on the Shenhu area of South China Sea. *J. Pet. Sci. Eng.* **2022**, *209*, 109855. [CrossRef]
28. Yin, F.; Gao, Y.; Chen, Y.; Sun, B.; Li, S.; Zhao, D. Numerical investigation on the long-term production behavior of horizontal well at the gas hydrate production site in South China Sea. *Appl. Energy* **2022**, *311*, 118603. [CrossRef]
29. Xue, K.; Liu, Y.; Yu, T.; Lv, J. Numerical Simulation of Optimized Step-Wise Depressurization for Enhanced Natural Gas Hydrate Production in the Nankai Trough of Japan. *Processes* **2023**, *11*, 1812. [CrossRef]
30. Ge, K.; Zhang, X.; Wang, J.; Cheng, C.; He, J. Optimization of the depressurization rate and stepwise strategy for hydrate exploitation using a genetic algorithm-based depressurization method. *Chem. Eng. Sci.* **2023**, *265*, 118218. [CrossRef]
31. Wang, J.; Zhang, X.; Ge, K.; Yu, W.; Wei, H.; Liu, J. Research on Natural Gas Production Behavior Using Stepwise Depressurization with a Horizontal Well. *Energy Fuels* **2023**, *37*, 10367–10382. [CrossRef]
32. Biastoch, A.; Treude, T.; Rüpke, L.; Riebesell, U.; Roth, C.; Burwicz, E.; Park, W.; Latif, M.; Böning, C.; Madec, G.; et al. Rising Arctic Ocean temperatures cause gas hydrate destabilization and ocean acidification. *Geophys. Res. Lett.* **2011**, *38*, L08602. [CrossRef]
33. Burton, Z.; Kroeger, K.; Hosford Scheirer, A.; Seol, Y.; Burgreen-Chan, B.; Graham, S.A. Tectonic uplift destabilizes subsea gas hydrate: A model example from Hikurangi margin, New Zealand. *Geophys. Res. Lett.* **2020**, *47*, e2020GL087150. [CrossRef]
34. Kim, B.; Zhang, Y. Methane hydrate dissociation across the Oligocene–Miocene boundary. *Nat. Geosci.* **2022**, *15*, 203–209. [CrossRef]
35. Burton, Z.; Dafov, L. Salt Diapir-Driven Recycling of Gas Hydrate. *Geochem. Geophys. Geosyst.* **2023**, *24*, e2022GC010704. [CrossRef]
36. Moridis, G.; Kowalsky, M.; Pruess, K. *TOUGH+ Hydrate V1.0 User's Manual*; Report LBNL-0149E; Lawrence Berkeley National Laboratory: Berkeley, CA, USA, 2008.
37. Piñero, E.; Hensen, C.; Haeckel, M.; Rottke, W.; Fuchs, T.; Wallmann, K. 3-D numerical modelling of methane hydrate accumulations using PetroMod. *Mar. Pet. Geol.* **2016**, *71*, 288–295. [CrossRef]
38. Burton, Z. Sediment organic contents required for gas hydrate formation: A survey of published basin and hydrocarbon system models. *Fuels* **2022**, *3*, 580–587. [CrossRef]

39. Moridis, G.; Reagan, M.; Boyle, K.; Zhang, K. Evaluation of the gas production potential of some particularly challenging types of oceanic hydrate deposits. *Transp. Porous Media* **2011**, *90*, 269–299. [CrossRef]
40. Moridis, G.; Kim, J.; Reagan, M.; Kim, S. Feasibility of gas production from a gas hydrate accumulation at the UBGH2-6 site of the Ulleung basin in the Korean East Sea. *J. Pet. Sci. Eng.* **2013**, *108*, 180–210. [CrossRef]
41. Moridis, G.; Reagan, M.; Queiruga, A.; Boswell, R. Evaluation of the performance of the oceanic hydrate accumulation at site NGHP-02-09 in the Krishna-Godavari Basin during a production test and during single and multi-well production scenarios. *Mar. Pet. Geol.* **2019**, *108*, 660–696. [CrossRef]
42. Su, Z.; Moridis, G.; Zhang, K.; Wu, N. A huff-and-puff production of gas hydrate deposits in Shenhu area of South China Sea through a vertical well. *J. Pet. Sci. Eng.* **2012**, *86–87*, 54–61. [CrossRef]
43. Yin, Z.; Moridis, G.; Chong, Z.; Linga, P. Effectiveness of multi-stage cooling processes in improving the CH4-hydrate saturation uniformity in sandy laboratory samples. *Appl. Energy* **2019**, *250*, 729–747. [CrossRef]
44. Kowalsky, M.; Moridis, G. Comparison of kinetic and equilibrium reaction models in simulating gas hydrate behavior in porous media. *Energy Convers. Manag.* **2007**, *48*, 1850–1863. [CrossRef]
45. Hao, Y.; Yang, F.; Wang, J.; Fan, M.; Li, S.; Yang, S.; Wang, C.; Xiao, X. Dynamic analysis of exploitation of different types of multilateral wells of a hydrate reservoir in the South China sea. *Energy Fuels* **2022**, *36*, 6083–6095. [CrossRef]
46. Zhang, W.; Liang, J.; Lu, J.; Wei, J.; Su, P.; Fang, Y.; Guo, Y.; Yang, S.; Zang, G. Accumulation features and mechanisms of high saturation natural gas hydrate in shenhu area, northern south China sea. *Pet. Explor. Dev.* **2017**, *44*, 708–719. [CrossRef]
47. Su, P.; He, J.; Liang, J.; Zhang, W. Natural gas hydrate migration and accumulation system and its controlling factors on northern deep water slope of the South China Sea. *Mar. Geol. Front.* **2017**, *33*, 1–10.
48. Moridis, G.; Kowalsky, M.; Pruess, K. Depressurization-induced gas production from class 1 hydrate deposits. *SPE Reserv. Eval. Eng.* **2007**, *10*, 458–481. [CrossRef]
49. Yu, T.; Guan, G.; Wang, D.; Song, Y.; Abudula, A. Numerical investigation on the long-term gas production behavior at the 2017 Shenhu methane hydrate production site. *Appl. Energy* **2021**, *285*, 116466. [CrossRef]
50. Sun, J.; Zhang, L.; Ning, F.; Lei, H.; Liu, T.; Hu, G.; Lu, H.; Lu, J.; Liu, C.; Jiang, G.; et al. Production potential and stability of hydrate-bearing sediments at the site GMGS3-W19 in the South China Sea: A preliminary feasibility study. *Mar. Pet. Geol.* **2017**, *86*, 447–473. [CrossRef]
51. Yuan, Y.; Xu, T.; Xin, X.; Xia, Y. Multiphase Flow Behavior of Layered Methane Hydrate Reservoir Induced by Gas Production. *Geofluids* **2017**, *2017*, 7851031. [CrossRef]
52. Zhang, J.; Yin, Z.; Li, Q.; Li, S.; Wang, Y.; Li, X. Comparison of fluid production between excess-gas and excess-water hydrate-bearing sediments under depressurization and its implication on energy recovery. *Energy* **2023**, *282*, 128315. [CrossRef]
53. Feng, Y.; Chen, L.; Suzuki, A.; Kogawa, T.; Okajima, J.; Komiya, A.; Maruyama, S. Enhancement of gas production from methane hydrate reservoirs by the combination of hydraulic fracturing and depressurization method. *Energy Convers. Manag.* **2019**, *184*, 194–204. [CrossRef]
54. Moridis, G.; Reagan, M.; Kim, S.; Seol, Y.; Zhang, K. Evaluation of the Gas Production Potential of Marine Hydrate Deposits in the Ulleung Basin of the Korean East Sea. *Spe J.* **2007**, *14*, 759–781. [CrossRef]
55. Li, G.; Li, X.; Zhang, K.; Li, B.; Zhang, Y. Effects of Impermeable Boundaries on Gas Production from Hydrate Accumulations in the Shenhu Area of the South China Sea. *Energies* **2013**, *6*, 4078–4096. [CrossRef]
56. Sun, Y.; Ma, X.; Guo, W.; Jia, R.; Li, B. Numerical simulation of the short- and long-term production behavior of the first offshore gas hydrate production test in the South China Sea. *J. Pet. Sci. Eng.* **2019**, *181*, 106196. [CrossRef]
57. Ma, X.; Sun, Y.; Liu, B.; Guo, W.; Jia, R.; Li, B.; Li, S. Numerical study of depressurization and hot water injection for gas hydrate production in China's first offshore test site. *J. Nat. Gas Sci. Eng.* **2020**, *83*, 103530. [CrossRef]
58. Cao, X.; Sun, J.; Qin, F.; Ning, F.; Mao, P.; Gu, Y.; Li, Y.; Zhang, H.; Yu, Y.; Wu, N. Numerical analysis on gas production performance by using a multilateral well system at the first offshore hydrate production test site in the Shenhu area. *Energy* **2023**, *270*, 126690. [CrossRef]
59. Qin, X.; Liang, Q.; Yang, L.; Qiu, H.; Xie, W.; Liang, J.; Lu, J.; Lu, C.; Lu, H.; Ma, B.; et al. The response of temperature and pressure of hydrate reservoirs in the first gas hydrate production test in South China Sea. *Appl. Energy* **2020**, *278*, 115649. [CrossRef]
60. Wang, H.; Liu, W.; Wu, P.; Pan, X.; You, Z.; Lu, J.; Li, Y. Gas recovery from marine hydrate reservoir: Experimental investigation on gas flow patterns considering pressure effect. *Energy* **2023**, *275*, 127482. [CrossRef]
61. Zhou, S.; Li, Q.; Zhu, J.; Gu, Q.; Wang, L.; Wu, Z. Experimental Study on Permeability of Methane Hydrate Clayey Interbedded Sediments Considering Effective Stress and Hydrate Dissociation. *Energy Fuels* **2023**, *37*, 11113–11124. [CrossRef]
62. Wu, Z.; Gu, Q.; Wang, L.; Li, G.; Shi, C.; He, Y.; Li, Q.; Li, Y. Experimental Study on Permeability and Gas Production Characteristics of Montmorillonite Hydrate Sediments Considering the Effective Stress and Gas Slippage Effect. *SPE J.* **2024**, *29*, 2525–2544. [CrossRef]
63. Uchida, S.; Klar, A.; Yamamoto, K. Sand production model in gas hydrate-bearing sediments. *Int. J. Rock Mech. Min. Sci.* **2016**, *86*, 303–316. [CrossRef]
64. Wu, N.; Li, Y.; Chen, Q.; Liu, C.; Jin, Y.; Tan, M.; Ding, L.; Hu, G. Sand production management during marine natural gas hydrate exploitation: Review and an innovative solution. *Energy Fuels* **2021**, *35*, 4617–4632. [CrossRef]
65. Deng, F.; Huang, B.; Li, X.; Liu, J.; Li, G.; Xu, Y.; Yin, B. Review of Sand Control and Sand Production in a Gas Hydrate Reservoir. *Energy Fuels* **2022**, *36*, 11712–11723. [CrossRef]

66. Zhu, H.; Xu, T.; Yuan, Y.; Feng, G.; Xia, Y.; Xin, X. Numerical analysis of sand production during natural gas extraction from unconsolidated hydrate-bearing sediments. *J. Nat. Gas Sci. Eng.* **2020**, *76*, 103229. [CrossRef]
67. Magoon, L.B.; Dow, W.G. The petroleum system. In *The Petroleum System—From Source to Trap*; Magoon, L.B., Dow, W.G., Eds.; American Association of Petroleum Geologists Memoir: Tulsa, OK, USA, 1994; Volume 60, pp. 3–24.
68. Burton, Z.; Moldowan, J.; Sykes, R.; Graham, S. Unraveling petroleum degradation, maturity, and mixing and addressing impact on petroleum prospectivity: Insights from frontier exploration regions in New Zealand. *Energy Fuels* **2018**, *32*, 1287–1296. [CrossRef]
69. Burton, Z.; Moldowan, J.; Magoon, L.; Sykes, R.; Graham, S. Interpretation of source rock depositional environment and age from seep oil, east coast of New Zealand. *Int. J. Earth Sci.* **2019**, *108*, 1079–1091. [CrossRef]
70. Jang, J.; Waite, W.; Stern, L. Gas hydrate petroleum systems: What constitutes the "seal"? *Interpretation* **2020**, *8*, T231–T248. [CrossRef]
71. Almashwali, A.; Bavoh, C.; Lal, B.; Khor, S.; Jin, Q.; Zaini, D. Gas hydrate in oil-dominant systems: A review. *ACS Omega* **2022**, *7*, 27021–27037. [CrossRef]

Journal of
Marine Science and Engineering

MDPI

Article

Gas–Water–Sand Inflow Patterns and Completion Optimization in Hydrate Wells with Different Sand Control Completions

Chenfeng Liu [1,2], Changyin Dong [1,2,*], Haoxian Shi [3], Yanjiang Yu [3] and Bin Yin [1,2]

1 Key Laboratory of Unconventional Oil & Gas Development, Ministry of Education, Qingdao 266580, China; liuchenfeng0312@163.com (C.L.); yinbin1678@163.com (B.Y.)

2 School of Petroleum Engineering, China University of Petroleum (East China), Qingdao 266580, China

3 Guangzhou Marine Geological Survey, China Geological Survey, Guangzhou 511466, China; shihaoxian00@163.com (H.S.); yuyanjiang2004@163.com (Y.Y.)

* Correspondence: dongcy@upc.edu.cn

Abstract: Sand production poses a significant problem for marine natural gas hydrate efficient production. However, the bottom hole gas–water–sand inflow pattern remains unclear, hindering the design of standalone screen and gravel packing sand control completions. Therefore, gas–water–sand inflow patterns were studied in horizontal and vertical wells with the two completions. The experimental results showed that gas–water stratification occurred in horizontal and vertical standalone screen wells. The gas–water interface changed dynamically, leading to an uneven screen plugging, with severe plugging at the bottom and high permeability at the top. The high sand production rate and low well deviation angle exacerbated screen plugging, resulting in a faster rising rate of the gas–water interface. The screen plugging degree initially decreased and then increased as the gas–water ratio increased, resulting in the corresponding variation in the gas–water interface rising rate. Conversely, gas–water stratification did not occur in the gravel packing well because of the pore throat formed between the packing gravels. However, the impact of gas and water led to gravel rearrangement and the formation of erosion holes, causing sand control failure. A higher gas–water ratio and lower packing degree could result in more severe destabilization. Therefore, for the standalone screen completion, sand control accuracy should be designed at different levels according to the uneven plugging degree of the screen. For the gravel packing completion, increase the gravel density without destabilizing the hydrate reservoir, and use the coated gravel with a cementing effect to improve the gravel layer stability. In addition, the screen sand control accuracy inside the gravel packing layer should be designed according to the sand size to keep long-term stable hydrate production.

Keywords: marine; natural gas hydrate; inflow pattern; standalone screen completion; gravel packing completion; gas–water–sand; sand control accuracy

1. Introduction

The ocean contains abundant renewable and non-renewable energy. Among the non-renewable energy sources in the ocean, natural gas hydrate (NGH) reserves are huge, and the recoverable natural gas reserves are almost twice the proven conventional oil and natural gas [1,2]. NGH is mainly distributed in deep sea slope areas with a water depth greater than 300 m and in the permafrost zone on land, among which marine NGH resources account for about 97% of the world's total resources [3,4]. However, the efficient development of marine NGH resources currently faces huge challenges.

The sand production problem in the process of hydrate development is one of the greatest challenges, which seriously limits the production capacity of marine hydrate wells [5–10]. Currently, dozens of NGH production tests have been conducted in China, America, Japan, and Canada [11–13], mainly using vertical and horizontal wells with the open-hole standalone screen (SAS), open-hole gravel packing (GP), cased-hole SAS, and cased-hole GP sand control completions (Figure 1). But most of the tests faced the problem of sand production, which limits NGH commercial production. To date, the longest period of hydrate production tests has only reached 60 days, and the maximum hydrate production has only reached 2.87×10^4 m^3/d [14–20], which is far lower than the critical commercial hydrate production of 5×10^5 m^3/d in the sea area [21]. To improve the production capacity of marine NGH wells, it is urgent to solve the sand production problem [22–25]. However, there is a contradiction between sand control and increasing production capacity, and the use of sand control completion technology will reduce the production capacity of hydrate wells [26,27]. Therefore, it is crucial to clarify the gas–water–sand inflow pattern in the sand control section of marine NGH wells and reasonably design sand control completions to maintain long-term and high NGH production.

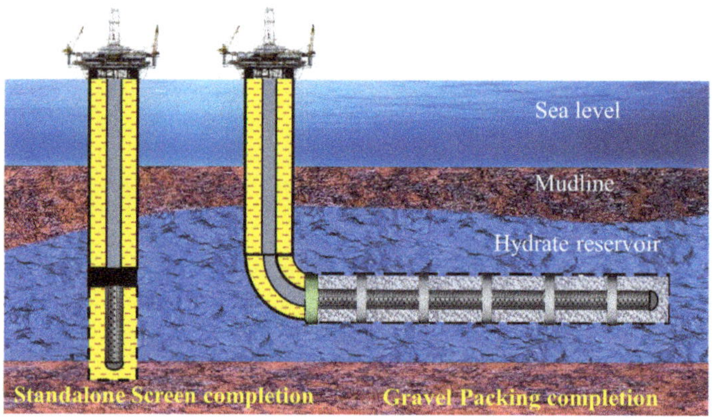

Figure 1. Schematic diagram of SAS and GP sand control completions.

In the past few decades, many research studies on sand control methods have been carried out to solve the sand production problem to improve marine NGH production and prolong the production period. In the evaluation of the sand control performance of different types of screens, the sand retention tests were conducted with wire wrap screens, metal sintered mesh screens, metal fiber screens, and prepacked gravel screens based on production data of the NGH production test in the South China Sea, which found that the permeability of screens generally showed three stages of slow decreasing, rapid increasing, and balance [28]. They analyzed the sand control effect of four screens with 20 μm, 40 μm, and 60 μm sand control accuracy and proved that the sand control in NGH reservoirs with highly argillaceous fine sands is practicable. Further research was conducted on the plugging mechanism and sand control effect of stainless-steel wire mesh screens, dutch wire screens, metal sintered mesh screens, and wire wrap screens from a microscopic point of view through sand retention tests from a micro-view [29]. The plugging pattern and mechanism of a non-consolidated prepacked gravel screen were investigated by using new sand retention tests, and they found that the anti-plugging ability of the screen is affected by the packing density and gravel size simultaneously [30]. The microscopic mechanism that influences the reservoir sand retention and retained permeability of the gravel pack layer was also further examined and interpreted by the CFD-DEM simulations [31]. In addition to the plugging of sand control media by sand in the hydrate reservoir, it was proved that the sand control media was also plugged by the reformation of hydrate through experiments [32].

In addition, in terms of the design of sand control parameters, a large number of laboratory tests were performed on various types of wire wrap screens with 6 to 16 sand control accuracy and premium mesh screens with 60 to 600 sand control accuracy [33]. They proposed that the screen should be initially selected based on sand retention performance, with the final selection confirmed based on flow capacity. The sand control effect of the wire wrap screen with 40 μm and 100 μm sand control accuracy was investigated during the depressurization production of hydrate indoors [34]. The results of the screen performance showed that sand production was not visibly observed in the screen sample, and the wire wrap screen with 100 μm had good application in sand control during hydrate reservoir exploitation. Hydrate reservoir sand control simulations were carried out using metal mesh screens with different pore sizes. The results showed that the design criterion of sand control accuracy for the metal mesh screen was finally obtained as D50 = 11 × d50 (where D50 is the median grain size of the gravel, and d50 is the median grain size of the formation sand) [35]. It was found that hydrate production was divided into three stages: water, gas with water drops, and gas [36]. Sand production only existed in the first two of the three stages, of which the first stage was fine sand and the second stage was sand particles. Based on the different stages of sand production, the concept of segmented graded sand prevention was proposed [36]. The Saucier gravel design method is mainly used in GP completion to develop conventional oil and gas reservoirs, but it is unsuitable for clayey silt hydrate reservoirs [37]. A new gravel sizing method was proposed for GP sand control completion named "Hold coarse while eliminating fine particles (HC and EF method)" for the clayey hydrate-bearing formations [38,39]. In this method, the formation sand was divided into coarse and fine components. Then, the gravel sizes for retaining coarse components and eliminating fine components were calculated, respectively. Finally, the intersection of these two gravel sizes was taken as the proper gravel size.

Existing research has enhanced the understanding of sand control of marine NGH reservoirs. At present, the research on hydrate sand control completions mainly focuses on the sand control mechanism of different types of sand control media, the comparison of sand control performance, and the design of sand control parameters. However, the current hydrate well sand control completions are generally statically designed based on parameters such as hydrate reservoir characteristics and the particle size of produced formation sand. However, the bottom of the hydrate well is a complex gas–water–sand three-phase fluid environment. Under the long-term flow impact of gas–water–sand, the stability of the sand control completion system, which is the only channel for gas–water–sand to enter the hydrate well, may change dynamically. Most of the research on hydrate sand control completions ignores the possibility of dynamic changes in the sand control completion system. At present, only Kaige Gao has found that the GP sand control completion has instability [40]. With the impact of gas–water–sand on the gravel layer, the gravel layer will lose stability and form erosion holes, resulting in the sand control failure of the gravel layer [40]. Inspired by his research, when designing sand control completions, we not only focused on the formation sand characteristics of hydrate reservoirs but also paid attention to the impact of the flow of gas–water–sand at the bottom of hydrate wells on sand control completions. Unfortunately, there is currently a blank in the research on the inflow pattern of gas–water–sand at the bottom of hydrate wells. The research related to the inflow pattern of gas–water–sand mainly focuses on the transport and deposition law of sand particles in the test pipeline during deep-water drilling or NGH production tests [41–44], the flow and deposition plugging law of hydrate in the deep-water wellbore, and the prediction of the location of hydrate production in the wellbore, etc. [45–47]. The unclear flow law of gas–water–sand at the bottom of hydrate wells limits the design of sand control completions under long-term production conditions.

To study the dynamic change characteristics of different sand control completion systems under the impact of long-term gas–water–sand inflow, to guide the design of sand control completions under long-term production conditions, this paper adopts indoor physical simulation experimental methods to study the inflow pattern of gas–water–sand at the bottom of hydrate wells for the SAS and GP sand control completions and analyzes the dynamic change characteristics of the sand control performance and flow performance of different completion systems under different flow parameters. Based on the experimental results, optimal design suggestions were proposed for SAS and GP sand control completions to prolong the effective period of sand control completions, increase natural gas production, and promote the commercial production of marine NGH.

2. Methodology

This paper adopts an indoor physical simulation experimental research method to study the inflow pattern of gas–water–sand in marine NGH vertical and horizontal wells with SAS and GP sand control completions and proposes optimization suggestions and inspirations for sand control completions based on the experimental results. Figure 2 shows the specific research idea and approach.

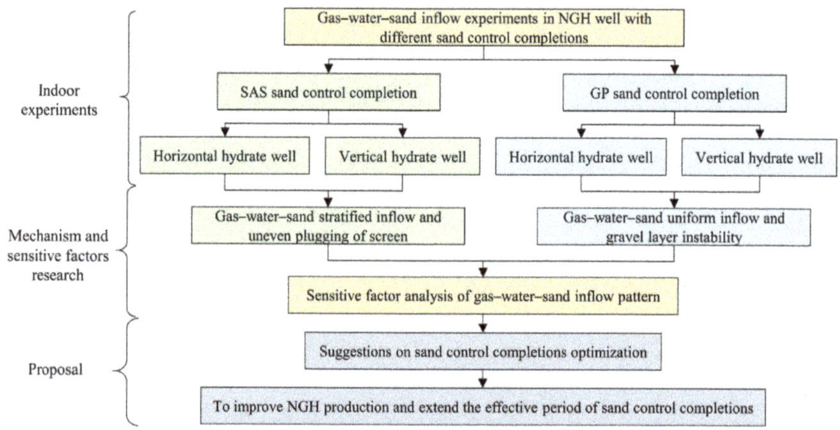

Figure 2. Flow chart of research idea and approach.

SAS and GP sand control completions are the mainstream sand control completion methods, and horizontal and vertical wells are mainly used in the current marine hydrate production tests in the world. Therefore, this paper conducts indoor physical experiments on two types of hydrate wells using SAS and GP sand control completions. First, the gas–water–sand inflow pattern in hydrate wells using the two sand control completions and the influence of gas–water–sand inflow on the sand control completions are explored. Then, the influencing factors of the gas–water–sand inflow pattern in the two sand control completions are studied. Finally, based on the experimental results, optimization suggestions for sand control completions are proposed to extend the effective period of completions, improve hydrate production, and help realize the marine NGH commercial production.

3. Materials and Methods

3.1. Apparatus and Materials

Hydrate decomposes to form natural gas and water during the production of the marine NGH. The strength of the hydrate reservoir decreased with the decomposition of NGH, leading to the collapse of the reservoir and shedding to produce a large amount of formation sand. Then, the formation of sand from the hydrate reservoir was carried by the gas–water fluid to the bottom of the hydrate well. Figure 3 shows the inflow process of gas–water–sand at the bottom of the hydrate well. The hydrate reservoir was divided into collapse, dissociated, and in the hydrate zone during the hydrate reservoir production. The gas–water–sand from the collapse zone first inflowed into the annular between the sand control screen and casing. Natural gas and water flowed upwards through the screen into the tubing. Formation sand with larger particle sizes was controlled by the screen and gradually plugged the screen, resulting in the gradual increase in formation sand deposited at the bottom of the wellbore annulus, and the water level gradually rose to the packer position with the plug of the screen, eventually filling the annulus. There are large differences in the gas–water–sand inflow patterns for hydrate horizontal and vertical wells with SAS and GP sand control completions. Exploring the gas–water–sand inflow pattern of hydrate wells with different completion methods is crucial to optimizing and designing sand control completions for different types of hydrate wells.

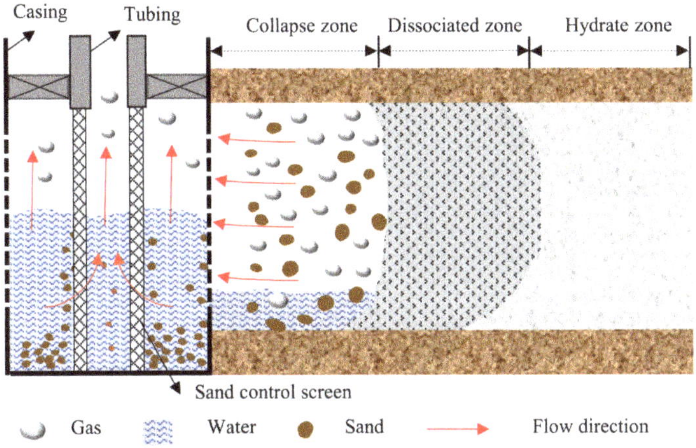

Figure 3. Schematic diagram of gas–water–sand inflow during the development of hydrate.

The marine hydrate well bottom inflow simulation device in this study is built to simulate the gas–water–sand inflow process in the sand control section of the hydrate well, which mainly consists of a screw pump, flow meter, simulated hydrate wellbore, data acquisition system, sand sink tank, return pump, liquid storage tank, automatic sand filler, air compressor, and so on (Figure 4a,c). The simulated hydrate wellbore is the core component of the device with an inner diameter of 220 mm, an outer diameter of 260 mm, and a height of 1000 mm, and it can accommodate a sand control screen with an outer diameter of 90–200 mm and a length of 900–998 mm. Six inlets were staggered on the well wall, and the flow outlets were located at the bottom of the wellbore. Six pressure sensors were equally spaced on the well wall to measure the pressure at different locations in the annulus, and four pairs of differential pressure sensors were evenly distributed at the bottom of the well to measure the differential pressure inside and outside the sand control screen (Figure 4b). The sand control screen was placed in the middle of the wellbore, and the angle of deviation of the wellbore can be adjusted in the range of 0–180° by adjusting the flip motor. The automatic sand filler realizes the automatic addition of formation sand to the pipeline and can precisely control the rate of sand addition. In addition, the data acquisition system can automatically collect flow, pressure, and differential pressure data during the experiment to calculate and analyze the screen permeability changes during the screen plug.

The NGH reservoir in the Shenhu area of the South China Sea is characterized by shallow burial and high mud content, and it belongs to a muddy silt sand hydrate reservoir with a mud content of 20–36% and a median grain size of 8–40 μm of which montmorillonite content is about 38%, illite content is about 32%, and kaolinite content is about 30% [11,18].

According to the characteristics of the formation sand of the NGH reservoir in the Shenhu area of the South China Sea, the sand with a median grain size of 17 μm and 20% mud content is compounded with quartz sand, montmorillonite, illite, and kaolinite (Figure 5). Figure 6 shows the grain size distribution curve of the compound formation sand. In addition, to simulate the GP sand control completion, the gravel used is 40–70 mesh conventional ceramsite.

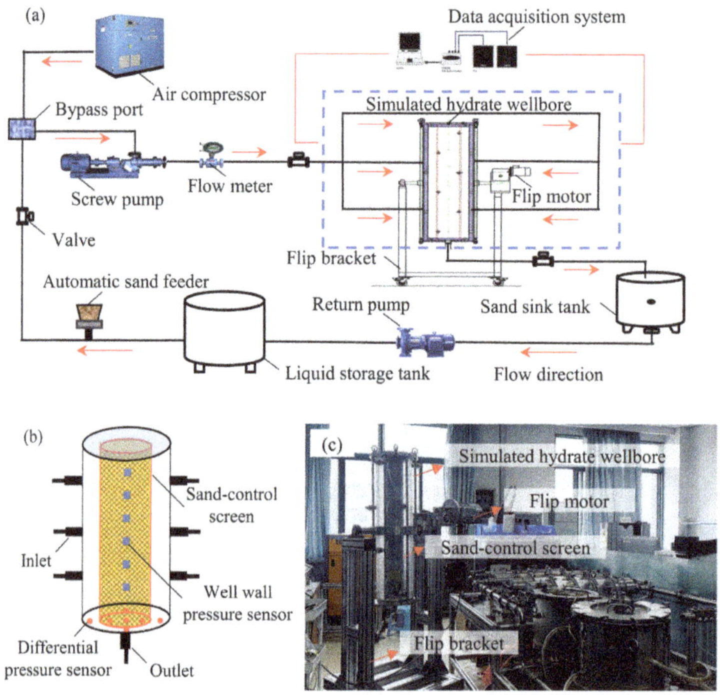

Figure 4. Physical and flowcharts of hydrate well bottom flow simulation device: (**a**) flowchart of device; (**b**) hydrate simulation wellbore; and (**c**) physical of device.

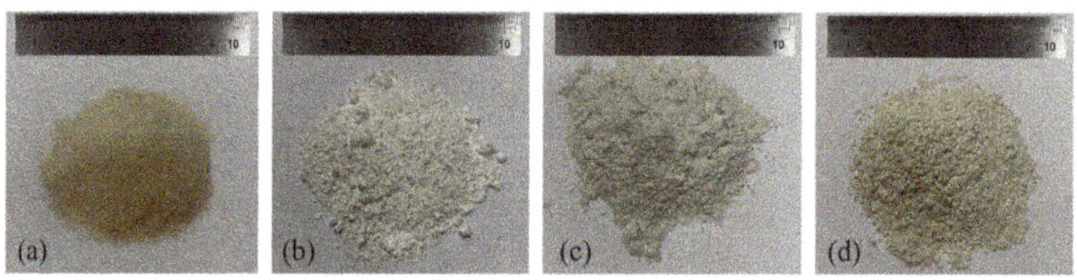

Figure 5. The composition of compound formation sand: (**a**) quartz sand; (**b**) montmorillonite; (**c**) illite; and (**d**) kaolinite.

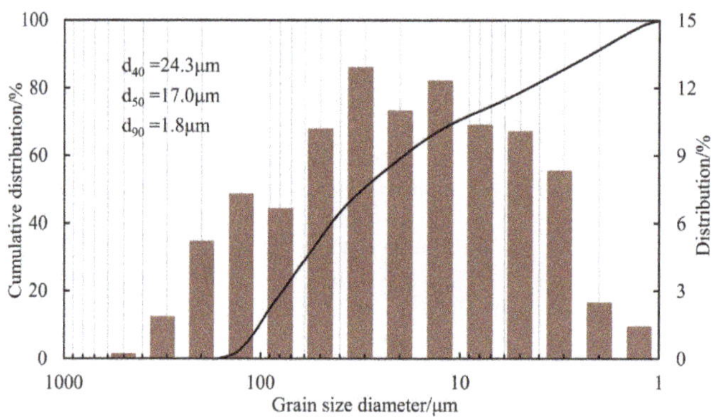

Figure 6. Grain size distribution of experimental sand.

The experimental flow rates of water and gas were calculated by using the dynamic similarity criterion to simulate the production conditions of the hydrate production test. Based on the first hydrate production test in the Shenhu area of the South China Sea, the experimental water flow rate was converted by the flow rate equivalence method. According to the gas–water ratio (GWR) and gas production at different moments of the production process, the experimental water flow rate was converted to 0.12–1.34 m^3/h. To cover the actual production conditions, the experimental flow rate of water was set to 1.5 m^3/h, which is higher than the converted flow rate of water. The GWR under reservoir conditions was converted by the gas volume coefficient based on experimental temperature and pressure conditions and hydrate reservoir temperature and pressure conditions. The experimental GWR ranges between 0.1 and 1.5 m^3/m^3, converted from the actual production GWR of hydrate wells. Therefore, the GWR in the basic experiment was set to 0.67 m^3/m^3. Another three GWRs of 0, 0.13, and 0.4 m^3/m^3 were set to research the effect of hydrate well production on the inflow pattern of gas–water–sand. In addition, the volumetric concentration of sand production (VCSP) or sand production rate in the NGH reservoir in the Shenhu area of the South China Sea is generally in the range of 0.05–1%. Therefore, four VCSPs (0.2%, 0.3%, 0.4%, and 0.5%) were set in this study to research the effect of the VCSP on the gas–water–sand inflow pattern, of which the basic experimental VCSP was 0.3%. All the experimental parameters of different sand control completion are summarized in Table 1.

Table 1. Experimental parameters of different sand control completion.

Number	Sand Control Completion	Deviation Angle/°	GWR/m^3/m^3	VCSP/%	Factor
1	SAS	0	0.67	0.2	VCSP
2	SAS	0	0.67	0.3	
3	SAS	0	0.67	0.4	
4	SAS	0	0.67	0.5	
5	SAS	0	0	0.3	GWR
6	SAS	0	0.13	0.3	
7	SAS	0	0.4	0.3	
8	SAS	0	0.67	0.3	
9	SAS	15	0.67	0.3	Deviation angle
10	SAS	30	0.67	0.3	
11	SAS	45	0.67	0.3	
12	SAS	60	0.67	0.3	
13	SAS	75	0.67	0.3	
14	SAS	90	0.67	0.3	
15	GP	0	0.67	0.3	GWR
16	GP	90	0.9	0.3	
17	GP	90	1.8	0.3	
18	GP	90	2.7	0.3	

3.2. Procedure and Methods

During the production of marine hydrate wells, hydrate may be reformatted in the hydrate reservoir near the wellbore and enter the wellbore along with the formation sand under the carrying of natural gas and water. However, the method of inhibitor or hot brine injection is generally used in the test production process to avoid hydrate reformation and prevent hydrate from plugging the sand control screen. Therefore, it is reasonable to disregard the hydrate dissociation and reformation process in this experiment and only simulate the inflow of gas–water–sand in the sand control section of the hydrate well. The experimental procedures include the following steps:

(1) Put the screen into the simulated wellbore and pack the gravel into the annular between the screen and casing, connect the experimental device, and check the air tightness of the device.
(2) Add the compound formation sand to the automatic sand filler, and fill the liquid storage tank with water.
(3) Open the screw pump to pump water into the pipeline, and check the sealing of the device.
(4) Set the target water and gas flow rate, adjust the VCSP, realize the mixing of gas, water, and sand through the gas–water mixer, turn on the screw pump, and return the pump to start the experiment.
(5) Collect the flow and pressure data in real time through the data acquisition system to calculate the permeability of the screen and gravel packing layer, record the height of the gas–water interface in the wellbore annular by a camera, and collect and measure the volume of the sand deposited in the annulus at the end of the experiment.

The flow pattern of gas–water–sand in the annulus of the sand control section of marine hydrate wells mainly includes the transition law of gas–water interface height, the deposition law of formation sand, and the changes in the structure of the gravel packing layer. To analyze the dynamic transition law of the gas–water interface, the decrease and rising rate of the gas–water interface height were proposed for characterization (Equations (1) and (2)). Considering the influence of the size of the wellbore and the screen on the dynamic change law of the gas–water interface in the annulus of the screen and casing, the gas–water interface height firstly performs a dimensionless (the ratio of the gas–water interface height to the annulus diameter) to make the experimentally obtained results universal. And then calculate the dimensionless gas–water interface height variation rate. In addition, a deposition volume ratio was proposed for the characterization to analyze the deposition pattern of formation sand in the annulus of the sand control well section (Equation (3)).

$$V_{\mathrm{D}} = \frac{H_{\mathrm{A}} - H_{\mathrm{B}}}{D_{\mathrm{A}} T_{\mathrm{B}}} \tag{1}$$

where V_{D} is the falling rate of the gas–water interface height, s^{-1}; H_{A} is the initial gas–water interface height in the annulus, m; H_{B} is the lowest gas–water interface height in the annulus, m; D_{A} is the size of the annulus between screen and casing, m; and T_{B} is the time when the height of the gas–water interface in the annulus reaches its lowest position, s.

$$V_{\mathrm{U}} = \frac{H_{\mathrm{C}} - H_{\mathrm{B}}}{D_{\mathrm{A}}(T_{\mathrm{C}} - T_{\mathrm{B}})} \tag{2}$$

where V_{U} is the rising rate of the gas–water interface height, s^{-1}; H_{C} is the highest gas–water interface height in the annulus, m; and T_{C} is the time when the height of the gas–water interface in the annulus reaches its maximum position, s.

$$R_{\mathrm{SV}} = \frac{V_1}{V_0} \tag{3}$$

where R_{SV} is the deposition volume ratio, dimensionless; V_1 is the total volume of the formation sand added during the experiment, mL; and V_0 is the volume of the formation sand deposited in the annulus of the wellbore after the experiment, ml.

The permeability was calculated by Darcy's formula using pressure and flow rate data to analyze the plug situation of the screen and gravel packing layer after the experiments. To maintain steady flow conditions to maintain the accuracy of the above parameter calculations, constant water and gas flow rates were maintained during the experiments through the use of variable frequency pumps and gas flow controllers. Because of the interference of electrical and magnetic signals, there are some slight fluctuations in both liquid and gas flow rates, but, overall, they are stable. The steady flow conditions ensure the accuracy of the above parameter calculations.

4. Results

4.1. Inflow Pattern in SAS Completion

4.1.1. Situation of Vertical SAS Well

(1) Gas–water interface

Figure 7 shows that gas–water stratification occurs in the annulus of the screen and casing in the NGH vertical well with SAS completion when the formation sand is carried by the water and gas into the wellbore and flows in the sand control well section. The gas is in the upper layer, and the water and sand are in the lower layer, with the water being the main sand-carrying phase. Various regions of the screen were gradually plugged as the experiment progressed, resulting in a dynamic transition in the height of the gas–water interface and the sand deposition in the bottom of the well.

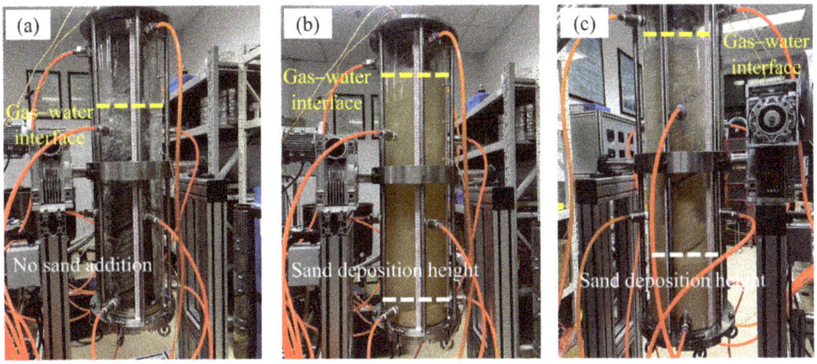

Figure 7. Dynamic variation in gas–water interface at different stages: (**a**) initial stage without sand addition; (**b**) initial plugging stage; and (**c**) final plugging stage.

Interestingly, experimental results show that the height of the gas–water interface decreases first and then increases with the screen plugging in some experimental conditions. Figure 8 shows that the variation law of the gas–water interface is significantly affected by the GWR. The annulus only contains gas and liquid without a sand addition at the beginning of the experiment. The resistance of the water and gas phase through the screen is equal, resulting in a dynamic balance of the gas–water interface and a constant initial height of the gas–water interface. A larger GWR strengthens the compression effect of the gas on the liquid, resulting in a lower initial height of the gas–water interface.

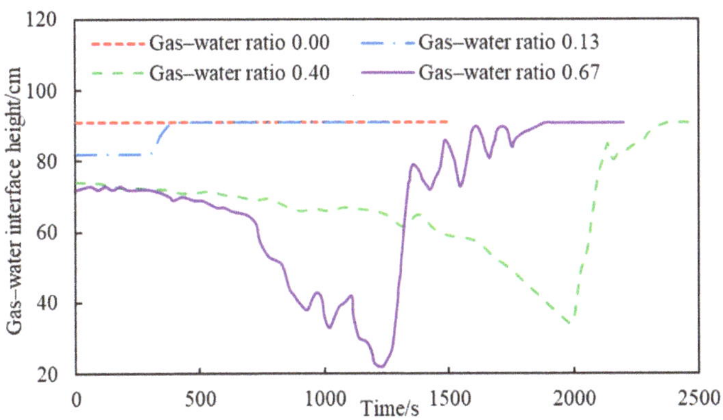

Figure 8. Variation in gas-water interface in different GWR conditions.

However, the balance state of the gas–water interface is destroyed because of gradual screen plugging when adding formation sand into the well, which leads to variation in gas and liquid flow resistance. Figure 8 shows that the height of the gas–water interface decreases and then increases when there is a high GWR. The rise and fall of the gas–water interface depends on the magnitude of the relative resistance of gas and water to passing through the screen. The gas flow resistance exceeds the water flow resistance with the plugging of the screen, requiring a larger flow space for gas flow. This action results in a fall of the gas–water interface, which weakens the degree of screen plugging and reduces gas flow resistance. However, the water requires more space to flow if the gas flow resistance is lower than that of water, resulting in a rise in the gas–water interface in the wellbore annulus. The compression of the gas in the water is weak when the GWR is low, resulting in a high initial gas–water interface. The water flow resistance was greater than the gas flow resistance with the plugging of the screen. The gas–water interface gradually rises, and there is no fall phenomenon of the gas–water interface during the experiment.

(2) Influencing factors

During NGH reservoir development, factors such as the VCSP in the NGH reservoir, production conditions (GWR), and the NGH well type significantly affect the degree of screen plugging, which in turn affects the inflow law of gas–water–sand in the sand control section of NGH well.

Figure 9 shows that the falling and rising rates of the gas–water interface are positively correlated with the VCSP, which both increase gradually with the increase in the VCSP. The quantity of formation sand entering the wellbore also increases as the VCSP increases, which increases the probability of formation sand deposition on the screen surface and may easily lead to screen plugging.

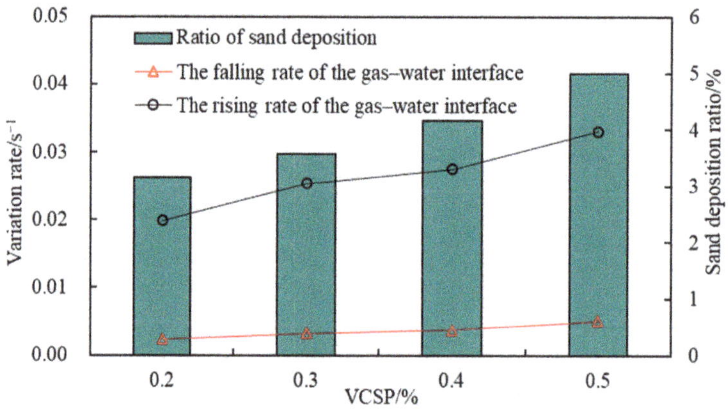

Figure 9. Variation rate of gas–water interface under different VCSP.

In the falling stage of the gas–water interface, the flow resistance of gas through the screen is greater than that of water through the screen. The high VCSP exacerbates the screen plugging. Consequently, an increase in the VCSP will result in an increase in the gas flow resistance through the screen, which will lead to a corresponding increase in the rate of gas–water interface decrease. In the rising stage of the gas–water interface, the flow resistance of water through the screen is larger than that of gas. The higher the VCSP is, the more space is needed for water to flow, which leads to a larger rate of gas–water interface increase.

In addition, Figure 9 also shows that higher VCSP hydrate reservoirs can exacerbate the deposition of formation sand in the annulus of the screen and casing in the hydrate well. Sand burial may occur in a hydrate well under prolonged production conditions, severely reducing gas production. Therefore, while ensuring that the screen has better sand control performance, the appropriate relaxation of sand control precision should be considered to avoid a sand burial phenomenon when designing the sand prevention parameters for hydrate reservoirs with a high VCSP.

The area and degree of screen plugging change with the variation in the gas–water interface. Figure 10 shows the relationship between the VCSP and screen permeability during gas–water–sand inflow. The screen permeability begins to decline earlier as the VCSP increases. The higher the VCSP, the more formation sand enters the wellbore at the same time, which leads to the easier plugging of the screen and the easier loss of screen permeability. Therefore, the VCSP is also a factor to be considered in the design of sand control parameters for hydrate reservoirs because the VCSP significantly affects the sand control effect of the screen. The accuracy of sand control should be appropriately reduced so that the screen can maintain high permeability performance for hydrate reservoirs with high sand concentration, which contributes to improving the hydrate well production.

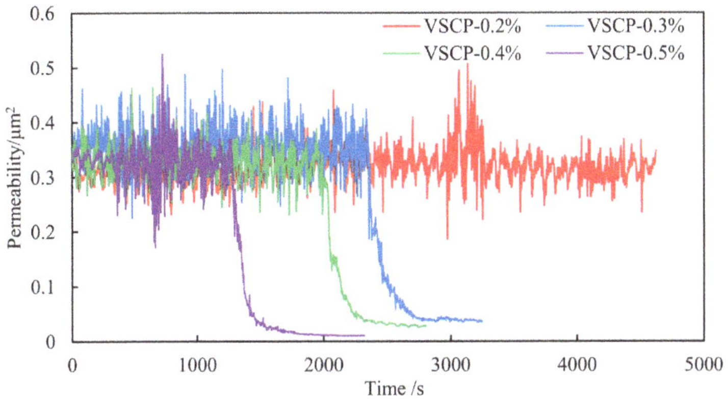

Figure 10. The curve of screen permeability variation with the VCSP.

When the gas–water–sand flows in the wellbore, the perturbing effect of the gas affects the variation in the gas–water interface and the formation sand deposition. Figure 11 shows the relationship between the GWR and the variation rate of the gas–water interface and the formation sand deposition. There is no gas–water stratification in the absence of gas injection into the wellbore. The gas-to-liquid compression is enhanced as the GWR increases, resulting in a larger falling rate of gas–water interface height. However, in the rising stage of gas–water interface height, the rising rate of the gas–water interface decreases first and then increases with the increase in GWR. The gas perturbation can lift part of the screen plugging when in a low GWR, reduce the sand deposition rate, and reduce the flow resistance of water through the screen, thus reducing the rising rate of the gas–water interface height. However, the gas–water interface rising rate will not always decrease with the increase in GWR. When the GWR is large, the gas carrying the sand effect gradually enhanced, resulting in strengthening the screen plugging degree and increasing the sand deposition rate and water flow resistance through the screen, which leads to increasing the gas–water interface rising rate.

A reasonable GWR can help to reduce screen plugging, and Figure 12 illustrates the relationship between the GWR and screen permeability. When the fluid entering the wellbore contains only formation sand and water, the screen is the first to plug, resulting in a permeability decrease in the screen. It takes longer for the screen permeability to decrease as the gas is injected, indicating that gas injection does have the effect of unplugging the screen. However, it is not the case that a higher GWR necessarily results in a more effective unplugging effect. Because, under the conditions of a GWR of 0.40 and 0.67, the time for permeability to decrease is 2450 s and 1600 s, respectively. Compared with the two conditions, under the condition of the GWR of 0.67, the screen is more likely to be plugged, resulting in an earlier decrease in permeability. When the ratio of gas to water reaches a certain threshold, the sand-carrying effect of the gas gradually intensifies, resulting in a gradual accumulation of formation sand on the surface of the screen, which leads to increased screen plugging and a greater loss of screen permeability. Therefore, controlling a reasonable hydrate production can help to improve the permeability performance of the screen tube, thereby increasing the hydrate production in actual NGH production tests.

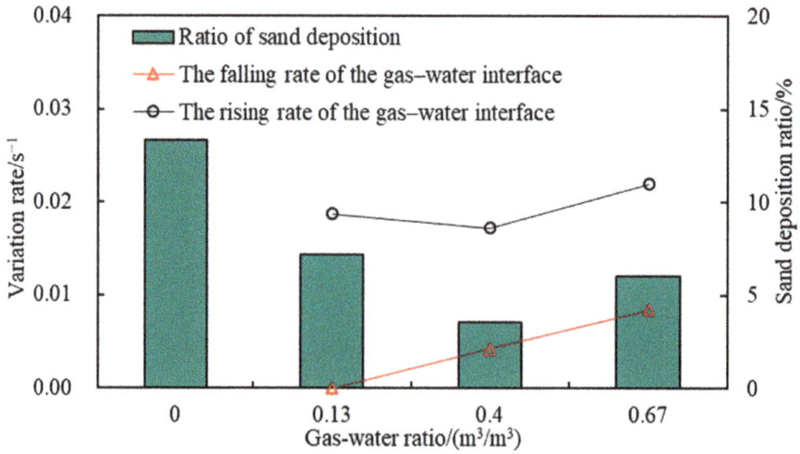

Figure 11. Variation in gas-water interface in different GWR conditions.

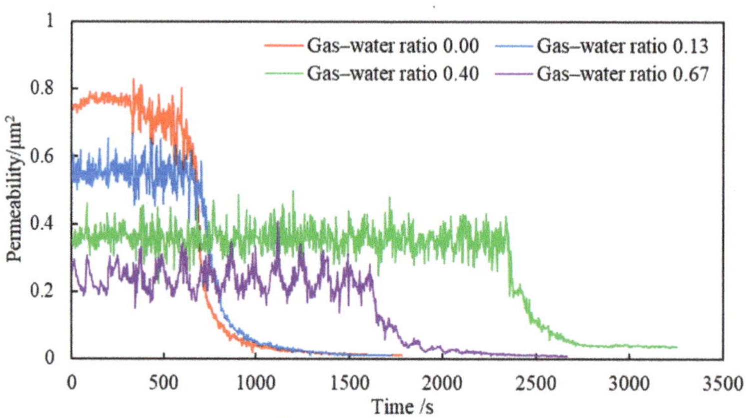

Figure 12. Variation in gas-water interface in different GWR conditions.

The pattern of gas–water–sand inflow varies significantly in different types of hydrate wells. Figure 13 shows the relationship between the deviation angle and variation rate of the gas–water interface and sand deposition ratio. Six deviation angles representing six types of hydrate wells were studied in this article. Both the gas–water interface rising rate and the gas–water interface falling rate decrease with the increase in the deviation angle, and the sand deposition rate increases with the rise in the deviation angle. As for the vertical hydrate well (deviation angle 0°), the screens located in the water phase are uniformly plugged by the formation sand. When the deviation angle increases, the formation sand is easily deposited at the bottom of the wellbore under the influence of gravity, resulting in the lower region of the screen maintaining a higher permeability and the overall degree of plugging of the screen being weakened. Therefore, the flow resistance of water through the screen decreases with the increase in the deviation angle in the rising stage of the gas–water interface, resulting in the rising rate of the gas–water interface decreasing with the rise in the deviation angle.

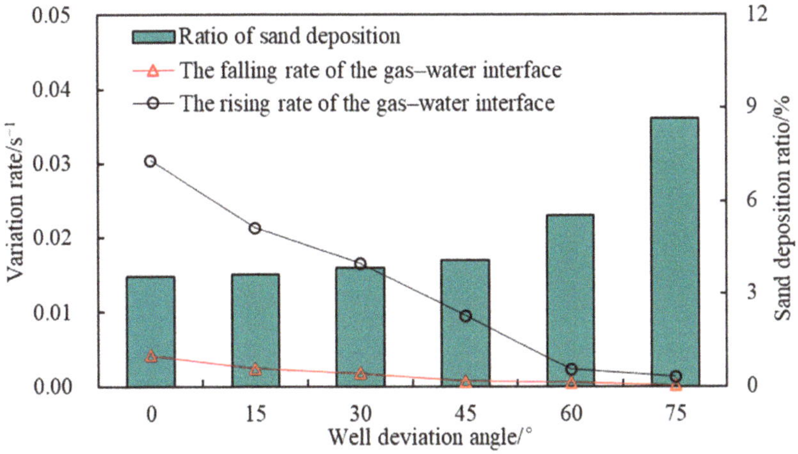

Figure 13. Variation in gas–water interface in different well deviation angle conditions.

The difference in the degree of screen plugging when gas–water–sand inflows in different types of hydrate wells is shown in Figure 14. The screen in a vertical well (at a 30° deviation angle) has the earliest decrease in permeability, indicating that the screen is most prone to plugging when the hydrate well is vertical. Theoretically, an increase in the deviation angle will cause the formation sand to be deposited at the bottom of the wellbore under the influence of gravity, reducing the degree of screen plugging. The larger the deviation angle, the weaker the degree of screen plugging. However, this is not the case in the experimental results. When the deviation angle is less than 30°, the decrease time of screen permeability gradually becomes longer with the increase in the deviation angle. However, the time becomes shorter with the increase in the deviation angle when the deviation angle is greater than 30°. This result suggests that there is a critical deviation angle in the range of 30–45°, which can weaken the degree of screen plugging, but the specific deviation angle needs further research.

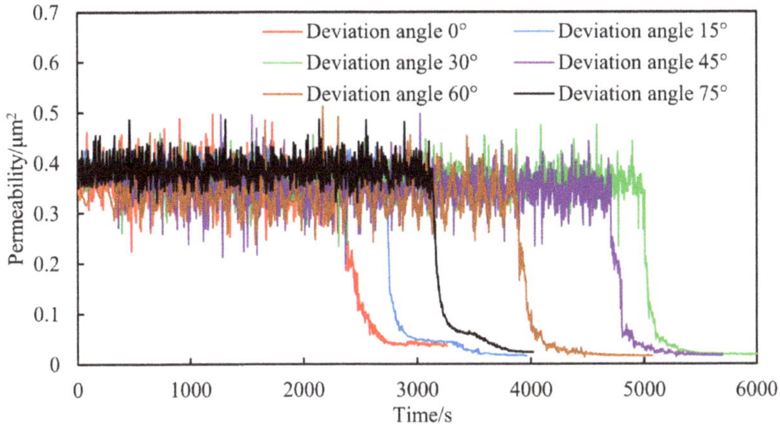

Figure 14. Variation in gas–water interface in different well deviation angle conditions.

4.1.2. Gas-Water–Sand Inflow in Horizontal SAS Well

For the horizontal NGH well with SAS completion, the same gas–water stratification phenomenon occurs when the gas–water–sand inflow enters the wellbore, with the gas located in the upper space of the wellbore, the liquid in the lower space of the wellbore, and the formation sand distributed in the water (Figure 15). During the flow of gas and water–sand in the wellbore, the lower part of the screen tube is first blocked by formation sand, which then leads to a gradual increase in the gas–water interface.

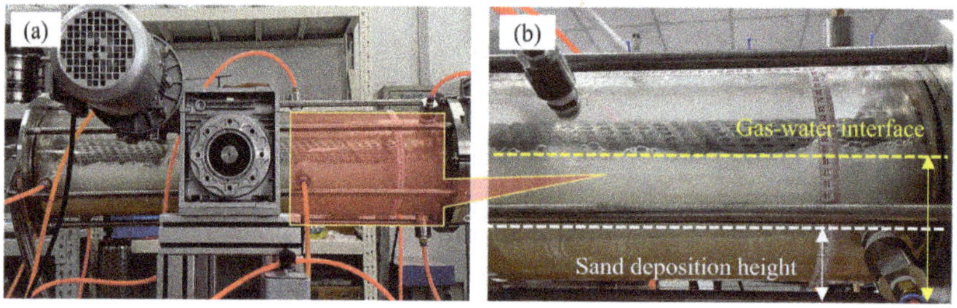

Figure 15. Gas–water stratification phenomenon: (**a**) macro view; and (**b**) zoom-in view.

For the horizontal and vertical NGH wells with SAS completion, although gas–water stratification occurs in both wells, the height of the gas–water interface in the wellbore shows a different pattern of variation. Figure 16 shows the varying law of the gas–water interface in a horizontal well. In horizontal wells with SAS completion, with the plugging of the screen, the height of the gas–water interface gradually rises until the liquid fills the annulus of the wellbore, and the height of the gas–water interface does not show any decreasing phenomenon. Because of the influence of gravity, the formation sand is not easy to be deposited on the surface of the lower part of the screen, which leads to the screen located in the water part of the screen being not easily clogged. Throughout the experiment, the liquid flow resistance is always greater than the gas flow resistance, resulting in a gradual increase in the height of the gas–water interface.

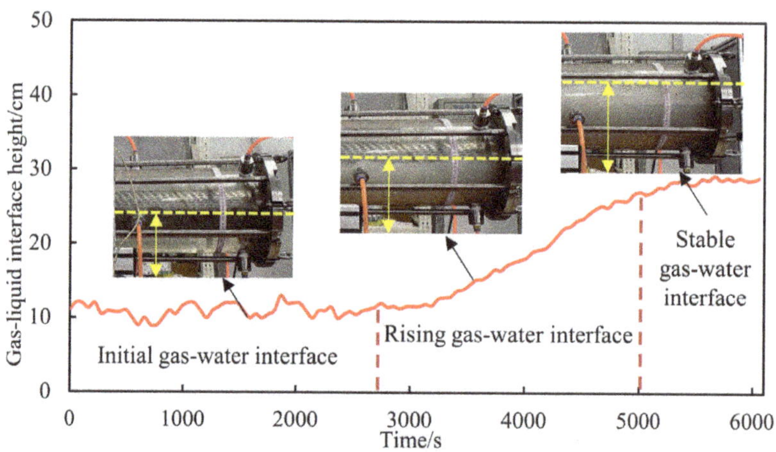

Figure 16. Varying law of gas–water interface in a horizontal well.

To further investigate the correlation between the screen plugging process and the gas–water interface change process, calculate and analyze the change rule of the screen permeability in the experimental process (Figure 17). The initial permeability of the screen is 0.33 D. When the experiment reaches 2817 s, the blockage of the screen begins to intensify, the pressure difference inside and outside the screen begins to increase, and the permeability begins to decrease. The phenomenon of phased changes in pressure and screen permeability was similarly demonstrated in other studies [30,48]. Figures 16 and 17 also show that, in the early stage of the experiment (0–2817 s), the height of the gas–water interface remained stable when the screen tube was not blocked. When the experiment progressed to 2817 s, the height of the gas–water interface began to rise after the screen tube was blocked, and the liquid eventually filled the annulus. The height of the gas–water interface reached its maximum value, and the final permeability of the screen tube was 0.013 D.

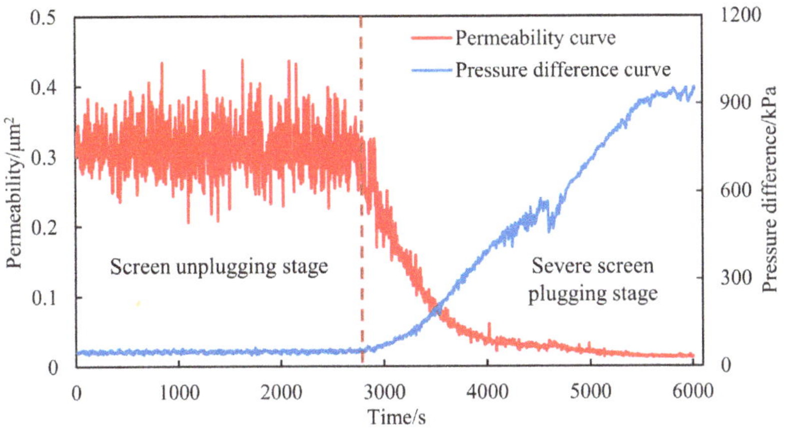

Figure 17. Varying law of screen permeability and pressure difference.

4.1.3. Mechanism of Gas–Water Interface Transition

When no sand is added at the beginning of the experiment, the height of the gas–water interface in the annulus remains unchanged. The relative resistance of liquid and gas passing through the screen is equal, and the gas–water interface is in dynamic equilibrium (Figure 18a,d). When the relative resistance of gas flow is greater than that of liquid flow, gas flow requires more flow space. The gas in the upper space of the wellbore annulus compresses the liquid, causing the height of the gas–water interface to begin to decrease (Figure 18b,e). When the relative resistance of gas flow is smaller than that of liquid flow, liquid flow requires a larger flow space, and the height of the gas–water interface in the wellbore annulus begins to rise (Figure 18c,f). When the GWR is low, the compression of the gas in the liquid is weak, resulting in a high initial gas–water interface. In the process of screen plugging, the liquid flow resistance has been greater than the gas flow resistance. The gas–water interface gradually increases as the experiment progresses, and there is no phenomenon of the gas–water interface decreasing during the experiment.

In horizontal wells with SAS completion, with the plugging of the screen, the height of the gas–water interface gradually rises until the liquid fills the annulus of the wellbore, and the height of the gas–water interface does not show any decreasing phenomenon (Figure 19). Because of the influence of gravity, the formation sand is not easily deposited on the surface of the lower part of the screen, which leads to the screen located in the water part of the screen being not easily clogged. Throughout the experiment, the liquid flow resistance is always greater than the gas flow resistance, resulting in a gradual increase in the height of the gas–water interface.

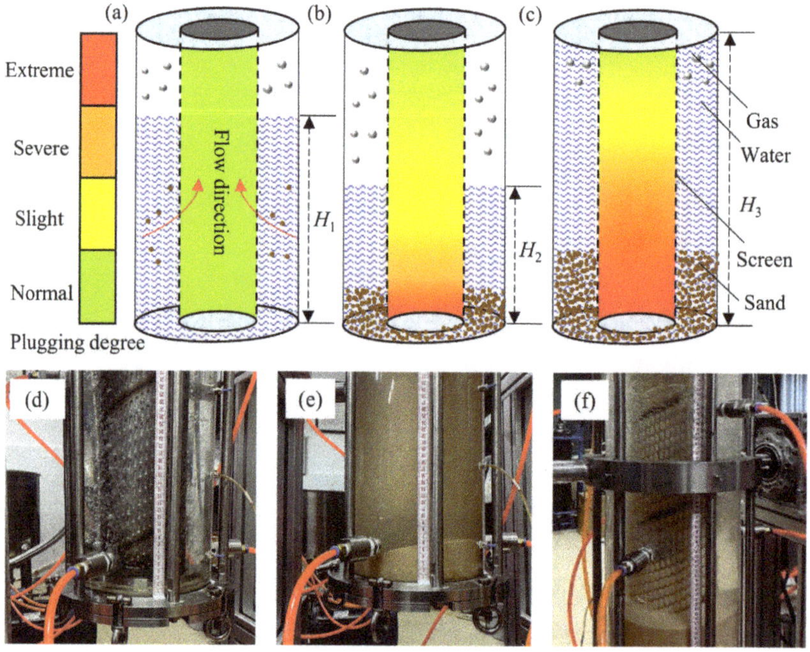

Figure 18. Schematic diagram of the mechanism of gas–water interface transition in the vertical well: (**a**,**d**) unplugged stage; (**b**,**e**) initial plugging stage; and (**c**,**f**) final plugging stage.

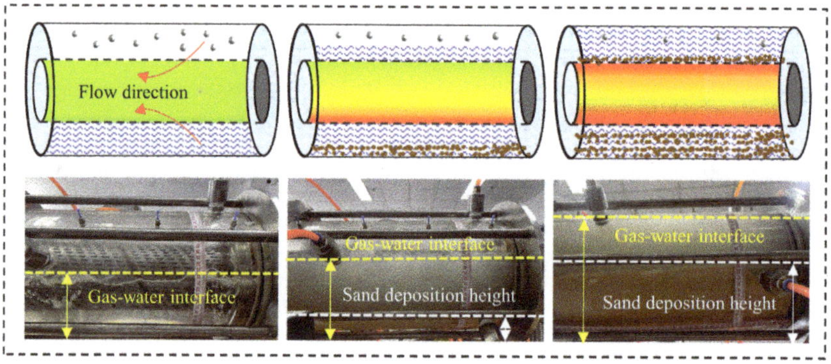

Figure 19. Diagram of the mechanism of gas–water interface transition in the horizontal well.

4.2. *Inflow Pattern Analysis of GP Well*

4.2.1. Uniform Inflow Pattern

The screen and casing annulus of the hydrate well with GP completion are packed with 40–70 mesh conventional ceramsite, resulting in a different gas–water–sand inflow pattern compared with SAS completion. Figure 20 shows the gas–water–sand inflow pattern in hydrate wells with GP completion. The gas–water–sand inflows uniformly both in the vertical and horizontal wells, and there is no gas–water stratification flow phenomenon.

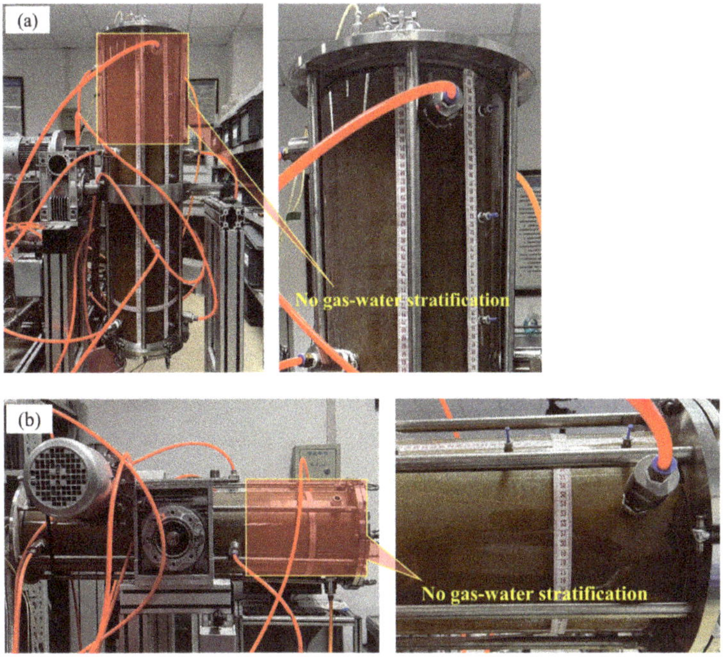

Figure 20. The uniform inflow of gas–water in vertical and horizontal wells with GP completion: (**a**) vertical well with GP completion; (**b**) horizontal well with GP completion.

Because of the 40–70 mesh conventional ceramsite filling in the wellbore, the gas–water–sand inflow pattern in the gravel layer was not photographed. Therefore, a schematic diagram was drawn to visualize the gas–water–sand inflow pattern in the NGH well with GP completion, as shown in Figure 21. Compared with the SAS completion, the gravels in the GP completion form a porous 3D structure, which reduces the flow area of gas–water–sand and leads to the rapid filling of water into the pores between the gravels, and the gas–water–sand flows uniformly in the gravel layer without gas–water stratification. When the gas–water–sand flows uniformly in the gravel layer, the gas is mainly in the form of bubbles in the gravel layer, and the fine sand gradually enters the inside of the screen through the gravel layer, and the coarse sand gradually invades and remains in the inside of the gravel layer.

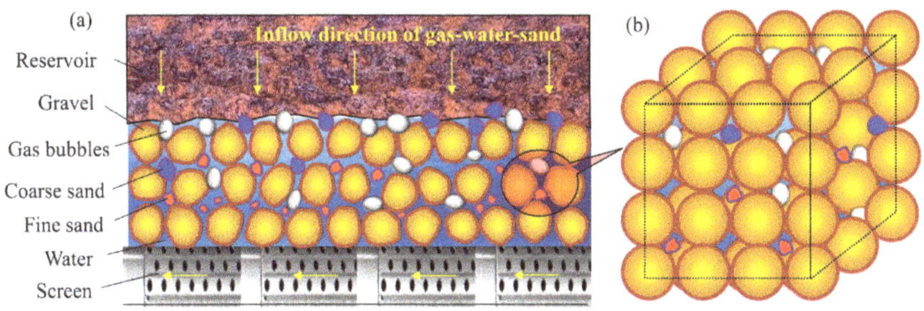

Figure 21. Schematic diagram of uniform inflow of gas–water–sand in GP completion: (**a**) macro view; and (**b**) zoom-in view.

4.2.2. Erosion Failure of Packed Gravel

Although gas–water stratification did not occur in the GP completion, it is interesting to note that the gravel layer structure was progressively damaged, and an erosion hole was formed as the gas–water–sand flowed in the gravel layer (Figure 22a).

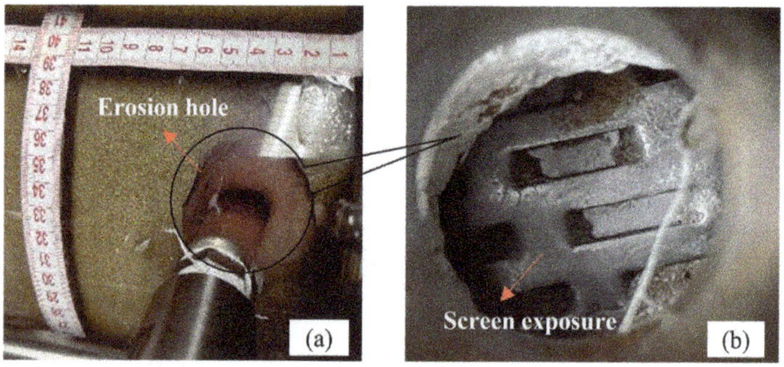

Figure 22. Formation and extension of erosion hole in the gravel layer: (**a**) formation of erosion hole; and (**b**) expansion of hole leads to screen exposure.

The formation sand is controlled on the outside of the gravel layer at the location of high gravel packing degree. However, the gravel layer is destabilized and reorganized by the impact of gas and water, and a hole is formed at the location of low gravel packing degree. Once an erosion hole is formed in a gravel layer, the gravel packing degree near the erosion hole decreases sharply. The volume (depth) of the erosion hole gradually increases under the continuous impact of air and water, weakening the sand control effect of the gravel layer. With the expansion of the erosion hole, the deficit volume of the gravel layer gradually increases, which gradually leads to the exposure of the screen inside the gravel layer (Figure 22b). Then, the gas–water–sand passes through the screen directly, and the sand control performance of the gravel layer decreases sharply.

The GWR significantly affects the formation of erosion holes in the gravel layer. Three GWRs (0.9, 1.8 and 2.7 m^3/m^3) were researched in this article. Figure 23 shows the relationship between the GWR and erosion hole in the gravel layer. It can be observed that an increase in the GWR results in a corresponding increase in the non-uniform inflow velocity of the gas–water–sand, which enhances the impact on the gravel filling layer and results in a more serious gravel destabilization and larger formation of erosion holes.

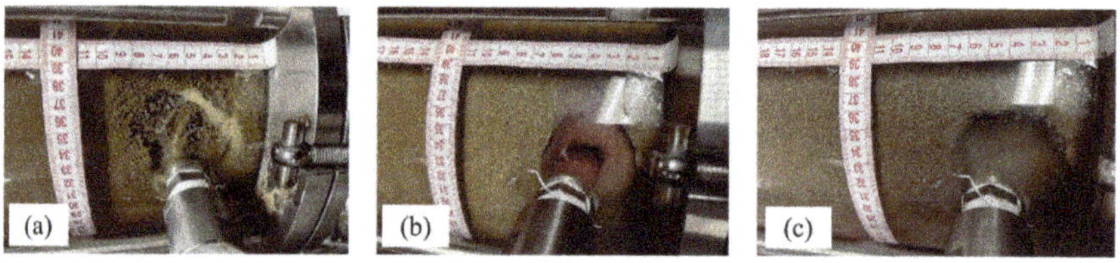

Figure 23. Variation in erosion hole in the gravel layer with GWR: (**a**) GWR is 0.9 m^3/m^3; (**b**) GWR is 1.8 m^3/m^3; and (c) GWR is 2.7 m^3/m^3.

The phenomenon of erosion holes in the instability of gravel layers was also confirmed in the studies of some scholars. The effect of gravel packing degree on the stability of the gravel layer was studied. Figure 24 shows the relationship between the gravel packing degree and the shape and depth of the erosion hole in the gravel layer. When using 40–70 mesh conventional ceramsite for experiments, the results show that the formation of erosion holes in the gravel layer is inevitable, even in the case of 100% packing degree (Figure 24a). Figure 24b also shows that the erosion holes gradually expanded and even broke through the gravel layer to the outer wall of the screen with a decrease in the packing degree [40].

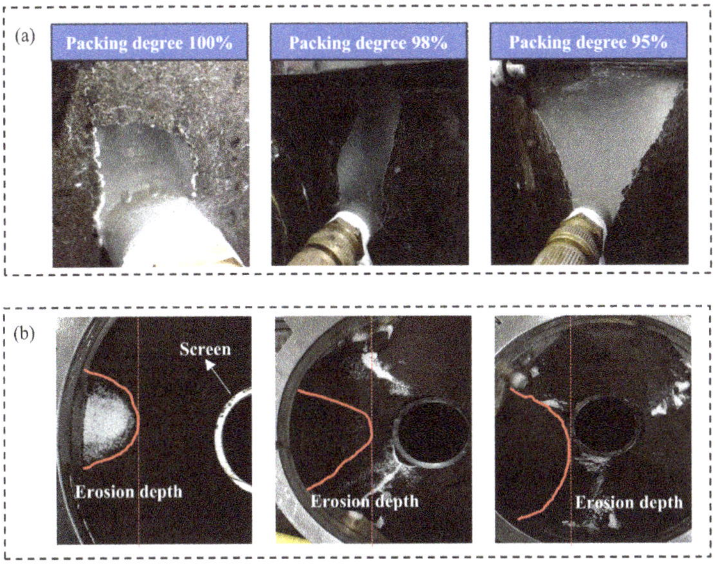

Figure 24. The shape and depth of the erosion hole of the gravel layer with different packing degrees: (**a**) the shape of the erosion hole; (**b**) the depth of the erosion hole. Reproduced from [40], with permission from North China Petroleum Administration Bureau/2018.

However, it is impossible to guarantee a 100% packing degree of gravel during the actual hydrate well development process. Furthermore, the presence of a high-speed inflow hotspot in the whole hydrate well will exacerbate the destabilization and damage of the gravel layer, which will lead to the sand control failure of the gravel layer. The screen should play a key role in controlling sand when the gravel layer fails. Therefore, when designing the GP completion, the sand control accuracy of the screen should be based on the principle that the screen directly controls the formation sand, rather than just supporting the gravel layer.

4.2.3. Plugging Performance

Figure 25 shows the variation in gravel layer permeability from unplugging to the destabilized state under different GWRs. When the gas–water–sand flows in the gravel layer, at the beginning of the experiment, the fine sand passes through the gravel layer directly (Unplugging stage). The gravel layer was gradually plugged, resulting in a decrease in the permeability of the gravel layer with the continuous intrusion of formation sand in the gravel layer (Plugging stage). The resistance to the flow of gas and water through the gravel layer increases after the plugging of the gravel layer. The gravel is displaced to form an erosion under the impact of gas and water, resulting in a gradual increase in the permeability of the gravel layer (Erosion hole formation stage).

The permeability of gravel layers at different stages under different GWRs, the rate of permeability loss, and the rate of permeability recovery after the formation of the erosion hole are summarized in Table 2. Table 2 shows that the permeability of the gravel layer after plugging is higher with the increase in GWR. Because the gas has a certain unplugging effect, the formation sand that enters the inside of the gravel layer is easily discharged under the disturbing effect of the gas. Therefore, the lower the permeability loss rate of the gravel layer as the GWR increases. However, the impact of gas and water on the gravel layer is strengthened as the GWR increases. A high GWR accelerates the formation of erosion holes in the gravel layer and intensifies the degree of erosion damage, leading to an increase in the permeability of the gravel layer and a weakening of the sand control performance. As the GWR increases, the recovery rate of permeability of the gravel layer rises (Figure 26), while the sand control performance of the gravel layer declines.

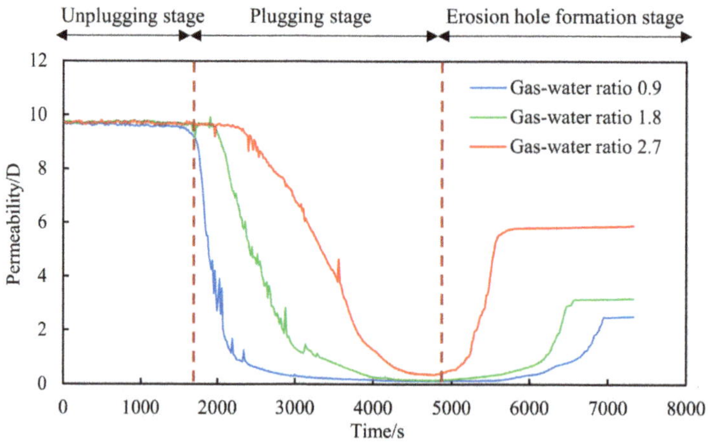

Figure 25. The curve of gravel permeability variation with the GWR.

Table 2. Permeability of gravel layer at different stages under different GWR.

GWR/(m^3/m^3)	Initial Permeability/D	Plugging Permeability/D	Permeability with Erosion Hole/D	Recovery Rate/%	Loss Rate/%
0.9	9.723	0.132	2.539	1823.48	98.64
1.8	9.719	0.154	3.196	1975.32	98.42
2.7	9.727	0.279	5.936	2027.60	97.13

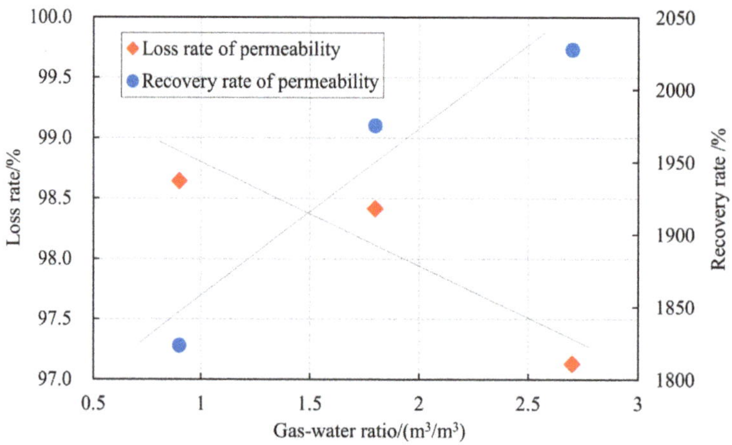

Figure 26. Variation in gravel layer permeability loss and recovery rates with GWR.

4.3. Optimization of Completion

Under the long-term marine NGH production, the sand control performance of different sand control completions changes dynamically under the impact of gas–water–sand inflow, which limits the development and utilization of marine NGH energy. These findings above about the inflow pattern of gas–water–sand in the sand control section of marine hydrate wells with different completions provide good inspiration for optimizing the design of completions for different types of marine NGH wells. Figure 27 shows an optimization diagram of different sand control completion methods.

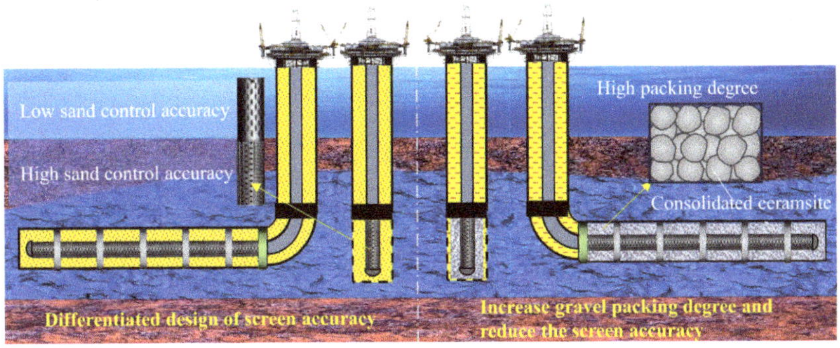

Figure 27. Optimization diagram of different sand control completions.

For vertical and horizontal wells with SAS completion, the gas–water stratification during the inflow of gas–water–sand in the wellbore leads to non-uniform plugging of the screen. After a long time of production of the hydrate well, the screen located in the water phase is plugged seriously, while the screen located in the gas phase maintains high permeability. However, the current SAS sand control completion for hydrate wells does not differentiate the sand control accuracy of the screen according to the difference in the plugging degree at different positions of the screen but adopts a unified sand control accuracy as a whole, which is costly and limits the increase in hydrate well production.

Therefore, for hydrate wells with SAS completion, it is recommended that the sand control screen is optimized to have two different sand control accuracies for the upper and lower parts of the screen, taking into account the non-uniform plugging phenomenon of the screen. The lower screen plays the role of sand control with high sand control accuracy. The upper screen is designed with a low sand control accuracy to enhance the gas conductivity, which contributes to improving the hydrate production.

Regarding the GP completion, the gravel layer destabilizes and forms an erosion hole at the high-speed inflow hotspots of the NGH reservoir, resulting in the exposure of the screen inside the gravel layer and the sand control failure of the gravel layer. Maintaining the stability of the gravel layer is beneficial to prolonging the validity of the GP completion. However, in the current hydrate production tests using GP sand control completion, to avoid high packing pressure from damaging the hydrate reservoir, lightweight ceramsite is used for gravel packing [49,50]. However, the low-density lightweight ceramsite will lead to a decrease in the gravel packing degree, which will sharply weaken the stability of the gravel packing layer.

Therefore, for hydrate wells with GP completion, the density of gravel should be increased as much as possible under the premise of ensuring the stability of the NGH reservoir. In addition, the coated ceramic with cementing effects can enhance the strength of the gravel layer to improve the stability of the gravel layer. Once the gravel layer is damaged, the screen inside the gravel layer will be the last sand control barrier of the GP completion. However, when designing the sand control accuracy of the screen in the GP completion, it is generally designed based on the ability to support gravel, and the sand control effect of the screen is not considered. The reasonable design of the sand control accuracy of the screen inside the gravel layer is conducive to prolonging the validity of the GP completion. Therefore, when designing the sand control accuracy of the screen, it is not only necessary to consider the supporting gravel, but also, more importantly, it should be considered that the screen can directly control the formation sand.

5. Discussion

The bottom of the marine NGH well is a complex three-phase fluid environment of gas, water, and sand. Complex fluids cause dynamic changes in different sand control completion systems. For SAS and GP sand control completion systems, the phenomena of uneven plugging of screens caused by gas–water stratified flow and the erosion failure of gravel layers caused by gas–water–sand impact observed in indoor experiments are objective. Because indoor physical experiments are affected by many factors such as experimental operators [51,52], signal interference, experimental time, and device scale, the laws of gas–water interface change, gravel layer instability may not be accurate, and the analysis metrics may have certain errors. Before the experiment, we conducted multiple test experiments to reduce the deviation caused by the signal interference and experimental device. During the experiment, we did not change the experimental operators to eliminate the deviation caused by the operators. In addition, outliers with large fluctuations caused by different factors were removed without affecting the regularity of the curves. The scale of indoor experiments is also a key factor that affects whether the experimental results can be applied to the actual field. Because indoor experiments cannot completely restore the actual field conditions, in indoor experimental research, based on the structural parameters of the experimental device and the actual hydrate well, the flow rate equivalent method is used to simulate the actual field production conditions, which enhances the correlation between indoor experimental results and the actual field. Therefore, the sand control completion optimization method based on experimental phenomena and experimental laws has a certain guiding significance for the actual field.

The gravel size optimization method proposed for GP sand control completion is different from the traditional gravel size design based only on the role of supporting screens [40,53] and is technically easier to achieve. In addition, many research studies have been conducted on how to improve the gravel-packing effect and avoid the formation of premature bridges during gravel packing [54–56]. The proposed gravel size design method will be more conducive to improving the validity period of gravel packing sand control completion in hydrate wells. For the SAS sand control completion, the traditional screen sand control accuracy design method generally does not consider the dynamic changes in the SAS completion system, and the screen sand control accuracy is designed based on static parameters such as the formation sand particle size [29]. The accuracy optimization method proposed in the paper is to design the screen sand control accuracy according to the plugging degree difference in different areas of the screen. Reduce the sand control accuracy in places where the screen plugging is light to increase the gas production. Although the idea is difficult to realize and will increase the operating cost, it is conducive to promoting the commercialization of NGH.

The research method in this paper is an indoor experimental study. Although we found interesting experimental phenomena such as the uneven plugging of screens caused by the gas–water stratified flow and the erosion failure of gravel layers caused by gas–water–sand impact, the mechanism of the above phenomena can be explained through qualitative analysis. However, there is a lack of quantitative analysis of the degree of uneven plugging of the screen and the erosion of gravel layers. In subsequent research, numerical simulation research can be considered to quantitatively analyze the changing laws of the above phenomena [57]. In addition, the optimized design method of the sand control completion system proposed based on the results of the indoor experiments will indeed increase the operation cost, but, at the same time, it will also increase the benefits of NGH production and promote the development and utilization of marine NGH resources. However, it is not clear whether the increased benefits will make up for the increased costs. In the future, it is necessary to carry out indoor research to evaluate the technical feasibility. Then, carry out actual field operation tests to evaluate the economic feasibility. Because field operations will cost a significant amount, indoor experiments and numerical simulation research on technical feasibility evaluation will be the focus of subsequent research.

6. Conclusions

This study investigated the inflow process of gas–water carrying formation sand into the sand control section of the marine NGH wellbore after NGH decomposition through physical experiments. The inflow laws and mechanisms of gas–water–sand in vertical and horizontal wells, respectively, using SAS completion and GP completion were analyzed, and a sensitivity factor analysis was conducted. Finally, suggestions are put forward for the sand control completion methods of the marine NGH well to promote the development of marine NGH energy. The following are the conclusions yielded from this study:

(1) The gas–water stratification occurs in horizontal and vertical wells with SAS completion. The gas is at the top of the wellbore, while the water carries sand at the bottom. The gas–water interface rises with the screen plugging. A high VCSP and low deviation angle exacerbated screen plugging, resulting in a faster rising rate of the gas–water interface. Moreover, the gas perturbation helps to unplug the screen when in a small GWR, but a high GWR increases the amount of sand suspended in the water, exacerbating the plugging degree of the screen. Therefore, the rising rate of the gas–water interface decreases first and then increases with the increase in the GWR. The screen is unevenly plugged due to the influence of the gas–water stratified flow, and the lower part of the screen is severely plugged, while the upper part maintains high permeability.

(2) Gas–water–sand flow uniformly without the gas–water stratification phenomenon due to the complex pore-throat structure formed between the gravels in the horizontal and vertical wells with GP completion. However, the gravel rearranges and forms erosion holes under the impact of gas and water, and the erosion holes gradually expand, ultimately leading to the internal screen leakage of the gravel layer and the sand control failure of the GP completion. The stability of the gravel layer is significantly affected by the gravel packing degree and the GWR. The lower the gravel packing degree and the higher the GWR, the larger the erosion holes formed in the gravel layer.

(3) For the SAS completion method, because the gas–water stratification flow causes uneven plugging of the screen, it is recommended to design the sand control accuracy of the screen in two sections. Increase the sand control accuracy of the lower part of the screen to enhance sand control performance. Conversely, decrease that of the upper part of the screen to maintain high permeability performance. For the GP completion method, the gravel layer is prone to instability, leading to sand control failure. Increase the gravel density under the premise of ensuring the stability of the hydrate reservoir to enhance the gravel packing degree, and use the coated ceramic with cementing effect to improve the stability of the gravel layer. In addition, when designing the sand control accuracy of the screen, it is not only necessary to consider the supporting gravel, but also, more importantly, it should be considered that the screen can directly control the formation sand.

Author Contributions: Conceptualization, C.L. and C.D.; methodology, C.L. and H.S.; validation, C.L. and B.Y.; formal analysis, C.L.; investigation, B.Y. and Y.Y.; resources, H.S. and Y.Y.; data curation, B.Y.; writing—original draft preparation, C.L.; writing—review and editing, C.L. and C.D.; supervision, C.D.; project administration, Y.Y.; funding acquisition, H.S. and Y.Y. All authors have read and agreed to the published version of the manuscript.

Funding: This research is financially supported by the Guangdong Major Project of Basic and Applied Basic Research (No. 2020B0301030003) and the National Key Research and Development Program of China (No. 2023YFC2811005).

Institutional Review Board Statement: Not applicable.

Informed Consent Statement: Not applicable.

Data Availability Statement: The data that support the findings of this study are available from the corresponding author.

Acknowledgments: The authors gratefully acknowledge the experimental support and technical support provided by China University of Petroleum (East China).

Conflicts of Interest: The authors declare no conflicts of interest.

References

1. Sloan, E.D. Fundamental principles and applications of natural gas hydrates. *Nature* **2003**, *426*, 353–359. [CrossRef] [PubMed]
2. Zhang, W.; Liang, J.; Wei, J.; Lu, J.; Su, P.; Lin, L.; Huang, W.; Guo, Y.; Deng, W.; Yang, X. Geological and geophysical features of and controls on occurrence and accumulation of gas hydrates in the first offshore gas-hydrate production test region in the Shenhu area, Northern South China Sea. *Mar. Pet. Geol.* **2020**, *114*, 104191. [CrossRef]
3. Liu, L.; Sun, Z.; Zhang, L.; Wu, N.; Yichao, Q.; Jiang, Z.; Geng, W.; Cao, H.; Zhang, X.; Zhai, B. Progress in Global Gas Hydrate Development and Production as a New Energy Resource. *Acta Geol. Sin. Engl. Ed.* **2019**, *93*, 731–755. [CrossRef]
4. Yang, M.; Zhao, J.; Zheng, J.; Song, Y. Hydrate reformation characteristics in natural gas hydrate dissociation process: A review. *Appl. Energy* **2019**, *256*, 113878. [CrossRef]
5. Dong, C.; Gao, K.; Dong, S.; Shang, X.; Wu, Y.; Zhong, Y. A new integrated method for comprehensive performance of mechanical sand control screens testing and evaluation. *J. Pet. Sci. Eng.* **2017**, *158*, 775–783. [CrossRef]
6. Dilimulati, S.; Dong, C.; Zhan, X.; Li, J.; Cui, G.; Liu, Q.; Bai, H. Experimental Investigation of Solids Production Mechanisms in a Hydraulic Screen-Through Fracturing Well in a Loose Reservoir and Its Control. *SPE J.* **2024**, *29*, 1818–1831. [CrossRef]
7. Li, Y.L.; Ning, F.L.; Xu, M.; Qi, M.H.; Sun, J.X.; Nouri, A.; Gao, D.L.; Wu, N.Y. Experimental study on solid particle migration and production behaviors during marine natural gas hydrate dissociation by depressurization. *Pet. Sci.* **2023**, *20*, 3610–3623. [CrossRef]
8. Lu, J.S.; Li, D.L.; He, Y.; Shi, L.L.; Liang, D.Q.; Xiong, Y.M. Experimental Study of Sand Production during Depressurization Exploitation in Hydrate Silty-Clay Sediments. *Energies* **2019**, *12*, 4268. [CrossRef]
9. Uchida, S.; Klar, A.; Yamamoto, K. Sand production model in gas hydrate-bearing sediments. *Int. J. Rock Mech. Min. Sci.* **2016**, *86*, 303–316. [CrossRef]
10. Uchida, S.; Klar, A.; Yamamoto, K. Sand Production Modeling of the 2013 Nankai Offshore Gas Production Test. In Proceedings of the Energy Geotechnics, Kiel, Germany, 29–31 August 2016. [CrossRef]
11. Li, J.F.; Ye, J.L.; Qin, X.W.; Qiu, H.J.; Wu, N.Y.; Lu, H.L.; Xie, W.W.; Lu, J.A.; Peng, F.; Xu, Z.Q. The first offshore natural gas hydrate production test in South China Sea. *China Geol.* **2018**, *1*, 5–16. [CrossRef]
12. Wu, N.; Li, Y.; Chen, Q.; Liu, C.; Jin, Y.; Tan, M.; Dong, L.; Hu, G. Sand Production Management during Marine Natural Gas Hydrate Exploitation: Review and an Innovative Solution. *Energy Fuels* **2021**, *35*, 4617–4632. [CrossRef]
13. Hunter, R.B.; Collett, T.S.; Boswell, R.; Anderson, B.J.; Digert, S.A.; Pospisil, G.; Baker, R.; Weeks, M. Mount Elbert Gas Hydrate Stratigraphic Test Well, Alaska North Slope: Overview of scientific and technical program. *Mar. Pet. Geol.* **2011**, *28*, 295–310. [CrossRef]
14. Lei, H.; Yang, Z.; Xia, Y.; Yuan, Y. Prospects of gas production from the vertically heterogeneous hydrate reservoirs through depressurization in the Mallik site of Canada. *Energy Rep.* **2022**, *8*, 2273–2287. [CrossRef]
15. Boswell, R.; Schoderbek, D.; Collett, T.S.; Ohtsuki, S.; White, M.; Anderson, B.J. The Iġnik Sikumi Field Experiment, Alaska North Slope: Design, Operations, and Implications for CO_2–CH_4 Exchange in Gas Hydrate Reservoirs. *Energy Fuels* **2017**, *31*, 140–153. [CrossRef]
16. Yamamoto, K.; Terao, Y.; Fujii, T.; Ikawa, T.; Seki, M.; Matsuzawa, M.; Kanno, T. Operational overview of the first offshore production test of methane hydrates in the Eastern Nankai Trough. In Proceedings of the Offshore Technology Conference, Houston, TX, USA, 5–8 May 2014. [CrossRef]
17. Liu, Z.; Wang, Z.; Sun, J.; Chen, L.; Wang, J.; Sun, B. Risk and preventive strategies of hydrate reformation in offshore gas hydrate production trials: A case study in the Eastern Nankai Trough. *J. Nat. Gas Sci. Eng.* **2022**, *103*, 104602. [CrossRef]
18. Ye, J.L.; Qin, X.W.; Xie, W.W.; Lu, H.L.; Ma, B.J.; Qiu, H.J.; Liang, J.Q.; Lu, J.A.; Kuang, Z.G.; Lu, C. The second natural gas hydrate production test in the South China Sea. *China Geol.* **2020**, *3*, 197–209. [CrossRef]
19. Feng, Y.; Chen, L.; Suzuki, A.; Kogawa, T.; Okajima, J.; Komiya, A.; Maruyama, S. Numerical analysis of gas production from reservoir-scale methane hydrate by depressurization with a horizontal well: The effect of permeability anisotropy. *Mar. Pet. Geol.* **2019**, *102*, 817–828. [CrossRef]
20. Xu, Z.; Hu, T.; Pang, X.Q.; Wang, E.Z.; Liu, X.H.; Wu, Z.Y.; Chen, D.; Li, C.R.; Zhang, X.W.; Wang, T. Research progress and challenges of natural gas hydrate resource evaluation in the South China Sea. *Pet. Sci.* **2022**, *19*, 13–25. [CrossRef]
21. Chen, X.; Lu, H.; Gu, L.; Shang, S.; Zhang, Y.; Huang, X.; Zhang, L. Preliminary evaluation of the economic potential of the technologies for gas hydrate exploitation. *Energy* **2022**, *243*, 123007. [CrossRef]
22. Deng, F.; Wang, Y.; Li, X.; Li, G.; Wang, Y.; Huang, B. A model-based study of the evolution of gravel layer permeability under the synergistic blockage effect of sand particle transport and secondary hydrate formation. *Appl. Energy* **2024**, *355*, 122209. [CrossRef]
23. Wang, Z.; Zhang, Y.; Peng, Z.; Shan, Z.; Sun, B.; Sun, J. Recent Advances in Methods of Gas Recovery from Hydrate-Bearing Sediments: A Review. *Energy Fuels* **2022**, *36*, 5550–5593. [CrossRef]
24. Deng, F.; Huang, B.; Li, X.; Liu, J.; Li, G.; Xu, Y.; Yin, B. Review of Sand Control and Sand Production in a Gas Hydrate Reservoir. *Energy Fuels* **2022**, *36*, 11712–11723. [CrossRef]

25. Miri, R.; Salimi, M.; Stewart, S.; Gonzalez, J.M.B.; Suárez, C.M.; Nouri, A. Sand control screen selection for cased dual-annulus gas offshore wells based on scaled laboratory tests. *Can. J. Chem. Eng.* **2024**, *102*, 1957–1969. [CrossRef]
26. Wang, Y.; Long, F.Q.; Li, X.; Xu, Y.T.; Kou, X. Experimental Study of Sand Migration under Distinct Sand Control Methods during Gas Hydrate Decomposition by Depressurization. *Energy Fuels* **2023**, *37*, 12966–12979. [CrossRef]
27. Uchida, S.; Seol, Y.; Yamamoto, K. Sand Migration Simulation during Gas Production from Gas Hydrate Reservoir at Kuparuk 7-11-12 site in the Prodhoe Bay Unit, Alaska. *Energy Fuels* **2022**, *36*, 7382–7390. [CrossRef]
28. Dong, C.; Wang, L.; Zhou, Y.; Huang, F.; Song, Y.; Zhou, B.; Xu, H. Microcosmic retaining mechanism and behavior of screen media with highly argillaceous fine sand from natural gas hydrate reservoir. *J. Nat. Gas Sci. Eng.* **2020**, *83*, 103618. [CrossRef]
29. Ding, J.P.; Cheng, Y.F.; Yan, C.L. Research on sand control effect and micro-plugging mechanism of sand control medium in the development of natural gas hydrate reservoir. *J. Pet. Sci. Eng.* **2022**, *215*, 110703. [CrossRef]
30. Ma, C.; Feng, Y.; Deng, J.; Wang, W.; Li, S.; Chu, M.; Gui, Y. Experimental investigation of the plugging mechanisms of non-consolidated prepacked gravel screens. *Int. J. Hydrog. Energy* **2021**, *46*, 34638–34651. [CrossRef]
31. Wang, Y.; Cheng, K.; Yang, Y.; Tao, Y.; Li, Y. Microscopic mechanical analysis of sand production using a new arbitrary resolved-unresolved CFD-DEM model. *Int. J. Multiph. Flow* **2022**, *149*, 103979. [CrossRef]
32. Li, Y.L.; Wu, N.Y.; Ning, F.L.; Gao, D.L.; Hao, X.L.; Chen, Q.; Liu, C.L.; Sun, J.Y. Hydrate-induced clogging of sand-control screen and its implication on hydrate production operation. *Energy* **2020**, *206*, 118030. [CrossRef]
33. Gillespie, G.; Deem, C.K.; Malbrel, C. Screen Selection for Sand Control Based on Laboratory Tests. In Proceedings of the SPE Asia Pacific Oil and Gas Conference and Exhibition, Brisbane, Australia, 16–18 October 2000. [CrossRef]
34. Lee, J.; Ahn, T.; Lee, J.Y.; Kim, S.J. Laboratory Test to Evaluate the Performance of Sand Control Screens During Hydrate Dissociation Process by Depressurization. In Proceedings of the Tenth ISOPE Ocean Mining and Gas Hydrates Symposium, Szczecin, Poland, 22–26 September 2013. Available online: https://onepetro.org/ISOPEOMS/proceedings/OMS13/All-OMS13/ISOPE-M-13-035/25372 (accessed on 20 August 2024).
35. Ding, J.; Cheng, Y.; Yan, C.; Song, B.; Sun, H.; Teng, F. Experimental study of sand control in a natural gas hydrate reservoir in the South China sea. *Int. J. Hydrogen Energy* **2019**, *44*, 23639–23648. [CrossRef]
36. Lu, J.S.; Xiong, Y.M.; Li, D.L.; Shen, X.D.; Wu, Q.; Liang, D.Q. Experimental Investigation of Characteristics of Sand Production in Wellbore during Hydrate Exploitation by the Depressurization Method. *Energies* **2018**, *11*, 1673. [CrossRef]
37. Saucier, R.J. Considerations in Gravel Pack Design. *J. Pet. Technol.* **1974**, *26*, 205–212. [CrossRef]
38. Li, Y.; Hu, G.; Liu, C.; Wu, N.; Chen, Q.; Liu, L.; Li, C. Gravel sizing method for sand control packing in hydrate production test wells. *Pet. Explor. Dev.* **2017**, *44*, 1016–1021. [CrossRef]
39. Li, Y.L.; Wu, N.Y.; Gao, D.L.; Chen, Q.; Liu, C.L.; Yang, D.Y.; Jin, Y.R.; Ning, F.L.; Tan, M.J.; Hu, G.W. Optimization and analysis of gravel packing parameters in horizontal wells for natural gas hydrate production. *Energy* **2021**, *219*, 119585. [CrossRef]
40. Dong, C.; Gao, K.; Xu, H.; Deng, J.; Hu, G.; Chen, Q. Experimental study on the influence of filling density on the sand blocking effect and stability of gravel. *Oil Drill. Prod. Technol.* **2018**, *40*, 123–130. [CrossRef]
41. Najmi, K.; McLaury, B.S.; Shirazi, S.A.; Cremaschi, S. Experimental study of low concentration sand transport in wet gas flow regime in horizontal pipes. *J. Nat. Gas Sci. Eng.* **2015**, *24*, 80–88. [CrossRef]
42. Chen, Y.; Sun, X.; Yan, T.; Yao, D.; Duan, R. Experimental study on micron-sized sand particles transport in the water flow path of hydrates production wellbore. *J. Nat. Gas Sci. Eng.* **2020**, *73*, 103088. [CrossRef]
43. Wang, J.; Meng, Y.; Han, B.; Liu, Z.; Zhang, L.; Yao, H.; Wu, Z.; Chu, J.; Yang, L.; Zhao, J. Hydrate blockage in subsea oil/gas flowlines: Prediction, prevention, and remediation. *Chem. Eng. J.* **2023**, *461*, 142020. [CrossRef]
44. Fajemidupe, O.T.; Aliyu, A.M.; Baba, Y.D.; Archibong, E.A.; Yeung, H. Sand minimum transport conditions in gas–solid–liquid three-phase stratified flow in a horizontal pipe at low particle concentrations. *Chem. Eng. Res. Des.* **2019**, *143*, 114–126. [CrossRef]
45. Kamyab, M.; Rasouli, V. Experimental and numerical simulation of cuttings transportation in coiled tubing drilling. *J. Nat. Gas Sci. Eng.* **2016**, *29*, 284–302. [CrossRef]
46. Wang, Z.Y.; Zhao, Y.; Zhang, J.B.; Pan, S.W.; Yu, J.; Sun, B.J. Flow assurance during deepwater gas well testing: Hydrate blockage prediction and prevention. *J. Pet. Sci. Eng.* **2018**, *163*, 211–216. [CrossRef]
47. Ding, J.; Cheng, Y.; Yan, C.; Lu, C.; Li, Y.; Xue, M. Numerical simulation on sand production in the exploitation process of gas hydrates based on skeleton failure of reservoirs. *J. Nat. Gas Sci. Eng.* **2021**, *94*, 104052. [CrossRef]
48. Dong, C.; Zhang, Q.; Gao, K.; Yang, K.; Feng, X.; Zhou, C. Screen sand retaining precision optimization experiment and a new empirical design model. *Pet. Explor. Dev.* **2016**, *43*, 1082–1088. [CrossRef]
49. Pedroso, C.A.; Sanches, E.D.; Oliveira, N.S.; Fernandes, L.H.; Gomes, M.; Farias, R.; Mendez, A.; Frata, F.M. Lightweight Proppants: Solution for Gravel Packing Horizontal Wells under Extreme Conditions. In Proceedings of the SPE International Symposium and Exhibition on Formation Damage Control, Lafayette, LA, USA, 15–17 February 2006. [CrossRef]
50. Pedroso, C.; Sanches, E.; Oliveira, N.; Fernandes, L.; Mickelburgh, I.J.; Coelho, G.S. New Solutions to Extend the Application Window of Horizontal Openhole Gravel Packing. In Proceedings of the SPE Annual Technical Conference and Exhibition, Dallas, TX, USA, 9–12 October 2005. [CrossRef]
51. Abbaszadeh Shahri, A.; Shan, C.; Larsson, S.; Johansson, F. Normalizing Large Scale Sensor-Based MWD Data: An Automated Method toward A Unified Database. *Sensors* **2024**, *24*, 1209. [CrossRef] [PubMed]
52. van Eldert, J.; Schunnesson, H.; Saiang, D.; Funehag, J. Improved filtering and normalizing of Measurement-While-Drilling (MWD) data in tunnel excavation. *Tunn. Undergr. Space Technol.* **2020**, *103*, 103467. [CrossRef]

53. Ma, C.; Dou, Y.; Deng, J.; Hui, C.; Zhao, K.; Feng, Y.; Dou, L. Numerical simulations of sand-screen performance in unconsolidated prepacked gravel screen. *Energy Sci. Eng.* **2024**, *12*, 983–1003. [CrossRef]
54. Farias, R.; Mendez, A.; Calderon, A. Best Practices and Lessons Learned in Open Brazil Hole Horizontal Gravel Packs Offshore. In Proceedings of the Offshore Technology Conference, Houston, TX, USA, 3–6 May 2004. [CrossRef]
55. Chen, Z.; Novotny, R.J.; Farias, R.; Mendez, A.; Pedroso, C.A.; Fernandes, L.H.C. Gravel Packing Deep Water Long Horizontal Wells under Low Fracture Gradient. In Proceedings of the SPE Annual Technical Conference and Exhibition, Houston, TX, USA, 26–29 September 2004. [CrossRef]
56. Bitencourt Zimmermann, J.H.; Haftani, M.; Wang, C.; Salimi, M.; Nouri, A. A novel laboratory technique for open-hole gravel-pack design. *J. Pet. Sci. Eng.* **2022**, *215*, 110713. [CrossRef]
57. Taheri Shakib, J.; Ghaderi, A.; Abbaszadeh Shahri, A. Analysis of hydraulic fracturing length and aperture on the production rate in fractured reservoir. *Life Sci. J.* **2012**, *9*, 1769–1777. [CrossRef]

Journal of
*Marine Science
and Engineering*

MDPI

Article

Selection Results of Solid Material for Horizontal and Highly-Deviated Well Completion Gravel-Packing: Experiments, Numerical Simulation and Proposal

Haoxian Shi [1,2], Changyin Dong [3,*], Xinjie Zhan [3], Chenfeng Liu [3], Lixia Li [1,2], Jianrong Ji [4], Yanjiang Yu [1,2,*] and Zhendong Li [3]

1 Guangzhou Marine Geological Survey, China Geological Survey, Guangzhou 511466, China; shihaoxian00@163.com (H.S.); gmgsllx2304@163.com (L.L.)
2 National Engineering Research Center of Gas Hydrate Exploration and Development, Guangzhou 511466, China
3 School of Petroleum Engineering, China University of Petroleum (East China), Qingdao 266580, China; zxjupc2022@163.com (X.Z.); liuchenfeng0312@163.com (C.L.); lizhendong20011112@163.com (Z.L.)
4 East China Petroleum Bureau of China Petroleum & Chemical Corporation, Nanjing 210019, China; 18505236880@163.com
* Correspondence: dongcy@upc.edu.cn (C.D.); yuyanjiang2004@163.com (Y.Y.)

Abstract: Lightweight and ultra-lightweight solid materials are being used in gravel packing for horizontal wells instead of traditional quartz and ceramsite to decrease the risk of premature plugging and improve packing efficiency. Physical and numerical simulation experiments of gravel packing were conducted to assess the effectiveness of reducing solid material density and investigate its impact on packing and sand control. Packed gravel destabilization experiments highlighted the importance of high-compaction degree packing for effective sand control. Further gravel packing experiments examined the packing performance of different solid materials, revealing that lightweight solids have minimal gravitational deposition effect because their density is similar to the gravel slurry, relying primarily on fluid flow for compaction. The numerical simulation indicated that lightweight ceramsite is unsuitable for horizontal and highly-deviated wells because of its poor compaction degree and sand control, especially with high-viscosity slurry. High-density particles enhance gravitational deposition, improving packing compaction and sand control. Lightweight materials are recommended only when advanced plugging of α wave packing cannot be avoided. In highly-deviated wells, high-density materials significantly improve packing stability and sand control. This study provides clear technical guidelines for selecting solid materials for gravel packing in horizontal and highly-deviated wells.

Keywords: gravel-packing; horizontal well; high-deviated well; numerical simulation; gravel destabilization

1. Introduction

Weakly consolidated sandstone reservoirs are one of the main types of hydrocarbon reservoirs that contribute a great amount of natural gas and crude to the energy industry [1–3]. Due to the relatively short geological age of this reservoir, the mechanical properties such as bond strength, permeability, and compressive strength of the rock are poor, especially in the development of deep-sea shallow oil and gas wells in this type of reservoir, tend to produce sand along with the fluid when the production parameters achieve a certain critical condition [4–6]. For decades, sand control has been the main solution to solve the problem of sand production in weakly-consolidated sandstone reservoirs [7–9]. Nowadays, gravel packing has been fully developed as an effective sand control method, not only in vertical wells but also in horizontal and highly-deviated wells, which concerns packing gravel into the annulus between the screen and the open-hole well wall or the perforated casing [10–12]. The screen is used to support the packed gravel; meanwhile, the packed gravel mainly undertakes the task of retaining the produced sand from the formation [13–15].

As shown in Figure 1, the gravel packing in horizontal wells involves the multiple solid–liquid two-phase coupled flow and sand-bed migration processes under complex conditions, such as inclined wellbore, formation filtration loss, and fluid mass exchange [16,17]. Aiming at the complex flow process, gravel filling mechanism, gravel filling law, and gravel filling design in horizontal wells and large inclined wells, many scholars have carried out a lot of research in the directions of experimental simulation and numerical simulation.

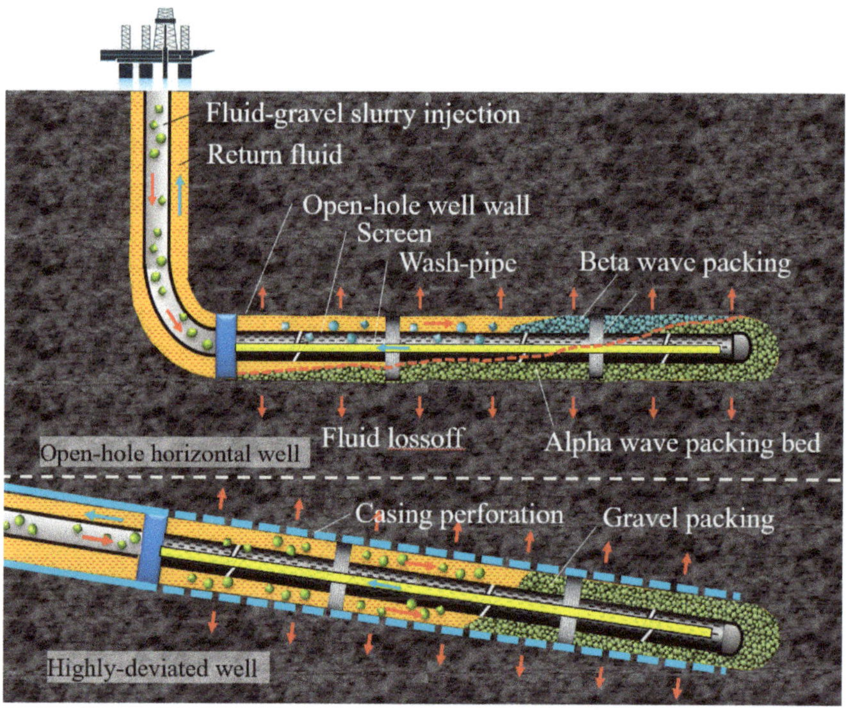

Figure 1. Diagrammatic sketch of gravel-packing process in horizontal and highly-deviated wells.

In the physical experimental simulations, Maly et al. found through in-house physical experiments that dunes form in the wellbore when the inclination is greater than 45° [18]. They believed that due to the continuous escape of sand-carrying fluids, the height of the dune gradually increased along the inclined section until the accumulation channel was blocked. They found the phenomenon of sand bridges during gravel filling, but no in-depth study was conducted on the causes and mechanisms of sand bridge formation. Gruesbeck et al. described the gravel-filling process in large inclined wells by carrying out gravel-filling experiments [19], further explained the experimental phenomena found in Maly's experimental study, and proposed the concept of a 'balanced dike' for the first time. Bigna et al. investigated the effects of screen offset, flow rate, and sand content on the equilibrium height of the wave sand layer by using a full-size horizontal well internal circulation gravel filling simulator to address the problems of solid–liquid two-phase flow and sand bed transport under complex conditions during the gravel filling process in horizontal wells or wells with large gradient [20]. Sanders et al. developed an alternative flow path system and verified the feasibility of the alternative flow path concept through small and large-scale experimental tests using equipment ranging in length from 1.524 to

304.8 m [21]. The alternative path allows the slurry to bypass the bridge if one is formed. In wells with low reservoir pressure, the safety window for gravel filling is narrow, and the dynamic pressure during the gravel filling operation can easily overcome the formation fracturing pressure and lead to reservoir fracture. Magalhães et al. proposed a new gravel-filling method by selecting two different proppants for filling [22]. In the α phase, when the filling pressure is low, a conventional density proppant is pumped, and in the β-wave phase, when a larger pressure increase is observed, a lightweight proppant will be pumped. Dong et al. simulated the process of gravel cyclic filling in horizontal wells and large inclined well tubes by using a horizontal well and large inclined well tubes gravel cyclic filling test simulator [23], analyzed the characteristics of α-wave and β-wave filling fronts, and researched the influences of the wellbore inclination angle, screen tubing offsets, flow rate, and sand volume fraction on the filling dynamics and the equilibrium height of the sand bed in α-wave and the mechanism.

Physical modeling studies have significant limitations at the experimental scale and are unable to simulate on-site gravel fill construction conditions accurately. Therefore, many more scholars have devoted great efforts to studying the mathematical description and simulation of the gravel packing process in horizontal wells [24–26].

In numerical simulations, since the 1970s, many scholars have begun to study the numerical modeling of gravel filling in horizontal wells. Gruesbeck et al. first put forward the concept of 'equilibrium dykes' based on the gravel filling process observed in the indoor test [19] and established a mathematical model for the inclined wells and horizontal wells in the case of complete filling. This model can be used to calculate the equilibrium height of the sand bed in a gravel-filled wellbore/screen tube annulus, but it cannot be used to plug in advance, nor can it simulate the whole filling process. Peden et al. extended the 'equilibrium sand bed' theory [27]. The semi-empirical formulae for predicting the filling rate of the screen tube/well simple annulus and shot hole aperture were obtained by using the method of magnitude analysis and fitting of experimental simulation results. The accuracy of the calculation is largely dependent on these experimental data and the limitation of experimental conditions, and the results of the calculation are not stable. A real-time numerical study of gravel filling in horizontal wells was carried out systematically [28]. They established a model for calculating the height of the equilibrium sand bed based on the probabilistic change in the limit state of particle movement on the sand bed surface. According to the theory of water turbulence, a set of theories and methods of stochastic analysis was proposed to study the particle motion state on the sand bed surface and the change rule of sand bed height. A three-dimensional mathematical model for gravel filling was established, considering the difference between gravel particles and sand-carrying liquid when solid–liquid flows in the annulus, and introduced the concept of deposition factor [29–32]. The model overcomes the shortcomings of one-dimensional and two-dimensional models, fully considers the effect of solid particle deposition when the mortar flows in the annulus, and reasonably predicts the concentration of in-situ gravel during the filling process so that the equilibrium height of the sand bed can be calculated more accurately. Bai et al. established a time-dependent mathematical model describing the α-filling process based on the equations of conservation of mass and momentum of sand and liquid in two independent flow systems of the wellbore and the rinse screen annulus [33] and the flow coupling equations between the systems. The entire filling process can be simulated visually using numerical model development software. Mimouna et al. achieved this by solving the mass and momentum conservation equations for the fluid [34] and gravel by interleaving them, taking into account the well, the downhole tool tubing column structure, the fluid and gravel properties, and the pumping time. The simulation predicts changes in fluid pressure as well as gravel concentration over time. Changes in gravel concentration provide reliable data on filling patterns and operational filling states. Sarraf Shirazi and Frigaard investigated and explained [11] the filling process of α-wave and β-wave by developing a mathematical model of gravel filling. The effects of important parameters such as mud flow rate, average solids concentration, flushing pipe diameter, and leakage rate on gravel filling flow were also investigated. Huang et al. proposed a design calculation method for α–β wave filling length considering the success of α-wave filling and the success of β-wave reverse filling [35]. The corresponding software was prepared to discuss and calculate the quantitative analysis of the factors affecting the α–β wave filling length, such as the density of sand-carrying liquid, gravel density, and wash pipe sieve ratio. Under specific conditions, certain criteria and methods can be used to design and optimize the gravel filling length of horizontal wells. Nie et al. made a detailed analysis of the whole process of gravel filling flow in horizontal wells [36], established a three-dimensional numerical model of gravel filling, and carried out several sets of gravel filling experiments to verify the improved three-dimensional numerical model.

These numerical simulation results enhance the understanding of the law of the influence of pumping parameters on filler performance and filler efficiency. Gravel packing efficiency is a key parameter that significantly affects the effectiveness of gravel packing against sand. However, in horizontal wells and large gradient wells, the high-density gravel material may lead to early sand plugging and reduce the gravel filling compactness because it tends to deposit solid particles in the fluid. Engineers are now gradually applying lightweight or ultra-lightweight gravels in horizontal packings [37–40], which reduces the risk of premature plugging during operations. The use of lightweight or ultra-lightweight materials as gravel for horizontal good gravel packing has almost become a trend to extend the length of horizontal good gravel packing, widen the safe operating window for gravel packing, and reduce the hazards to the reservoir. However, there is a lack of effective evidence on whether the use of light or ultra-light materials can really improve the effect of gravel packing. Therefore, this paper will study a selection of gravel packing materials for horizontal wells and highly deviated wells.

2. Methodology

This paper adopts a method combining physical experimental simulation and numerical simulation to carry out research. Figure 2 shows the methodology and approach. Numerical simulation experiments overcome the problem of small-scale indoor physical simulation experiments. Indoor physical simulation experiments can supplement and verify the results of numerical simulations and improve the accuracy and reliability of numerical simulation results.

(1) The density of gravel packing affects the instability morphology of the gravel layer. First, an indoor physical simulation experiment of gravel layer instability is carried out to study the law of gravel layer instability under different gravel packing density conditions.
(2) However, the density of gravel packing is affected by the type of gravel packing. Therefore, an indoor physical simulation experiment of horizontal well gravel packing is carried out. Based on the results of indoor physical simulation experiments, the type of gravel packing material is preliminarily selected to improve the density of gravel packing.
(3) The scale of indoor physical simulation experiments is small and cannot simulate the gravel packing process in actual oil wells. Therefore, based on the actual oil wells in a certain oil field, a numerical simulation study of gravel packing is carried out. Based on the results of numerical simulation experiments, the types of gravel filling materials are selected for horizontal wells and highly deviated wells, respectively.
(4) Finally, the types of gravel filling materials are selected by combining the results of physical experimental simulation and numerical simulation, and construction suggestions are put forward for on-site gravel filling of horizontal wells and highly deviated wells.

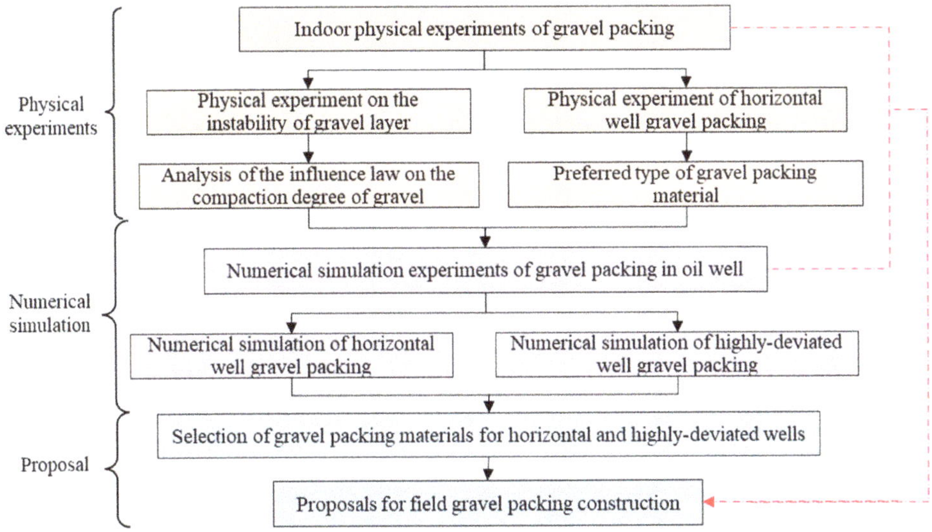

Figure 2. Flow chart of research methodology and approach.

3. Packing Gravel Instability Experiment

3.1. Methods and Materials

To accurately replicate the production conditions of gravel-packed wells, a comprehensive visualized gravel-packed production simulation experiment system has been carefully constructed (Jiangsu Tuochuang Technology Co., Ltd., Nantong, China). The experimental system is mainly used to simulate the morphological changes in the gravel layer after filling operation in the production process after being carried by fluid impact, and its flow is shown in Figure 3a. The experimental apparatus comprises a fluid tank, a screw pump, a primary container, as well as a comprehensive data acquisition and control system. The experimental setup can be flexibly assembled with different main containers to conduct different experiments. Containers AB are two different main devices; container A is a simulated vertical wellbore, and container B is a simulated horizontal wellbore. The main container A has an inner diameter of 0.3 m, which can accommodate a short section of the screen with a length of 0.245 m; Unit B, with an inner diameter of 0.22 m, is designed to house a sand-proof screen pipe that is 1 m long and measures 0.1 m diameter. Additionally, a clamping mechanism is installed at both ends to prevent any radial displacement of the screen when it is inserted into the main apparatus.

For the experiments, samples from screens commonly used in oilfield construction were placed into the main device, making sure that both ends of the screen were sealed to both sides of the device. Different types of filler particles were used to fill the cavities formed by the screen and the outer wall of the device (simulated casing) according to the densities set in the experimental conditions. The filling materials used in the experiment are normal quartz sand and ceramsite with a diameter of 0.424–0.85 m. Fluid was injected into multiple inlet ports along the sidewall of the device using the same discharge volume, and the pressure, flow rate, and particle accumulation pattern over time were recorded and photographed in real-time by a data acquisition system to simulate the production process. Based on the equivalent flow rate method, the experimental flow rate is converted into 2 m^3/h according to the actual production of the oil well. All test data and images were used for further analysis of the formation process and morphology of the erosion holes in the gravel layer.

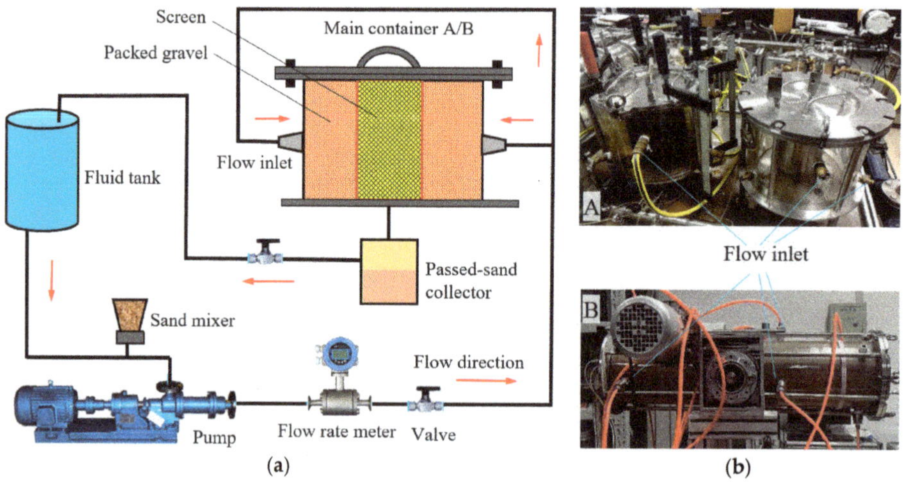

Figure 3. Experimental apparatus of packed-gravel destabilization. (**a**) Flow chart. (**b**) Picture of main container A (Simulate vertical well)/B (Simulated horizontal well).

3.2. Results and Discussion

Using the above apparatus and methods, the simulation experiments were carried out around the gravel-packed layers formed by different types of particles, and the gravel-packed densities were set to 95%, 96%, 98%, and 100%. Figures 4 and 5 show the erosion and sediment intrusion patterns of the quartz sand and ceramsite particle-filled layer under the impact of local high-flow rate load fluids.

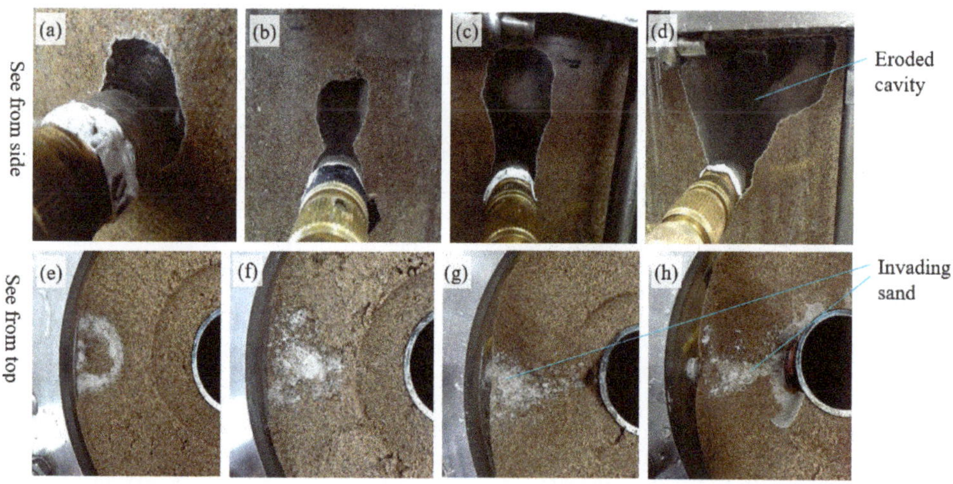

Figure 4. Observed erosion and sand invasion patterns in packed quartz sand with different compaction degrees. (**a**,**e**) Compaction degree of 100%. (**b**,**f**) Compaction degree of 98%. (**c**,**g**) Compaction degree of 96%. (**d**,**h**) Compaction degree of 95%.

As evident from Figure 4, the degree of filling compactness significantly influences both the stability of the gravel layer and the patterns of intrusion and plugging exhibited by the formation sand when subjected to local high-flow velocity fluid impact conditions. When the compactness is 100%, as shown in Figure 4a, the space for sliding and displacement between quartz sand grains is very limited. The impact and carrying action of the fluid are counteracted by the tightly arranged grains, and only a small portion of the gravel layer is displaced to form holes. When the formation sand intrudes with the fluid, it can only be transported between the pores of the quartz sand body, and no new transport channel can be generated. When the intrusion reaches a certain depth, with a reduction in the sand-carrying capacity of the fluid and the accumulation of sand clogging, resulting in the subsequent strata sand being difficult to intrude further, the intrusion behavior is close to stagnation.

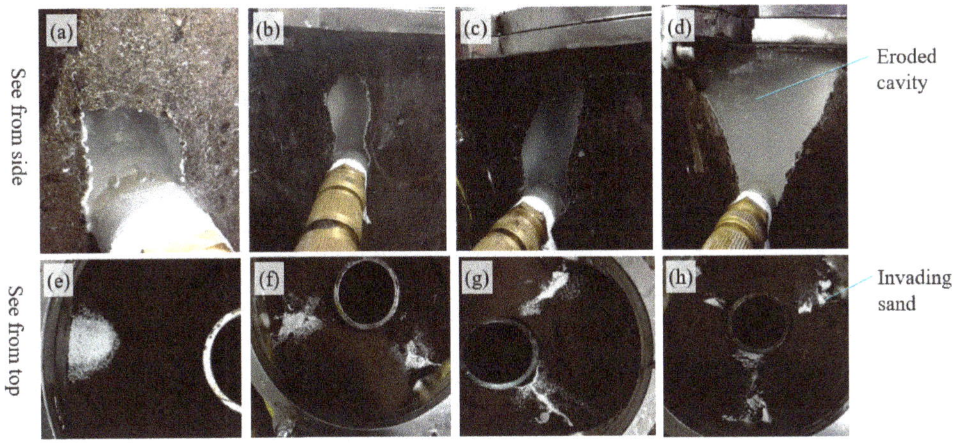

Figure 5. Observed erosion and sand invasion patterns in packed ceramsite with different compaction degrees. (**a**,**e**) Compaction degree of 100%. (**b**,**f**) Compaction degree of 98%. (**c**,**g**) Compaction degree of 96%. (**d**,**h**) Compaction degree of 95%.

Moreover, with a decrease in filling compactness, as shown in Figure 4b–d, the quartz sand particles are no longer tightly stacked with each other. The particles are displaced by the impact of high-flow fluid, which produces large holes near the inlet, and the holes eventually extend to the screen with the prolongation of production time.

In addition, when the fluid carries the formation of sand intrusion, as the pressure of the intruding fluid increases, when it exceeds the original stabilizing force between the quartz sand grains, the quartz sand grains are disturbed and thus rearranged. The pressure change causes the formation sand to squeeze out new flow channels and allows deeper intrusion, which is why the formation sand is able to contact the screen in Figure 4d–h.

It can be found that the same pattern is observed when ceramsite grains are used as filling materials. It is worth noting that when the fluid carries the formation of sand to invade the packed layer, the fluid always preferentially breaks along the stress weak point of the filling zone. In the experiment shown in Figure 5, the weak stress direction is axial, so the erosion holes along the axial direction formed first. Subsequently, the holes gradually developed radially toward the screen until the edges of the holes contacted the screen and expanded.

Using the main container B in Figure 3 to simulate the gravel-packed layer of a horizontal well to carry out simulation experiments, it can be found that under different inflow angle conditions, the noncompact filling layer produces erosion holes. Figure 6 shows the changes in the erosion cavity of gravel packing layers in horizontal wells under different levels of gravel compaction degree.

The erosion dynamics in the lower and middle part of the simulated wellbore near the lateral inlet are the same as that of the vertical well, which first expands along the weak stress area and then extends in the radial direction after formation. The cavity near the inlet at the top of the simulated wellbore is in the shape of a regular trumpet, which is affected by the gravity effect of ceramsite grains, and it is difficult for the local high-flow rate fluid to break through the gravel layer directly, and the formation sand can still be blocked by the gravel layer after erosion.

(**a**) (**b**) (**c**)

Figure 6. Observed erosion cavity in packed ceramsite using main container B. (**a**) Compaction degree of 100%. (**b**) Compaction degree of 95%. (**c**) Compaction degree of 90%.

4. Gravel-Packing Experiments

4.1. Method and Materials

In order to simulate the dynamic process of gravel packing in horizontal wells and large inclined wells, a large-scale visual gravel packing simulation experiment system was built (Jiangsu Hua'an Technology Co., Ltd., Nantong, China). The experimental system mainly clarifies the transportation and accumulation patterns of different types of particles inside the annular space between the screen and casing, as well as the final filling effect, and its process and device objects are shown in Figure 7. The apparatus mainly includes a liquid storage tank, screw pump, sand filler, sand collector, main device unit, data acquisition and control system, and so on. The simulated wellbore has an inner diameter of 250 mm and a length of 10 m. It is capable of accommodating standard screen tubing actually used in oilfield production sites. In this study, a 5 in screen pipe with a length of 9.5 m and a flushing pipe with an outside diameter of 63 mm and a length of 8.5 m were used to simulate the gravel filling process by allowing the load fluid to enter through an inlet at the root end of the wellbore, located below the screen pipe.

The flow rate is set as 21 m^3/h, and the sand-carrying fluid is clean water with a viscosity of 1 mPa·s and a sand ratio of 5% in the gravel packing experimental. The gravel packing simulation is carried out using different types of particles according to the set discharge volume and sand ratio. The gravel particles are carried and mixed uniformly by the experimental fluid into the annulus of the screen sleeve, the gravel particles are blocked by the screen and retained in the annulus, and the liquid is returned to the tank through the wash pipe, forming a cyclic packing process. During the experiment, the movement state and overall morphology of the sand bed formed by gravel deposition were recorded. The experimental pressures and discharges were collected to analyze the filling dynamics of different types of gravels and the final filling effect.

Three types of gravel particles were used in this paper (Beijing Qixiang Ceramic Materials Co., Ltd., Beijing, China), including conventional ceramsite particles (Figure 8a), lightweight ceramsite particles (Figure 8b), and ultra-lightweight ceramsite particles (Figure 8c), with apparent densities of ceramicite particles of 2.76, 1.92, and 1.00 g/cm^3, and stacking densities of 1.49, 1.31, and 0.62 g/cm^3, respectively. The diameters of the three types of particles were 0.425–0.850 mm (Table 1).

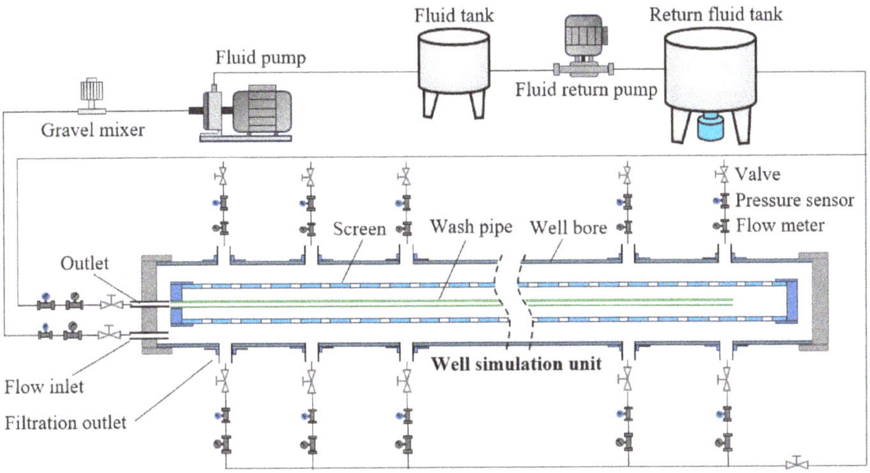

(**a**) Flow chart

(**b**) Picture of the experimental installation

Figure 7. Experimental system of horizontal gravel-pacing.

(**a**) (**b**) (**c**)

Figure 8. Used packing solid materials in the experiments and their apparent densities. (**a**) Normal ceramicite. (**b**) Light ceramicite. (**c**) Ultra-Light ceramicite.

Table 1. The parameters differences of different packing materials.

Type	Apparent Density/g/cm^3	Stacking Density/g/cm^3	Diameter/mm
Normal Ceramicite	2.76	1.49	0.425–0.850
Light Ceramicite	1.92	1.31	0.425–0.850
Ultra-Light Ceramicite	1.00	0.62	0.425–0.850

4.2. Results and Discussion

4.2.1. Basic Packing Process Analysis

In the simulation experiment of the gravel packing process in horizontal wells, the filling process of conventional and lightweight ceramicite grains all showed obvious α and β waves. During the filling process, the sedimentary sand bed extends along the direction of fluid flow, but the closer the inlet is, the stronger the effect of fluid turbulence is, which leads to the inability of ceramicite grains to accumulate. Therefore, during the α-filling stage, there is a section of unfilled area between the inlet and the sedimentary sand bed, as shown in Figure 9a. The density of lightweight ceramicite grains is closer to that of the load fluid, so they are more easily carried by the load fluid of the same viscosity and displacement during the filling process. The fast transportation speed of the lightweight ceramicite grains α-wave deposited sand bed makes the angle between the front section of the sand bed and the horizontal plane larger, as shown in Figure 9b,d.

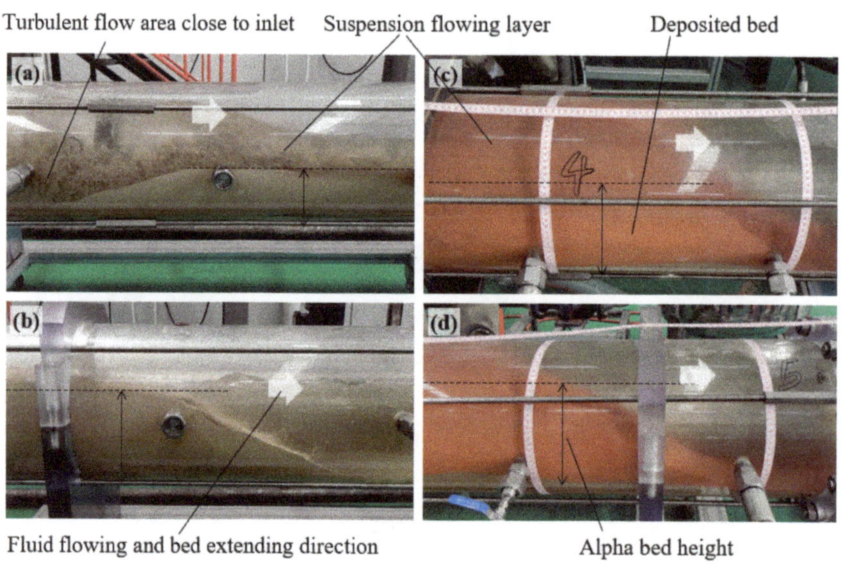

Figure 9. Picture of α wave packing with different ceramicite. (**a**,**b**) Normal ceramicite. (**c**,**d**) Light ceramicite.

Under the same discharge and sand-carrying fluid conditions, the low-density ceramicite grains were transported over a long distance, showing an overall transport but not accumulation filling trend.

Influenced by the density difference between particles and load fluid, more particles are transported or floated into the same filling space during the filling process of lightweight ceramicite grains. A stronger fluid-carrying effect and weaker gravity deposition prevent lightweight ceramicite grains from forming dense deposits during the filling process. In the formation of β-wave filling, the height of the conventional ceramicite grains deposited sand bed is higher, the angle of the β-wave front is small, the backfilling speed is fast, and the fluid impact and compaction effects are strong, which can further enhance the dense degree of gravel packing, as shown in Figure 10c,d. Among them, Figure 10b is the filling state near the end of the β-wave filling, so the sand bed has a larger angle along the front edge, which is not contrary to the above analysis.

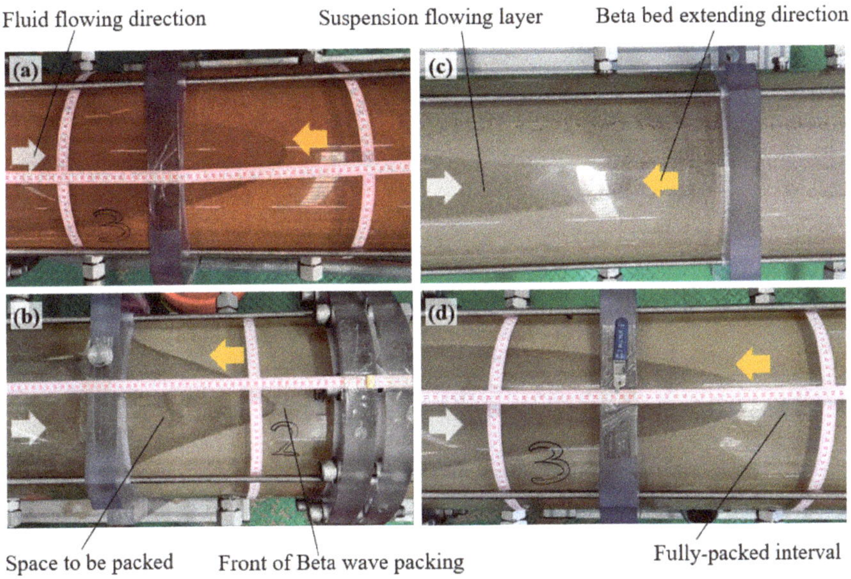

Figure 10. Picture of β wave packing with different ceramicite. (**a**) Light ceramicite. (**b–d**) Normal ceramicite.

Figure 11 shows the changing curves of packing pressure and flow rate during normal ceramsite packing. The test process shows that in the α-wave filling stage, the filling flow rate and filling pressure are relatively stable as the filling sediment bed advances. When entering the β-wave filling stage, as the distal end of the filling port gradually becomes dense, the channel for sand-carrying liquid to return to the bed gradually decreases, and the flow resistance increases, which is manifested as a significant increase in the pressure and a decrease in filling displacement. When the β-wave is about to reach the filling port, in order to prevent the damage of the filling tool caused by excessive pressure and the disturbance of the filled area, the final desanding is achieved by decreasing the displacement, and with an obvious increase in the filling pressure, it is proved that the grains carried by the β-wave have reached the filling port, and the filling test is finished.

In the horizontal well circulation filling process, the load fluid and the filled ceramicite particles enter the annular cavity formed by the screen and the well wall through the filling port and are transported to the outside of the screen. The filled ceramicite particles are blocked by the screen pipe from entering the annular cavity of the screen pipe and wash pipe, while the load fluid without ceramicite particles enters the annular cavity of the screen and wash pipe and finally enters the inlet at the bottom of the wash pipe, and then flows back to the wellhead.

The filling process is generally divided into α-wave filling (Figure 12a) and β-wave filling (Figure 12b). During the α-wave filling, the ceramicite grains preferentially accumulate at the bottom of the annulus between the screen and the well wall and gradually push the sedimentary sand bed along the axial direction toward the end of the well bottom. After the front edge of the sand bed touches the bottom of the well, β-wave filling follows, and the ceramicite particles in this process will fill the cavity between the deposited sand bed and the top portion of the well wall until the end of the desanding pressure is reached when the ceramicite particles pile up in the vicinity of the filling port.

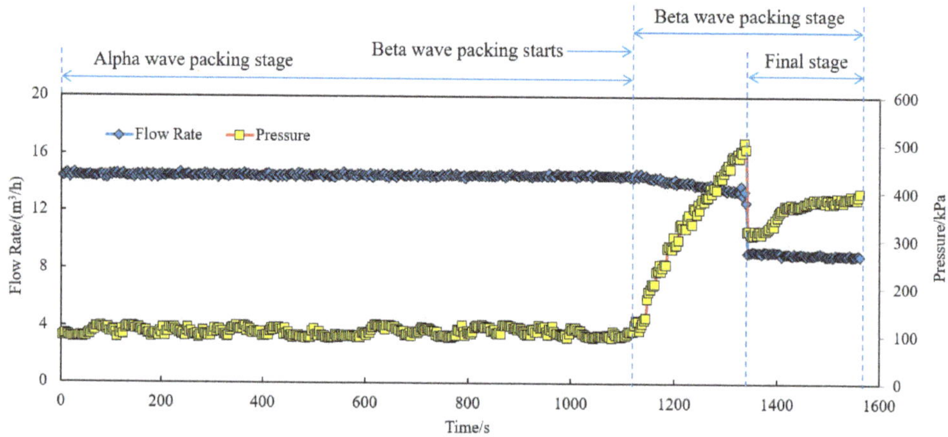

Figure 11. Tested flow rate and pressure performance of gravel-packing.

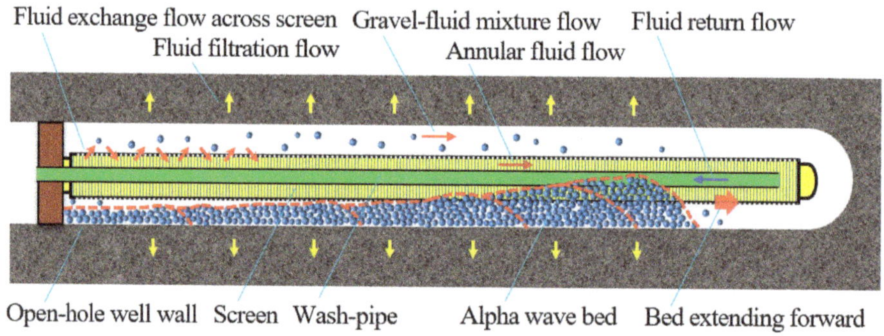

(**a**) α wave packing process.

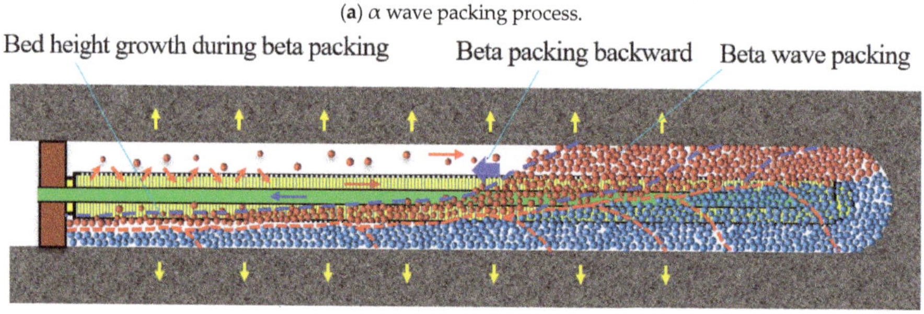

(**b**) β wave packing process.

Figure 12. Diagrammatic sketch of horizontal gravel-packing mechanism and process.

4.2.2. Packing Performance with Different Solid Materials

The apparent density of the filled ceramic grains has a significant effect on the filling effect. As can be seen in Figure 13, taking the α-wave filling process as an example, as the density of the filled ceramic grains decreases, their gravitational deposition effect is relatively weak or even absent, so it takes longer time to reach the closed-loop pressure of the α-wave, which is also confirmed by the pressure curve in Figure 13b. Compared with the conventional ceramic grains with a density of 1.49 g/cm^3 that reach the closed-loop pressure of the α-wave in 800 s, the ultralight ceramic grains need a time of 1700 s, which more than doubles the time and reduces the construction efficiency. Especially in the high-viscosity sand-carrying liquid, the ultra-lightweight ceramic particles are easily carried and disturbed, the original filling structure is destroyed, resulting in the filling compactness can not be guaranteed.

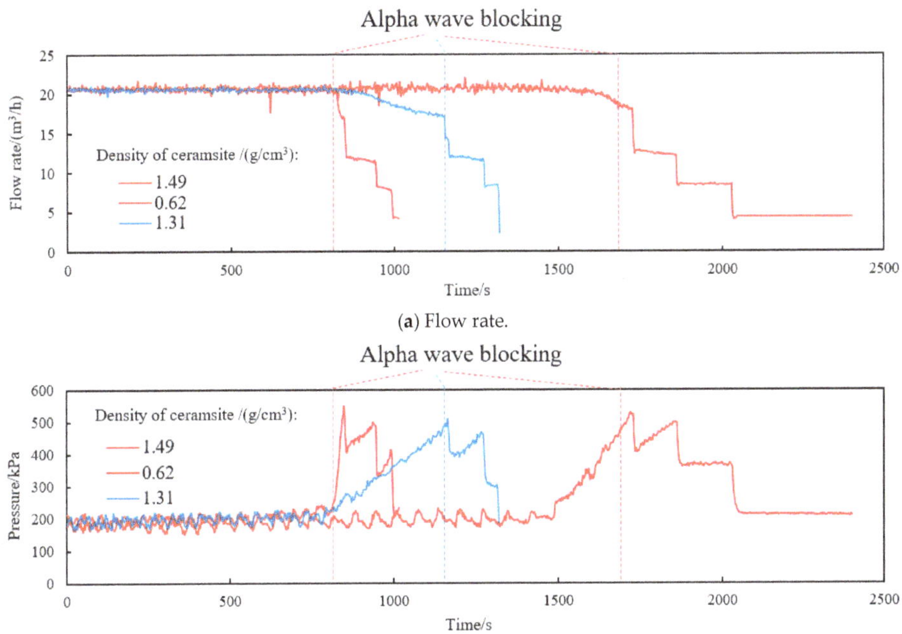

(**b**) Pressure.

Figure 13. Packing performance with different ceramicite.

Comparing the phenomena during the filling process of the three different densities of vitrified gravel, it can be found that the ultra-lightweight vitrified grains are similar to the density of clear water because the apparent density is close to 1 g/cm^3. Therefore, the stratification is obvious in the α-wave filling stage (Figure 14), with a suspended layer at the bottom and a sedimentary layer at the top with a flow zone in the middle, which is opposite to the phenomenon of conventional ceramicite gravel packing.

Since the accumulation of gravel is only supported by the buoyancy of the loaded liquid, and the filling effect cannot be further enhanced by the impact compaction of the liquid and gravity, the accumulation of gravel is less dense than that of the other two densities of gravel. The gravel layer with low density has poor stability and is prone to erosion holes, so the formation sand invades the screen directly, which is not conducive to the stable production and control of sand for a long period of time.

When the filling process of β-waves is carried out, the sedimentary surface of the sand bed of ultra-lightweight ceramicite grains is extremely irregular under the scouring effect of the impact, while the surface of the sand bed of lightweight, conventional ceramicite grains is more regular, which is conducive to the improvement of the final filling and densification degree (Figure 15).

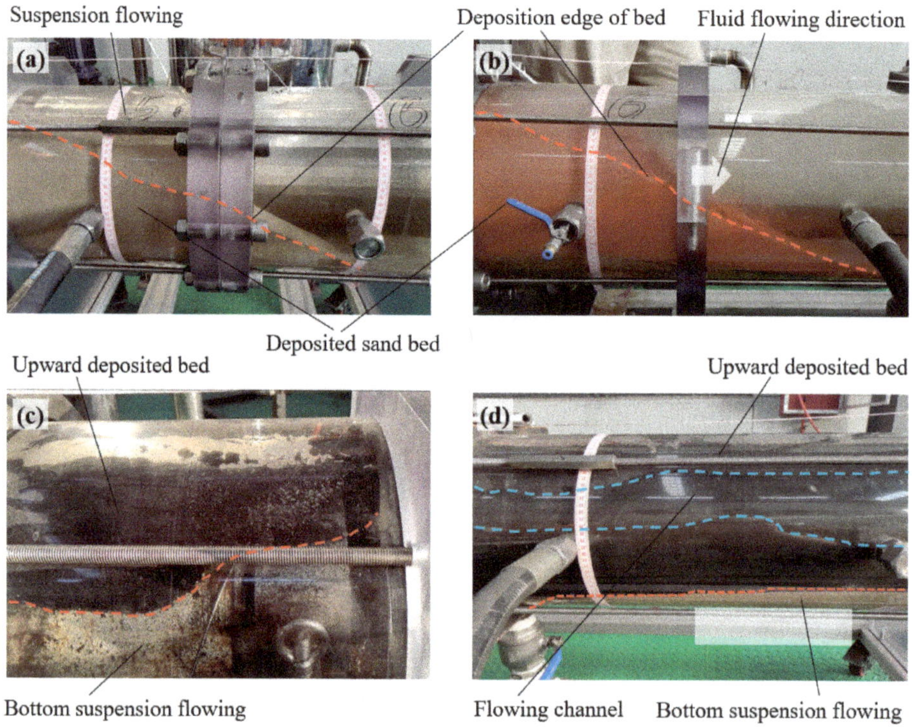

Figure 14. α wave packing patterns with normal and lightweight solid materials. (**a**) Normal ceramicite. (**b**) Light ceramicite. (**c**,**d**) Ultra-light ceramicite.

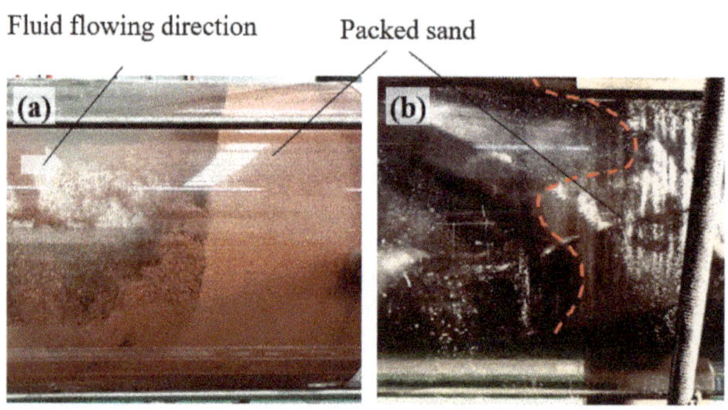

Figure 15. β wave packing patterns of light and ultra-light ceramicite. (**a**) Light ceramicite. (**b**) Ultra-light ceramicite.

Comparing the phenomena during the filling process of three different densities of vitrified gravel, it can be found that when using normal ceramsite and lightweight ceramsite for gravel filling experiments, the density of ceramsite is greater than the density of water. During the α wave filling process, most of the ceramsite is deposited at the bottom of the wellbore under the action of gravity and gradually forms a sand bed. Under the carrying effect of the mortar, the sand bed gradually moves forward (Figure 14a,b).

The ultra-lightweight vitrified grains are similar to the density of clear water because the apparent density is close to 1 g/cm^3. Therefore, the stratification is obvious in the α-wave filling stage (Figure 14c,d), with a suspended layer at the bottom and a sedimentary layer at the top with a flow zone in the middle, which is opposite to the phenomenon of conventional ceramicite gravel packing. Since the accumulation of gravel is only supported by the buoyancy of the loaded liquid, and the filling effect cannot be further enhanced by the impact compaction of the liquid and gravity, the accumulation of gravel is less dense than that of the other two densities of gravel (Figure 16). The gravel layer with low density has poor stability and is prone to erosion holes so that the formation sand directly invades the screen, which is not conducive to the stable production and control of sand for a long period of time.

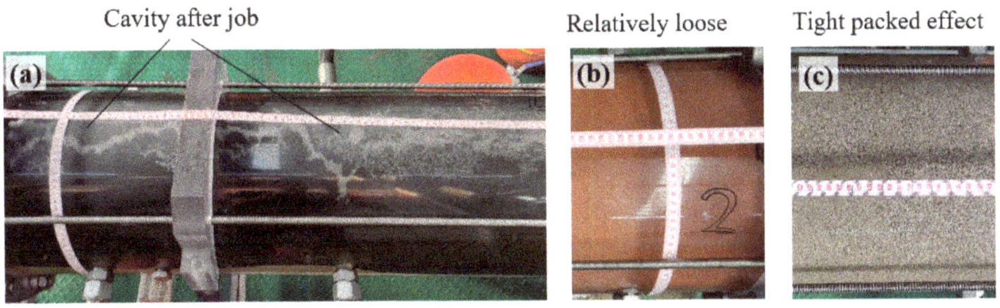

Figure 16. Packing result with different ceramicite. (**a**) Ultra-light ceramicite. (**b**) Light ceramicite. (**c**) Normal ceramicite.

The dynamic filling process of ultra-lightweight ceramicite is different compared with conventional ceramicite filling because of its weak gravity effect (Figure 17). In the α-wave filling stage, most of the ultra-lightweight particles are in suspension, making it difficult to achieve effective sand control with low-density filling. Only in the β-wave filling stage, under the action of fluid pressure, the reverse accumulation and the accumulation surface are gradually pushed to the filling mouth, showing the full β-wave filling pattern.

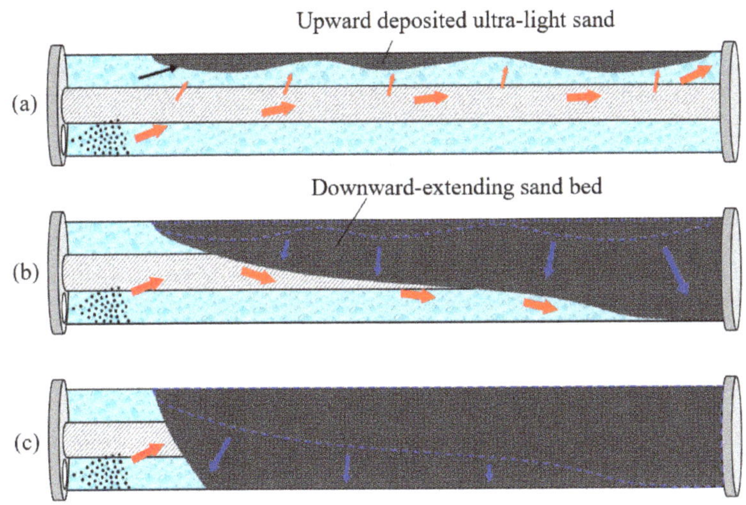

Figure 17. Packing pattern and mechanism of ultra-light solid material. (**a**) Early stage of gravel packing. (**b**) Mid stage of gravel packing. (**c**) End stage of gravel packing.

5. Numerical Simulation Analysis

5.1. Numerical Model and Simulator

Well A is an open-hole horizontal well in the CB field with a reservoir of loose sandstone that requires a gravel-filled, sand-proof completion. The main borehole size of Well

A is 152.4 mm, the drilling depth is 2060 m, and the maximum inclination angle is 91.0°. The starting depth of the filling tool is 1784 m, and the filling length is 276 m. The screen size is 124 mm, and the wash pipe size is 73.0 mm. A circulating gravel filling simulation calculation is carried out for Well A. The filling material is 0.4–0.8 mm ceramic grains, the apparent density is 2.5 g/cm^3, the viscosity of load liquid is 1.5 mPa·s, and the density is 1.0 g/cm^3. Based on basic data from Well A, the calculation is made by a simulator. Visualization of gravel filling dynamic process in horizontal wells as shown in Figure 18.

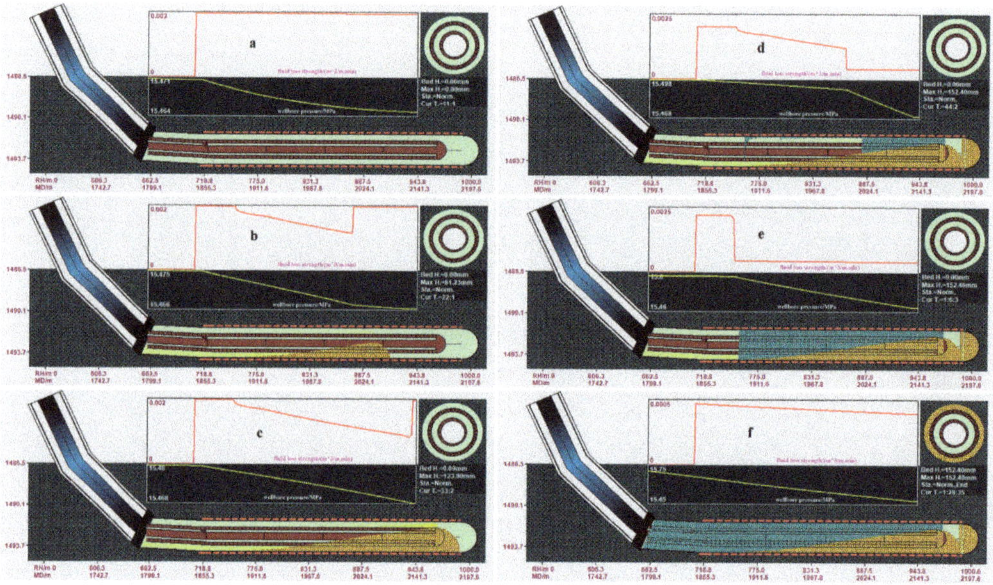

Figure 18. Simulation image of packing process of horizontal Well A. (**a**) Early stage of α-wave gravel packing. (**b**) Mid stage of α-wave gravel packing. (**c**) End stage of α-wave gravel packing. (**d**) Early stage of β-wave gravel packing. (**e**) Mid stage of β-wave gravel packing. (**f**) End stage of β-wave gravel packing.

The simulation results show that complete filling can be achieved using a combination of ceramicite with an apparent density of 2.5 g/cm^3 and 1.5 mPa·s load liquid. No early blockage occurred during the filling process, and the filling rate was 100%.

However, from Figure 18c, it can be found that when the α-wave filling stage is about to end, and the leading edge of the deposited sand bed contacts the bottom of the artificial well, the height of the deposited sand bed is too high, the overall slope of the sand bed is large, and there is a risk of forming an early blockage, and the safety of the filling process is low. In addition, from Figure 18b, it can be found that the formation location of the sedimentary sand bed in the α-wave filling stage is far away from the inlet, about 85 m, which leads to a longer travel time in the final filling stage of the β-wave, as shown in Figure 18e,f. In this stage, the ceramicite is carried by the fluid and impact to complete the filling, and the gravity effect is difficult to play a real role, which hinders the further improvement of the filling degree of compactness. The filling parameters of Well A still need to be optimized.

5.2. Sensitivity Analysis

5.2.1. Solid Material Density

Based on the above calculation results, by changing the apparent density of ceramicite, the filling rate index was analyzed to determine the filling process stability and safety index. A comprehensive analysis was conducted to obtain the comprehensive filling effect evaluation index and establish different filling particle densities under the filling evaluation index plate, as shown in Figure 19.

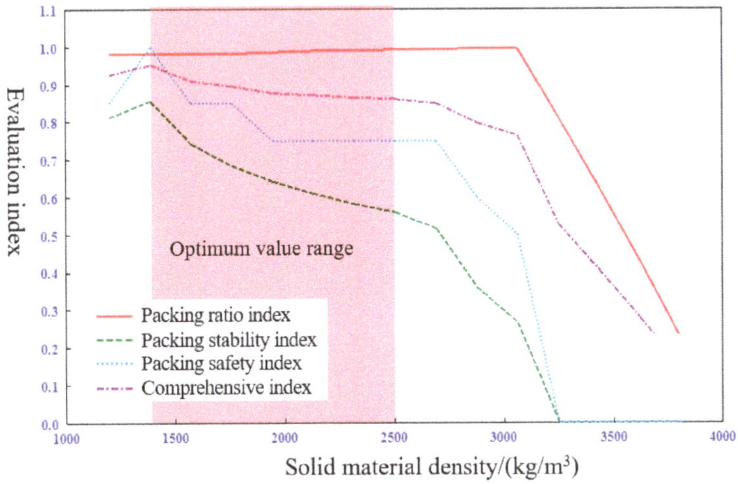

Figure 19. Packing effects with different solid material densities.

Packing ratio index: the ratio of the actual filling volume to the theoretical filling space volume, which evaluates the degree of annular filling at the end of filling.

$$\Psi_p = \frac{V_A}{V_T} \tag{1}$$

where Ψ_p is the packing ratio index, dimensionless; V_A is the actual filling volume, m^3; V_T is the theoretical filling volume, m^3.

Packing stability index: used to evaluate the safety and stability of the α wave packing process. Specifically, the change in sand bed height per unit length of sedimentary sand bed extension indicates that the higher the safety evaluation index of the filling process, the better the filling effect.

$$\Psi_{Hb} = 1 - \frac{R_W - H_b}{R_W} \tag{2}$$

where Ψ_{Hb} is the packing stability index, dimensionless; R_W is the inner diameter of the casing or wellbore, m; H_b is the final equilibrium height of the sand bed, m.

Comprehensive index: comprehensively evaluates the final effect of gravel packing. It is calculated by weighting the packing process safety and packing rate evaluation indicators.

$$V = \alpha_{Hb}\Psi_{Hb} + \alpha_p\Psi_p \tag{3}$$

$$\alpha_p + \alpha_{Hb} = 1 \tag{4}$$

where V is the comprehensive index, dimensionless; α_{Hb} is the weighted coefficient of balanced height evaluation indicators, dimensionless; α_p is the weighted coefficient of the filling rate evaluation index, dimensionless.

According to the calculation results, it was found that the filling evaluation indexes decreased when the density was less than 1.4 g/cm^3, which indicated that ultra-lightweight ceramicite was not necessary for conventional gravel filling of horizontal wells.

Adjusting the density of ceramicite in a reasonable range enables the formulation of an optimal load fluid to obtain a higher sand control capacity after filling while ensuring the filling effect. Therefore, the optimal density of filled particles is recommended by the platen to be between 1.4 and 2.5 g/cm^3.

5.2.2. Fluid Viscosity

The viscosity of the load fluid was taken as the sensitivity factor and the change rule of each index was analyzed.

Through sensitivity analysis, the filling evaluation index plate under the influence of sand-carrying fluid viscosity was established, showing that when the viscosity of sand-carrying fluid is 6~10 mpa·s, the balance between the sand-carrying effect and sand-sedimentation effect is reached, which meets the construction requirements of high filling rate for the long horizontal well section, and the comprehensive filling evaluation effect is the best (Figure 20).

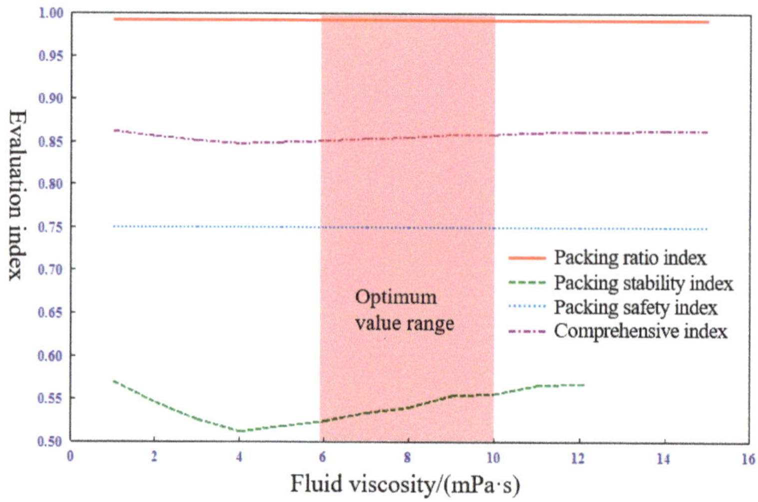

Figure 20. Packing effects with different fluid viscosity.

5.2.3. Discussion

Through simulation experiments, the changes in the erosion pattern and the dynamics of intrusion of formation sand particles under the action of local high-flow velocity fluids during the production process of gravel layers of different densities are analyzed. The dynamic process of gravel filling with different densities of ceramicite and its differences are clarified. Based on the experimental results, a numerical model and simulation software were developed to reveal further the sensitivity law of solid and load fluid properties.

In order to obtain a high sand control effect, it is an important condition to improve the compactness of the gravel layer. Under the impact and disturbance of local high-flow velocity fluid, large holes are easily eroded in the uncompacted gravel layer. After the holes are created, the sand particles produced by the formation will invade the screen along the holes quickly, causing serious blockage of the screen, or even pass through the screen directly and produce to the ground with the fluid. The gravel layer with a high degree of compactness, whether it is localized high-flow velocity fluid or low-velocity fluid, can not erode and penetrate the gravel layer, forming large holes. Fluid scouring can only slightly change the accumulation pattern on the surface of the gravel layer, and a small portion of the sand produced from the stratum intrudes into the interior of the gravel layer, while most of the sand particles are blocked on the surface of the gravel layer. The high-density gravel layer maintains a good sand control effect during the whole production process.

For different densities of ceramicite, the filling process of horizontal well gravel is simulated experimentally. For lightweight and conventional ceramicite, there are clear α-wave and β-wave filling processes in the filling process, while this process is not obvious in the experiments of ultra-lightweight ceramicite. In the filling process, the α-wave filling can make full use of the gravity deposition effect to improve the density of the whole filling process, and affected by the density difference between the particles and the load fluid, more particles were transported or floated in the same filling space during the filling process of lightweight and ultra-lightweight ceramicite. The stronger fluid-carrying effect and the weaker gravity deposition effect make the lower-density ceramicite unable to form dense deposition during the filling process. This problem is particularly evident when filling with ultra-lightweight ceramicite, which, at the end of the α-wave filling, differs from other ceramicites in that the ceramicite forms a suspended sand bed. At the β-wave filling stage, the front surface of the sand bed of the ultra-lightweight ceramicite is extremely irregular under the scouring effect of the fluid. In contrast, the sand bed surface of the lightweight, conventional ceramicite was more regular, which was favorable to improving the final filling densification. Under the displacement range of the simulation experiments, the degree of filling compaction decreased with a decrease in ceramicite density.

For high-angle wells, since the wellbore has a certain downward deflection angle, the gravel particles can reach the bottom of the well smoothly with the fluid, and it is almost impossible to form a sedimentary sand bed, and basically, there is no α-wave filling. Early plugging rarely occurs in gravel filling operations in high-angle wells, so the use of gravel with higher density can further enhance the gravel layer densification to obtain a high sand control effect.

Based on the experimental results, a numerical model and simulation software were developed to calculate the process and effect of gravel filling based on data from actual production wells and to analyze the sensitivity between ceramicite density and the viscosity of load fluid. The simulation results show that adjusting the gravel density within a reasonable range can safely and stably complete the gravel filling operation so that the filling rate reaches 100%. The ultra-lightweight ceramicite was not within the recommended range. The use of ultra-lightweight ceramicite is rare and only applies to the situation where premature clogging of α-wave filling cannot be avoided even after all construction parameters are optimized.

Although this article conducted simulations based on data from actual production wells and carried out large-scale physical modeling experiments, the relevant understanding has not yet been validated in actual production. Subsequently, it is necessary to combine filling data from multiple different well types to optimize understanding gradually.

6. Recommendations and Conclusions

In the case of high velocity of inflow from formation to the wellbore, the restructuration destabilization of packed gravel is easy to take place, leading to the form of the cavity, which emasculates the sand retention function of the packing gravel layer. Therefore, how to improve and ensure the compaction degree of packing gravel is absolutely a considerable issue to achieve the desired sand control effect. Through experiments and simulations, the results indicate that:

(1) For gravel-packing in highly deviated wells, due to the deviated wellbore and direction of downward flow, α wave packing can hardly happen. The premature plugging almost won't happen during the job. Therefore, the usage of high-density solid material can obviously improve the compaction of packing and help to ensure the stability of the packing layer and the effect of sand retention.

(2) In horizontal gravel-packing, which involves α and β wave packing, the gravitational effect of solid material really helps to improve the compaction of packing. So, the utilization of high-density particles favors strengthening the effect of gravity deposition obviously and resultantly improves the effect of sand retention during production.

(3) The density of light or ultra-lightweight solid ceramsite is close to, even smaller than, the density of normal-used carrier fluid. In this case, the effect of gravitational deposition is extremely weak even negligible, which does no favor to the compaction during packing. The compaction degree depends on only the compact of fluid flowing. From the perspective of compaction and sand retention, light and ultra-lightweight ceramsite are not a good choice for horizontal and highly-deviated gravel packing,

especially in the case of using high-viscosity fluid. The lightweight solid material will be recommended for use only in one situation, under which the premature plugging of α wave packing can not be avoided under all of the parameters optimization.

Author Contributions: Conceptualization, H.S. and C.D.; methodology, H.S., Y.Y., X.Z. and C.L.; software, Z.L. and L.L.; validation, H.S., C.D. and Y.Y.; formal analysis, H.S., C.D., Y.Y. and J.J.; investigation, X.Z. and Z.L.; resources, C.D. and Y.Y.; data curation, H.S., X.Z., C.L., Z.L. and J.J.; writing—original draft preparation, H.S., X.Z., C.L., Z.L. and L.L.; writing—review and editing, C.D. and Y.Y.; project administration, Y.Y. and J.J.; funding acquisition, C.D. and Y.Y. All authors have read and agreed to the published version of the manuscript.

Funding: This research was funded by the National Natural Science Foundation of China (NSFC) (Grant No. 52074331), National Key Research and Development Program of China (No. 2023YFC2811005) and Guangdong Major project of Basic and Applied Basic Research (No. 2020B0301030003), the Key Areas of Science and Technology Programme Project (No. 2023ZD017), the Marine Geological Survey Program of China Geological Survey (No. DD20230063).

Institutional Review Board Statement: Not applicable.

Informed Consent Statement: Not applicable.

Data Availability Statement: The data that support the findings of this study are available from the corresponding author.

Acknowledgments: The authors wish to thank Qiuping Lu and Wenwei Xie from the GMGS for their help in experimental methods, data handling and effectiveness analysis.

Conflicts of Interest: Author Jianrong Ji was employed by the company East China Petroleum Bureau of China Petroleum & Chemical Corporation. The remaining authors declare that the research was conducted in the absence of any commercial or financial relationships that could be construed as a potential conflict of interest.

References

1. Paköz, U.; Ceyhan, A.G.; Aktepe, S.; Cruden, N.; Patey, I.; McLaughlin, R. Formation damage challenges and solutions in cased hole gravel pack completions in deep offshore unconsolidated laminated sandstone formations in the Sakarya Field. *Geoenergy Sci. Eng.* **2024**, *233*, 212527. [CrossRef]
2. Nie, S.; Li, H.; Hu, Z.; Wen, M.; Gao, S.; Zhang, H.; Luo, H.; Zhang, L. A review of the research status and development prospects for gravel packing sand control in horizontal wells. *Geoenergy Sci. Eng.* **2023**, *229*, 212152. [CrossRef]
3. Bhowmick, A.; Amritkar, A.; Ehlig-Economides, C. Determining Gravel Pack Fluidization Velocity. In Proceedings of the SPE Annual Technical Conference and Exhibition, Dallas, TX, USA, 24–26 September 2018. [CrossRef]
4. Li, Y.; Li, M.; Wang, L.; Yao, Z.; Meng, W. A Productivity Prediction Model for the Gravel-packed Horizontal Well. *Pet. Sci. Technol.* **2013**, *31*, 633–642. [CrossRef]
5. Firth, J.A.; Hamid, H.; Saldungary, P.M. Novel Techniques Enable Successful Gravel Packing of Long Horizontal North Sea Wells. In Proceedings of the SPE Europec, Online, 1–3 December 2020. [CrossRef]
6. Marques, L.C.; Pedroso, C.A. Cardinal Points to Achieve the Excellence in Horizontal Open-hole Gravel Packing Operations in Deep- and Ultra-deepwaters. In Proceedings of the SPE Annual Technical Conference and Exhibition, Denver, CO, USA, 30 October–2 November 2011. [CrossRef]
7. Martins, A.L.; de Magalhaes, J.V.M.; Ferreira, M.V.D.; Calderon, A.; de Sa, A.N. Sand Control in Long Horizontal Section Wells. In Proceedings of the Offshore Technology Conference, Houston, TX, USA, 4–7 May 2009. [CrossRef]
8. Wassouf, P.; Jain, S.; Dannish, G.; Ramnarine, A. Evaluation and Improvement of Gravel Pack Treatments Using Advanced Downhole Pressure Analysis. In Proceedings of the IPTC 2012: International Petroleum Technology Conference, Bangkok, Thailand, 7–9 February 2012. [CrossRef]
9. Colbert, F.C.; Costa, A.L.; Gachet, R.F.; Garcia, F.M.M.; Neto, A.T. Sand Control Technology for Extended-Reach Wells: Best Practices from the Longest Gravel Packs Performed Offshore Brazil. In Proceedings of the Abu Dhabi International Petroleum Exhibition & Conference, Abu Dhabi, United Arab Emerites, 13–16 November 2017. [CrossRef]
10. Parlar, M.; Tibbles, R.J.; Gadiyar, B.; Stamm, B. A New Approach for Selecting Sand-Control Technique in Horizontal Openhole Completions. *Spe Drill. Complet.* **2015**, *31*, 004–015. [CrossRef]
11. Sarraf Shirazi, A.; Frigaard, I.A. Gravel packing: How does it work? *Phys. Fluids* **2020**, *32*, 053308. [CrossRef]
12. Tan, M.; Li, Y.; Qi, M.; Wang, H.; Wang, Y.; Lu, J.; Chen, M.; Wu, H. A novel multi-path sand-control screen and its application in gravel packing of deepwater horizontal gas wells. *Nat. Gas Ind. B* **2022**, *9*, 376–382. [CrossRef]
13. Dong, C.; Zhang, Q. Real-time numerical simulation of gravel_packing process in horizontal wells. *Acta Pet. Sin.* **2004**, *25*, 96–100.

14. Mendez, A.; Curtis, J.; Evans, B.; Farias, R. A Quantum Leap in Horizontal Gravel-Pack Technology. In Proceedings of the SPE Latin American and Caribbean Petroleum Engineering Conference, Rio de Janeiro, Brazil, 20–23 June 2005. [CrossRef]
15. Zhao, X. Research and application of gravel-packing flow-regulation water-control screen completion technique in horizontal wells. *IOP Conf. Ser. Earth Environ. Sci.* **2019**, *227*, 042034. [CrossRef]
16. Changyin, D.; Jiajia, L.; Yanlong, L.; Huaiwen, L.; Lifei, S. Experimental and Visual Simulation of Gravel Packing in Horizontal and Highly Deviated Wells. In Proceedings of the SPE Latin America and Caribbean Petroleum Engineering Conference, Maracaibo, Venezuela, 21–23 May 2014. [CrossRef]
17. Chen, Z. Horizontal Well Gravel Packing: Dynamic Alpha Wave Dune Height Calculation and Its Impact on Gravel Placement Job Execution. In Proceedings of the SPE Annual Technical Conference and Exhibition, Anaheim, CA, USA, 11–14 November 2007. [CrossRef]
18. Maly, G.P.; Robinson, J.P.; Laurie, A.M. New Gravel Pack Tool for Improving Pack Placement. *J. Pet. Technol.* **1974**, *26*, 19–24. [CrossRef]
19. Gruesbeck, C.; Salathiel, W.M.; Echols, E.E. Design of Gravel Packs in Deviated Wellbores. *J. Pet. Technol.* **1979**, *31*, 109–115. [CrossRef]
20. Bigna, Y.; Oyeneyin, M.S.; Peden, J.M. Investigation of Pore-Blocking Mechanism in Gravel Packs in the Management and Control of Fines Migration. In Proceedings of the SPE Formation Damage Control Symposium, Lafayette, LO, USA, 7–10 February 1994. [CrossRef]
21. Sanders, M.W.; Klein, H.H.; Nguyen, P.D.; Lord, D.L. Gravel Pack Designs of Highly-Deviated Wells with an Alternative Flow-Path Concept. In Proceedings of the International Symposium and Exhibition on Formation Damage Control, Lafayette, LO, USA, 20–21 February 2002. [CrossRef]
22. de Magalhães, J.V.M.; Calderon, A.; de Oliveira, T.J.L.; Pires, I.J.; Martins, A.L. A New Gravel-Pack Approach Using Two Different Proppants in Long Horizontal Well Completions. In Proceedings of the SPE International Symposium and Exhibition on Formation Damage Control, Lafayette, LO, USA, 13–15 February 2008. [CrossRef]
23. Dong, C.; Wu, L.; Wang, A.; Liu, C.; Zhang, Q. Experimental simulation of gravel-packing in horizontal and highly deviated wells. *J. China Univ. Pet.* **2010**, *34*, 74–77.
24. Sarraf Shirazi, A.; Frigaard, I. A New Three-Layer Model for Gravel Packing Applications in Horizontal Wells. In Proceedings of the ASME 2019 38th International Conference on Ocean, Offshore and Arctic Engineering, Glasgow, Scotland, 9–14 June 2019. [CrossRef]
25. Xu, F.; Zhou, S.; Zhang, C.; Yu, Y.; Dong, Z. Analysis of Shunted Screen Gravel Pack Process and Calculation of Friction in Deepwater Horizontal Wells. *Geofluids* **2021**, *2021*, 4651199. [CrossRef]
26. Wen, M.; Huang, H.; Zhou, S.; Fan, B.; Qiu, H. Friction calculation and packing effect analysis for gravel packing in deepwater horizontal wells. *Energy Sci. Eng.* **2021**, *9*, 2380–2387. [CrossRef]
27. Peden, J.M.; Russell, J.; Oyeneyin, M.B. A Numerical Approach to the Design of a Gravel Pack for Effective Sand Control in Deviated Wells. In Proceedings of the SPE Annual Technical Conference and Exhibition, Houston, TX, USA, 16–19 September 1984. [CrossRef]
28. Dong, C.; Zhang, Q. A new probability model for predication of equilibrium sand-bed height for liquid-solid two-phase flowin horizontal pipe. *J. China Univ. Petroleum.* **2004**, *28*, 46–48.
29. Ojo, K.P.; Osisanya, S.O.; Ayeni, K.B. Factors Affecting Horizontal Well Gravel-Pack Efficiency. In Proceedings of the SPE Annual Technical Conference and Exhibition, San Antonio, TX, USA, 24–27 September 2006. [CrossRef]
30. Ojo, K.P.; Osisanya, S.O.; Ayeni, K.B. 3D Numerical Simulator for Horizontal-Well Gravel Pack. In Proceedings of the SPE Annual Technical Conference and Exhibition, San Antonio, TX, USA, 24–27 September 2006. [CrossRef]
31. Ojo, K.P.; Osisanya, S.O.; Ayeni, K.B. Development of a 3D Numerical Simulator of Horizontal Well Gravel Pack. In Proceedings of the Canadian International Petroleum Conference, Calgary, AB, Canada, 19–23 June 2006. [CrossRef]
32. Ojo, K.P.; Osisanya, S.O.; Ayeni, K. Factors Affecting Horizontal Well Gravel Pack Efficiency. *J. Can. Pet. Technol.* **2008**, *47*. [CrossRef]
33. Bai, Y.; Dong, C.; Ren, M. Visual Simulation of Horizontal Gravel-packing and Its Effect Evaluation. *Sci. Technol. Eng.* **2012**, *12*, 4149–4153.
34. Mimouna, A.; Verma, A.; Guddati, M.N. Generalized Framework to Simulate Gravel Packing in Wellbore Completions. In Proceedings of the Offshore Technology Conference, Houston, TX, USA, 20–23 March 2018. [CrossRef]
35. Huang, H.; Wen, M.; Xing, X.; Qiu, H.; Hou, Z.; Zhou, S. The Gravel Packing Length Determination Method and Influencing Factors Analysis in Deepwater Horizontal Wells. *Geofluids* **2022**, *2022*, 2912652. [CrossRef]
36. Nie, S.; Li, H.; Gao, S.; Hu, Z.; Luo, H.; Li, Q.; Ma, X.; Cui, X.; Liu, Z.; Zhang, L. Numerical simulation and experimental study on the whole process of gravel packing in horizontal wells. *Geoenergy Sci. Eng.* **2023**, *224*, 211603. [CrossRef]
37. Pedroso, C.A.; Sanches, E.D.; Oliveira, N.S.; Fernandes, L.H.; Gomes, M.; Farias, R.; Mendez, A.; Frata, F.M. Lightweight Proppants: Solution for Gravel Packing Horizontal Wells Under Extreme Conditions. In Proceedings of the SPE International Symposium and Exhibition on Formation Damage Control, Lafayette, LO, USA, 15–17 February 2006. [CrossRef]
38. Trujillo, H.; Tengono, J.A.; Hernández, A.; Castaño, R.; Ortiz, E.; Charry, W.; Anaya, L.A.; Portela, F.; Castillo, R.; López, M.; et al. Long Horizontal Gravel Pack with the Lightest Gravel Ever Used. In Proceedings of the SPE Latin American and Caribbean Petroleum Engineering Conference, Lima, Peru, 1–3 December 2010. [CrossRef]

39. Chen, Z. The Application of Light and Ultra-Light Weight Proppant in Horizontal Well Sand Control: Unified Model and Case Histories. In Proceedings of the SPE Deepwater Drilling and Completions Conference, Galveston, TX, USA, 20–21 June 2012. [CrossRef]
40. Denney, D. Ultralightweight Proppants for Long Horizontal Gravel Packs. *J. Pet. Technol.* **2012**, *64*, 152–156. [CrossRef]

Journal of
Marine Science and Engineering

MDPI

Article

Seafloor Subsidence Evaluation Due to Hydrate Depressurization Recovery in the Shenhu Area, South China Sea

Benjian Song [1] and Qingping Zou [2,*]

[1] Maxwell Institute for Mathematical Sciences, Heriot-Watt University, Edinburgh EH14 4AS, UK; bs2028@hw.ac.uk

[2] The Lyell Centre for Earth and Marine Science and Technology, Institute for Infrastructure and Environment, Heriot-Watt University, Edinburgh EH14 4AS, UK

* Correspondence: q.zou@hw.ac.uk

Abstract: Submarine hydrate mining can trigger geological disasters, including submarine landslides and seafloor subsidence due to excess pore pressure and weakened layers, which may potentially lead to the reactivation of faults and increased seismic activity. However, current research encounters challenges in assessing geotechnical issues associated with long-term and large-scale production from well grids located in sloped areas. Limited by the complexity of the hydrate sediment, a multifield coupled numerical model of hydrate slope in the Shenhu area was established. Utilizing the modified Mohr–Coulomb model as the constitutive model for hydrate-bearing sediments to track the dynamic reduction in strength and employing the shear strength method to assess submarine slope stability, a series of depressurization strategies are applied to evaluate the risks associated with submarine landslides and seafloor subsidence. Results show that the hydrate dissociation tends to stagnate after a period of mining. The strength of the hydrate decomposed area is severely reduced, and a volume deficit occurs in this area, causing formation displacement. The peripheral region of the decomposed area is compacted by high stress, resulting in a serious decrease in permeability and porosity, which limits the continued decomposition of hydrates. The large-scale submarine landslides with hydrates decomposition will not appear in this block. However, several meters' seafloor subsidence over a wide range risks engineering safety significantly. The amount of seafloor subsidence in the first 50 days is approximately half of the final settlement. A higher production pressure drop can speed up the recovery rate while resulting in more significant seafloor subsidence and slippage. Therefore, the balance between mining speed and formation stability needs more research work.

Keywords: natural gas hydrates; hydrate decomposition; numerical modeling; stratum response; seafloor subsidence; pore network destruction; deep sea mining; compaction

1. Introduction

Natural gas hydrate (NGH) is a crystalline substance with an ice-like appearance, consisting of water and hydrocarbon gas, which forms under conditions of low temperature and high pressure [1]. Typically, it is found in the upper strata of permafrost regions and in the sedimentary layers of deep ocean floors. The energy stored in a hydrate reservoir is more than twice that in traditional fossil fuels [2,3]. Although constrained by temperature and pressure conditions, meaning natural gas hydrates are not as ubiquitously distributed as traditional oil and gas reserves, they are still considered a significant potential alternative energy source soon due to their extensive reserves, broad distribution, minimal environmental impact, and high energy density [4,5].

Presently, there are five primary techniques for extracting NGH: depressurization, thermal stimulation, chemical injection, carbon dioxide replacement, and solid fluidization [6,7]. In 2007 and 2008, Canada tried to exploit land hydrates twice using the depressurization method and produced 2000–4000 m^3 hydrate after six days of continuous gas production [8]. In 2013, Japan conducted its inaugural offshore hydrate extraction trial using the depressurization method in the Nankai Trough. Over a period of six days, the operation successfully produced more than 12,000 m^3 of gas, and eventually ended with severe sand production [9]. In 2017 and 2020, the China Geological Survey conducted on-site hydrate extractions in the Shenhu area using the depressurization method, initially with vertical wells and subsequently with horizontal wells. The productions from these methods were 300,000 m^3 and 575,000 m^3, respectively [10]. The outcomes of these field experiments suggest that depressurization is practical for ocean NGH extraction. In contrast, there is a lack of field tests and resources to use other methods for large-scale production of hydrates. For instance, thermal stimulation may serve as a supportive technique in this process for the depressurization method but not the leading method alone for hydrate production at present [6,10–12]. In 2017, the solid fluidization method was first implemented in the Shenhu area. However, the results were not as successful as anticipated [7].

Initial research on hydrates primarily concentrated on the efficiency of hydrate dissociation and multiphase flow dynamics [13–16]. More recently, there has been a growing focus on the geomechanical issues that arise from hydrate dissociation [17–21]. When the hydrate is stable in seafloor sediments, it cements the formation particles together to form an integrated structure. During the dissociation of hydrates, the cemented structure would dissolve, and the huge volume of gas generated from NGH dissociation would further damage the skeleton structure of the seabed sediment [22–24]. The mechanical strength of hydrate-bearing zones significantly diminishes, resulting in geomechanical challenges such as sand production, seabed subsidence, and wellbore instability [25–28].

Depressurization reduces sediment pore pressure, destabilizing the marine substrate and potentially triggering submarine landslides. This process, by weakening sediment cohesion and reducing effective stress, can lead to large-scale seabed movements, especially on sloped terrains. When discussing the mechanical properties of sedimentary layers in their natural state, and assessing potential subsidence and landslides, it is crucial to determine the composition and type of the sediments, their sources, and whether features indicative of earthquakes and submarine landslides occurred during the sedimentation process. Jmail conducted detailed research on these problems in the South China Sea, including the past subsidence and slides induced by seismic activity [29–31]. These factors are equally essential in characterizing the nature of submarine sedimentary layers. Considering that the first NGH exploration site in the Shenhu area is situated on the northern slope, it is essential to evaluate the impact of hydrate dissociation on the stability of the seabed slope. However, currently there is a lack of systematic analysis, and pilot demonstration about the likelihood of large-scale dissociation of hydrates caused by humans will lead to large-scale submarine landslides in this area and the dynamic effect of hydrate dissociation on slope stability [32–34].

Hydrate decomposition is a highly complex multiphase problem involving gas–liquid–solid coupling, phase transformation, heat conduction, and geomechanical processes. The change in structure strength during hydrate dissociation further complicates the problem. Numerical simulation is the most robust and economic method to capture the coupled multiphase dynamic processes such as yield, efficiency, mechanical, and thermal response during hydrate dissociation. Currently, numerical simulations of hydrates have progressed from initial fluid–solid coupling to more complex thermo–gas–liquid–solid–chemical–mechanical coupling models [23,33].

In the earliest studies, around the 1980s, hydrate simulation was primarily based on the TOUGH software, focusing on the transport processes of multiphase fluids in porous media. Around 2002, Reagan developed TOUGH + Hydrate, which modeled the thermodynamics and fluid dynamics of hydrates, though it did not include geomechanical behavior [11]. By around 2010, the incorporation of Biot's theory of poroelasticity led to the development of TOUGH + Hydrate + Biot, which was among the first to evaluate the geological impacts of hydrate-related activities. This approach has been used to the present day, as exemplified by Jin et al.'s study of seabed subsidence behaviors resulting from NGH extraction via depressurization in horizontal wells, exploring various formation failure modes during hydrate dissociation [6].

Around 2015, the introduction of Flac3D into Hydrate + Tough allowed for a more detailed capture of the mechanical behavior during the decomposition of hydrate reservoirs, as demonstrated by Dong et al.'s simulation of hydrate sediment deformation behavior during hydrate recovery [35]. Additionally, Gupta et al. developed a C++ software package to simulate thermo–chemo–hydro–geomechanical coupling during hydrate dissociation, utilizing a linear elastic model, and analyzed the dynamic deformation response of hydrate sediment sample cores through indoor experiments [22].

With the advancement and refinement of large-scale commercial finite element software, such as ABAQUS and COMSOL, more sophisticated descriptions of multiphysical coupling problems in porous media have become possible. ABAQUS, in particular, has shown accuracy and reliability in geomechanical simulations and can be extended via subroutines [1,5,25]. Importantly, ABAQUS allows for the implementation of the strength reduction method through field variables, which is crucial for analyzing issues related to the stability of submarine slopes caused by hydrates [1,25].

It is generally acknowledged that NGH dissociation can lead to seabed subsidence, potentially destabilizing submarine slopes and triggering submarine landslides [36–38]. However, at present, there is a lack of systematic and comprehensive studies of submarine slope subsidence and submarine landslides under hydrate-dissociation conditions. Moreover, to the best of our knowledge, no studies have been conducted to examine the likelihood of landslides with a set of hydrate production wells. By taking this into consideration, this study aims to explore the effects of submarine slopes on NGH recovering in a series of vertical wells by simulating the whole stratum response of submarine slopes using the finite element software ABAQUS 2022 secondary development.

2. Methodology

2.1. Solution Strategy

There are three major challenges in the thermo–chemo–hydro–geomechanical coupled simulation of hydrate dissociation: (1) Kinetics of methane hydrate dissociation and chemical reaction, (2) Changes in formation strength due to hydrate dissociation, (3) Thermal–multiphase coupling in porous media. Based on existing studies, we have simplified these problems so that they can be coded into ABAQUS to perform coupling analysis. For the item (1), a relationship between formation temperature, pressure, permeability, time, and the rate of hydrate decomposition was established. This relationship was used to update the distribution of hydrate saturation within the formation. For the item (2), an enhanced hydrate constitutive model was developed based on the modified Mohr–Coulomb criterion, incorporating the effects of hydrate saturation, is constructed to describe the dynamic changes in structure strength during hydrate dissociation [36]. For the item (3), ABAQUS integrates an extensive well-established hypothesis, formula, and governing equations to analyze the thermal–multiphase–mechanical coupling behaviors in porous media and has been validated against benchmark tests [37].

The hydrate sediments consist of gaseous, liquid, and solid phases (sand and hydrate). We assume that the first three phases fully fill the void space, and the solid phase comprises the bearing structure of total sediment [22,38,39]. The non-dominant factors in practical sediment processes, such as dissolution, evaporation, and attach effect, are neglected. It is also assumed that the liquid phase does not contain any soluble gas and once the hydrate dissociates, no water and gas is attached to the surface of NGH. The saturation of gas, water, and NGH in a porous medium is described by the following equation:

$$s_i = \frac{V_i}{V_p} \tag{1}$$

where, s_i represents the saturation of phase i, g for gas, w for water, and h for hydrate.

2.2. Hydrate Model

The methane hydrate phase equilibrium relationship curve is given by Moridis [15]:

$$CH_4 \cdot nH_2O + heat \leftrightarrow CH_4(gas) + nH_2O(water) \tag{2}$$

In this study, the value of n is 5.5~6, with a chosen value of $n = 6$ for the analysis according to Moridis [15] to simplify the model although it may be changed with P-T conditions.

The phase boundary of methane hydrate is described by the following expression [40].

$$P_{eq} = \exp\left(e_1 - \frac{e_2}{T}\right) \tag{3}$$

where P_{eq} represents the phase equilibrium pressure at temperature T, e_1 and e_2 are the regression parameters, e_1 = 39.08, and $e_2 = 8533$ for temperatures above 0 °C.

Sun pointed out that a complete description of the hydrate decomposition kinetics would involve complex physical and chemical theories, but it can be simplified to include the formation temperature and pressure conditions, the dynamic permeability of the porous medium, and the time function [41,42]. Hydrate-dissociation kinetics can be described as follows:

$$\begin{gathered} -g^h = M_h k_{reac} \Gamma_r A_s \left(P_{eq} - f_g\right) \\ g^g = g^h \frac{M_g}{M_h} \\ g^w = n\, g^h \frac{M_w}{M_h} \\ A_s = \sqrt{\frac{\varphi_e^3}{2k}} \quad \Gamma_r = \left(s_g s_w s_h\right)^{\frac{2}{3}} \end{gathered} \tag{4}$$

where g^i(phase $i = h, g, w$) is the hydrate, gas, and water generation rate, in kg/m^{3}s. k_{reac} is the hydrate decomposition kinetic constant, mol/m^{2}Pa s. f_g is the methane gas fugacity, which can be derived from Peng–Robinson EOS, Pa. When the temperature changes, P_{eq} and f_g will change accordingly, thus affecting the rate of hydrate decomposition [42]. A_s quantifies the total surface area where the hydrate phase is in contact with both liquid and gas phases, Γ_r measures the proportion of the surface area actively involved in hydrate decomposition relative to the total contact area. Sun and Mohanty modeled and calculated A_s and Γ_r [42].

The saturation of NGH markedly influences the permeability within bearing formations. When the NGH exists to stabilize the sediment, it does not contribute to multiphase flow processes in the pores or pore fraction. After the hydrate is decomposed, however, as effective porosity increases, so too does formation permeability. Therefore, the permeability of the NGH bearing sediment equation is given by [15,23]:

$$\begin{gathered} k = k_0 \left[\frac{\varphi_e}{\varphi_0}\right]^5 \times \left[\frac{1-\varphi_0}{1-\varphi_e}\right]^2 \\ \phi_e = \phi_0 \cdot \left(s_g + s_w\right) \end{gathered} \tag{5}$$

where k is the permeability of NGH bearing sediment, which is influenced by both start NGH saturation and the decomposed rate. k_0 is the initial permeability of the NGH bearing sediment. ϕ_e is defined as the effective porosity. ϕ_0 refers to the start porosity.

2.3. Hydrate Constitutive Model

The constitutive model of hydrate-bearing sediments is fundamental to addressing geomechanical issues related to hydrate formations. When hydrates are present in the strata, they not only exhibit high intrinsic strength but also provide cementation and support to the formation. As hydrates dissociate, these reinforcing effects diminish, leading to a significant reduction in the overall strength of the formation [43]. Results from triaxial tests indicate that the constitutive models traditionally used for describing rocks or soils, such as the Duncan–Chang model, the Cam–Clay model, and the Mohr–Coulomb model, cannot be directly applied to hydrate-bearing formations. However, these models can be modified to develop a suitable constitutive model for hydrate-bearing sediments [39].

On the basis of the modified Mohr–Coulomb model, the hydrate formation strength equation with consideration of hydrate saturation is given by [39]:

$$c = c_s + c_h = c_1\left(1 - e^{-\frac{\sigma_3'}{\sigma_t'}}\right) + c_2 s_h^{c_3} \tag{6}$$

$$sin\varphi = sin\varphi_s + sin\varphi_h = sin\left(\delta_1 - \delta_2 \ln\left(\frac{\sigma_3'}{1_{MPa}}\right)\right) + \delta_3 s_h\left(1 - e^{-\frac{\gamma}{\delta_4}}\right) \tag{7}$$

$$E = E_s + E_h = e_1\left(\frac{\sigma_3'}{1_{MPa}}\right)^{e_2} + e_3 s_h \tag{8}$$

where the c_s and c_h are cohesiveness of sand and hydrate, respectively. φ, φ_s, φ_h denote the dilation angle of overall formation, sand and hydrate, respectively. Likelihood E, E_s, E_h represent the Young's modulus of overall formation, sand and hydrate, respectively. All other parameters in Equations (6)–(8), are hydrate-bearing formation skeleton coefficient. Different hydrate sediment skeleton coefficients are used for different areas and block [9]. According to Pinkert et al. [39], we take a typical hydrate sediment sample from Nankai Trough as the study object [44]. And all of the variables and constants in Table 1 are obtained through the optimization process [36].

Table 1. Skeleton coefficient of hydrate-bearing sediments.

Parameter	Value	Unit	Parameter	Value	Unit
c_1	293	kPa	e_2	0.642	
c_2	1960	kPa	e_3	450,000	kPa
c_3	1.7		σ_3'	1962	kPa
δ_1	0.4		σ_t'	382	kPa
δ_2	0.1		δ_3	0.22	
e	2.72		δ_4	0.04	
e_1	241,200	kPa	γ	9	

The relationship between the strength of hydrates and their saturation is delineated in Figure 1, according to the formula.

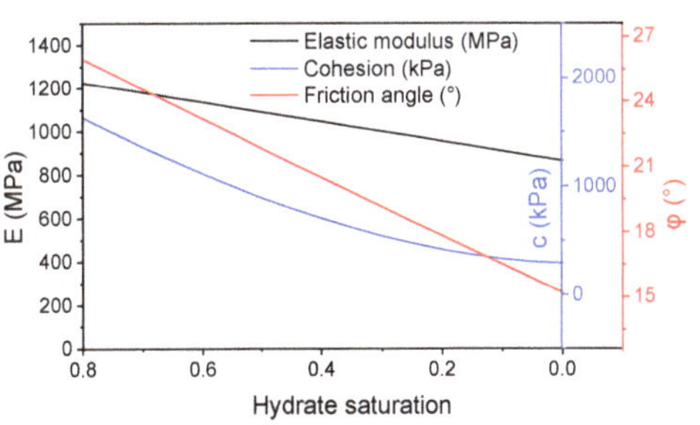

Figure 1. Relationship between hydrate saturation and hydrate formation strength parameters: black curve and corresponding axis for elastic modulus, blue curve and axis for cohesion, and red curve for friction angle.

2.4. Analysis of Porous Media

Hydrate-bearing formation is modeled as porous medium in ABAQUS by using multifield fully coupling solution. ABAQUS integrates extensive hypotheses, formula, and governing equations to describe the porous media behaviors in soil and rock [5,26,45]. The equations involved can be integrated into ABAQUS by embedding them through the

USDFLD subroutine. Adhering to the principle of effective stress, the model solution is derived in four sequential steps:

Step 1: Discretization of the porous medium, achieving a state of discrete equilibrium.

Step 2: Calculation of the constitutive behavior of the fluid and the skeletal structure of the discretized material.

Step 3: Fulfillment of the continuity equation.

Step 4: Coupling of the temperature and pressure fields with the diffusion and transport behavior of substances.

In the solution process, it is also necessary to solve the effective stress criterion, discrete equations, control equations, continuity equations, constitutive equations, and thermodynamic equilibrium equations.

3. Numerical Model Set Up

3.1. Site Location and Description

Located on the north slope of the South China Sea, the Shenhu area has slopes over 10 km long and 5 km wide, with successful hydrate recovery operations conducted by the China Geological Survey in 2017 and 2020. Seabed temperatures range from 2.3 to 3.7 °C, and water depths exceed 1000 m. The area features a stratum temperature gradient of 30–45 °C/km, hydrate sediment thickness of 10–120 m, and porosity of 33–48%. Hydrate saturation varies significantly, ranging from 0.2 to 0.9. Over 96% of the dissociation products are methane, with the rest being ethane and propane. The mechanical characteristics of the surrounding sediments are akin to those of fully saturated hydrates [6].

3.2. Model Domains

Based on exploration survey data from the Shenhu area and typical parameters of a hydrate reservoir, the numerical model of the stratum is structured into three layers: the top layer consists of overburden sediment with a thickness of 300 m, the middle layer comprises a hydrate zone of 100 m thickness, and the bottom layer includes underburden sediment ranging from 600 to 1100 m in thickness. As shown in Figure 2, this model is set up for a gentle slope of 3.18°. Two plateaus on both sides of the slope have the same length of 500 m in x direction. Meanwhile, the slope is 9000 m long in x direction. The top and toe of the slope left side is 1500 m and 1000 m above the zero-displacement boundary. The apex of the slope is situated 1000 m below sea level, and the top width extends 2000 m in the y direction [14,46]. It is assumed that each layer is made of homogeneous and isotropic material. There is a series of vertical wells with the 0.1 m radius on the x–z plane and the number of wells and the space between wells is determined by the recovery scenario [47,48].

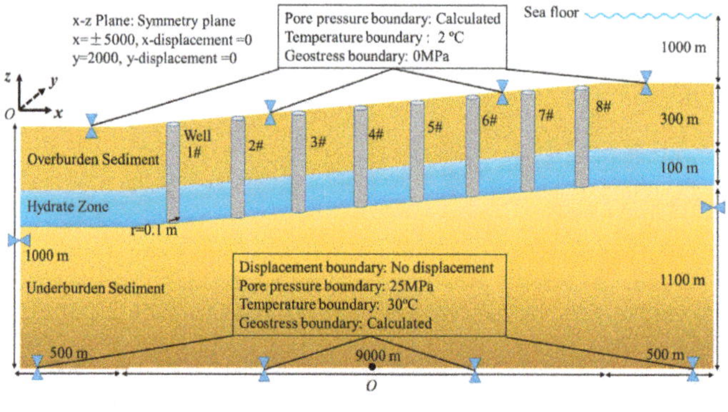

Figure 2. Model diagram and initial conditions, including well positions and numbers, seabed domain, hydrate zone distribution, and boundary conditions.

It is expected that there will be many ways to produce hydrates commercially in the future. It is not practical to predict the method which will be used in the future and analyze the corresponding stratum stability. Therefore, the recovery scheme adopted in this paper is assumed to be the most likely used method for primary hydrate productions. The potential geological problems in the block and the extent and magnitude of its instability can then be predicted.

3.3. Initial State and Boundary Conditions

Referring to Figure 2, the top and bottom surface are the important boundaries in the numerical model. The top surface is the seafloor surface with a temperature equal to the seawater and a pore pressure equal to the hydrostatic pressure from overlying water. And the bottom surface is considered as the non-influenced boundary. Therefore, the displacements of these two surfaces are fixed in the vertical direction. The temperature, pore pressure, and geostatic pressure are considered as functions of depth. Assuming the wellbore will not be destroyed, the wellhole is fixed in the horizontal direction. Meanwhile, the wellhole is the boundary of depressurization pressure and production surface. Other parameters such as hydrate saturation, porosity, and permeability are set as initial conditions.

The overburden stress increases with depth and densities which are assumed to be constants in the overburden and underburden sediment layers and calculated in the hydrate zone. The top and bottom domain sides are temperature boundaries at 2 °C and 30 °C, respectively [6,14]. The model developed in this study simulates the thermal variations induced by hydrate dissociation and regeneration, resulting from changes in formation pore pressure, which consequently alters the temperature field of the formation. It also captures the dynamics of formation pore pressure and fluid migration during the depressurization of formation hydrates. Key seepage and thermodynamic parameters are detailed in Table 2.

Table 2. Thermal and physical properties of numerical model.

Parameter	Value	Parameter	Value
Porosity	0.4	Sand instinct permeability	6 mD
NGH density	920 kg/m^3	NGH instinct permeability	2 mD
NGH thermal conductivity	0.5 W/m K	Liquid density	1040 kg/m^3
NGH specific heat	2.1 kJ/kg K	Liquid specific heat	4.2 kJ/kg K
Sand density	2600 kg/m^3	Liquid thermal conductivity	0.5 W/m K
Sand thermal conductivity	1.0 W/m K	Liquid saturation of Sand	1
Sand specific heat	1.0 kJ/kg K	Liquid saturation of sediment	0.2

For the initial states of model domain, the initial porosity is 0.4, while the permeability for non-hydrate-bearing formations is 0.6 mD. Specifically, the initial permeability of the hydrate formation is fixed at 0.2 mD, with post-dissociation permeability calculated using Equation (5). Additionally, thermal properties of the materials are dependent on temperature conditions. To simplify calculations, the heat transfer coefficient is maintained at a constant value.

When all these boundary conditions and the initial parameters are input and drive the Geostatic Step in ABAQUS, the geostatic balance state of the stratum will be output at the end of the step. This state can be considered as the origin state of the stratum with the preliminary distribution of multifield, including stress, strain, hydrostatic, and pore pressure, etc. This preprocessing analysis helps the model predictions to achieve desirable accuracy without changing the initial load.

3.4. Recovery Scenarios

The depressurization method is extensively employed in NGH recovery experiments and represents one of the most economical and practical approaches to extracting NGH [13]. Figure 3a shows the hydrate equilibrium curve and formation and hydrate reservoirs P-T conditions. The x axis is temperature. The x coordinate of the left and right edge of the green area is −3500 m and 3500 m. There are eight vertical wells and the distance between wells is 1000 m. Figure 3b shows the recovery scenario tested in the simulation. In the diagram, the x axis represents a constant temperature for each individual production well, while the y axis indicates the production pressure at the well bottom. Five scenarios of production pressure are considered, creating five cases per well, depicted as green rectangles representing the production conditions. When the P-T conditions exceed the equilibrium curve, NGH decomposes into gas and water. The black points in Figure 3b illustrate the P-T conditions at the well bottom. When the production pressure is 8 MPa or 9 MPa, there are two or three wells that would not have hydrate decomposition. Therefore, the comparison between the model results for the well with and without hydrate decomposition provided the evaluation of the effect of hydrate dissociation on seafloor subsidence.

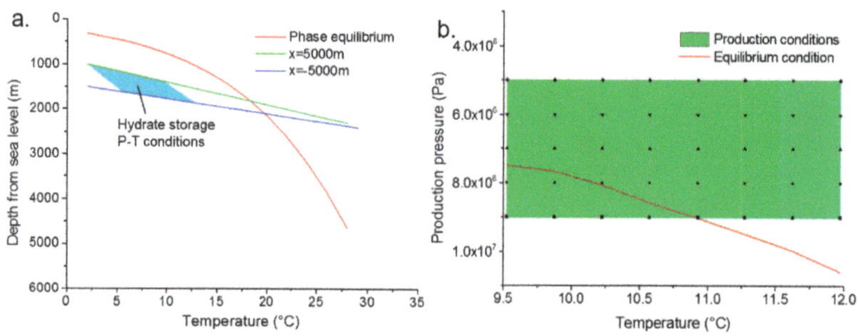

Figure 3. (**a**) Hydrate phase equilibrium conditions and strata P-T conditions and hydrate existence conditions (**b**) recovery scenario conditions.

4. Results and Discussion

4.1. Procedure of Hydrate Dissociation

Figure 4 illustrates the cumulative gas volume across wells 1 to 8, highlighting the substantial influence of formation conditions on the degree of hydrate dissociation [22,23]. The gas production process is segmented into four distinct stages: initial decomposition, rapid production, production decline, and the stable stage. The duration from the onset to the conclusion of the third stage, along with the cumulative gas production, is significantly shaped by the recovery conditions. As depicted in Figure 3b, a greater deviation from the recovery condition to the phase equilibrium curve correlates with higher cumulative gas production and a shorter transition to the stable stage.

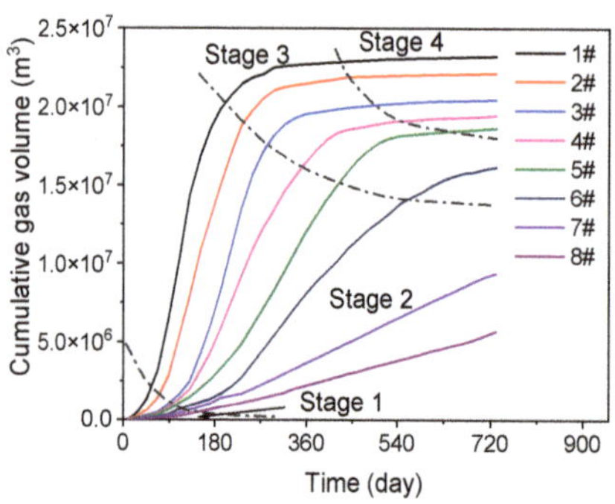

Figure 4. Accumulated gas production in different wells with regards to different stages.

Hydrate decomposition results in dynamic alterations in the mechanical behaviors of the formation, as depicted in Figure 5, which displays the spatial distribution of hydrate saturation at various times under a production pressure of 7 MPa. The rate of hydrate decomposition is markedly higher in the deeper parts of the formation compared to the shallower regions, influenced by the formation temperature, pore pressure conditions, and production pressure. Notably, Figure 5 shows that the periphery of the decomposed area forms a continuous dark blue ring where the hydrate saturation exceeds the initial conditions. Furthermore, as hydrate decomposition progresses, the saturation in this region increases and eventually stabilizes.

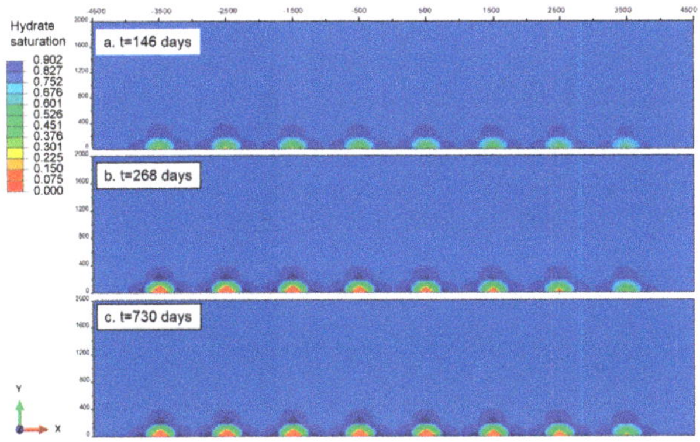

Figure 5. Time evolution of hydrate saturation distributions around wells and different times.

The depressurization method enhances the effective stress within the formation, prompting hydrate breakdown. Concurrently, the strength of the decomposed zone deteriorates significantly, rendering it unable to withstand the overburden pressure. Therefore, the surrounding area must bear the part of the extra geostress which was previously born by the hydrate decomposition zone. Correspondingly, decomposed gas and water may reform into hydrate in this area due to higher stress, leading to a slight increasing of hydrate saturation. Subsequently, the increase in effective stress leads to compaction of the surrounding region, resulting in a decrease in both effective porosity and permeability. This would lead to a drastic reduction in seepage properties of the surrounding area and pore pressure trapping. Further hydrate dissociation would then be trapped in this area so that the hydrate dissociation would stagnate when the hydrate is exhausted.

4.2. Stratum Geomechanics Response

4.2.1. Pore Pressure Distribution

The reduced bottom hole flowing pressure leads to a redistribution of the formation's pore pressure field. Given that variations in bottom hole temperature are not accounted for in this scenario, changes in pressure become the primary driving force for hydrate decomposition. Figure 6 illustrates the distribution of pore pressure and pressure drop across the slope, captured under a production pressure of 7 MPa at 730 days. It shows that the reduction in pore pressure is not evenly distributed in the formation. At this time, the left half stratum has reached the stable stage so that the hydrate saturation almost remains stable in this area while it slowly changes in the right half stratum. The closer to the wellbore, the larger the pressure drops at a certain distance. The high pressure drop (more than 7 MPa) only occurs in a narrow area next to the wellbore, which is the only region where the hydrate may decompose (Figure 6b,c). Due to the poor seepage capacity of the overlying and underlying rock formations, as depicted in Figure 6a, the low pressure within the hydrate formation exerts a minimal impact on adjacent formations. Ultimately, influenced by the pore pressure at the boundaries, the rate of pore pressure drop within the formation decelerates and generally stabilizes.

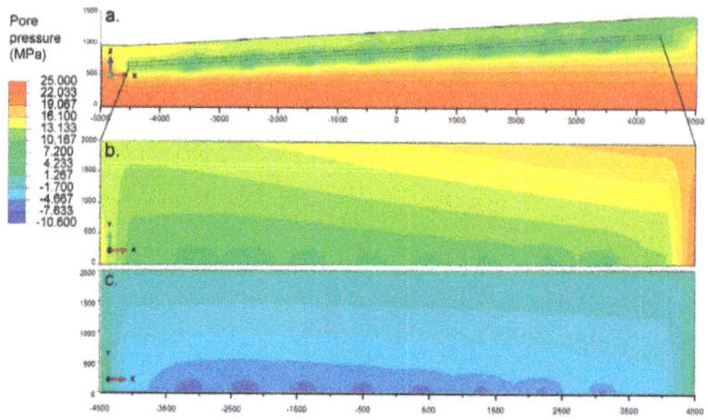

Figure 6. Pore pressure distributions (**a**) overview; (**b**) hydrate sediment profile view for the rectangular box indicated in (**a**); (**c**) zoom in hydrate sediment profile view of pore pressure drop distribution for 730 days.

4.2.2. Stress Distribution

The stress that the formation skeleton withstands is the effective stress, and the formation would strain when the effective stress changes. Figure 7 shows the minimum principle stress and strain in hydrate formation after two years of production. Apparently, the minimum principal stress distribution is spatially consistent with the hydrate saturation distribution (Figures 5 and 7b). If the skeleton strength remained constant, a reduction in pore pressure would correspondingly increase the effective stress. However, the decomposition of hydrates significantly weakens the formation's strength, altering this dynamic. In the formation, high strength corresponds to high stress and vice versa. Therefore, in a hydrate decomposed or decomposing area, the shorter distance to the wellhole, the lower the effective stress. Meanwhile, the effective stress in the peripheral region of the hydrate decomposition area is considerably increased because the overburden sediments would not completely collapse into the hydrate decomposition and collapse region due to their mechanical strength. Thus, part of the overburden pressure initially borne by the hydrate decomposed zone is transferred to the undecomposed area. Therefore, the support for the overlying layer and the compaction for the underlying layer of the hydrate decomposition zone is reduced. Additionally, the decomposition of hydrates leads to weakened compaction in the underlying layer, causing the underlying formation to experience strain as pronounced as that in the overlying formation, as illustrated in Figure 7c. This phenomenon results from the interplay of pore pressure, effective stress, and the compacting effects exerted by both overlying and underlying formations; the hydrate saturation would eventually reach an equilibrium stable state. The formation area changes differently in response to different formation strength.

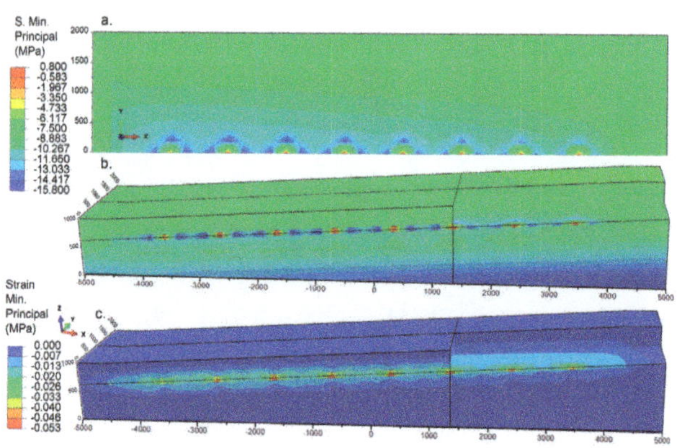

Figure 7. (**a**) Profile view of hydrate zone; (**b**) stereoscopic view of Minimum principle stress distributions; (**c**) minimum principle strain distribution.

4.2.3. Seepage Response

Hydrate saturation, porosity, permeability, pore pressure, and minimum principal stress distribution have similar spatial patterns. Figure 8 shows the top view of the effective porosity and the absolute permeability of hydrate formations in eight wells after two years of production. The degree of hydrate decomposition decreases gradually from well 1# to well 8# spatially. Coinciding with hydrate decomposition degree, the maximum porosity and permeability increase accordingly. As shown in Figure 8a, the effective porosity in the hydrate decomposed area decreases with distance from the wellbore. In the surrounding region, the effective porosity is compacted to 0.008, which is under a large negative impact on seepage. The permeability in this area also reduced from 0.4 md to 0.069 md as the hydrate decomposed area and surrounding area is compacted due to low strength and the stress transfer. Therefore, the hydrate decomposed peripheral region becomes a hypotonic zone which prevents the lower pressure from wells from propagating far away. This zone traps more than half of the pressure drop in a small area so that the hydrates cannot decompose continuously. At the same time, permeability determines whether the products of hydrate dissociation can be rapidly transported, significantly affecting the relative concentrations of gas and liquid phases, as well as the hydrate phase, around the site of dissociation. This, in turn, influences the subsequent rate of hydrate dissociation. Therefore, seepage plays a crucial role in the hydrate-dissociation processes and significantly impacts the entire system.

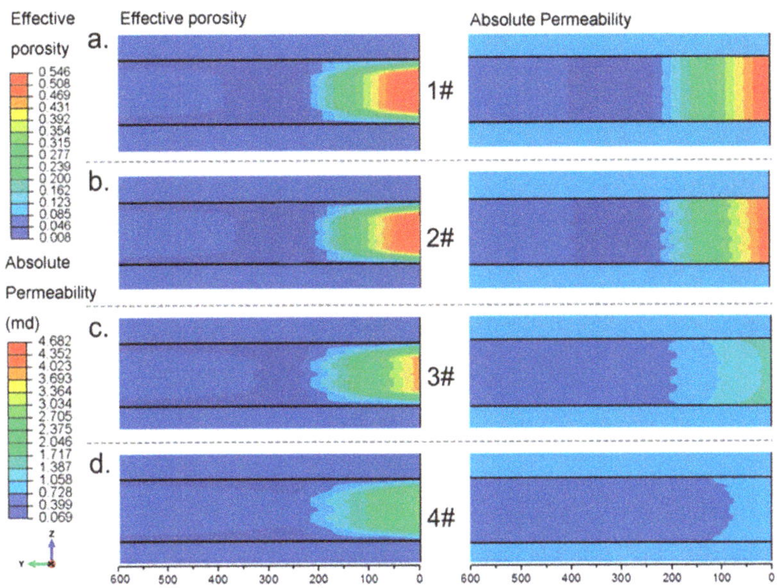

Figure 8. Effective porosity and permeability distributions in different wells after two years of production (**a**) well 1#, (**b**) well 2#, (**c**) well 3#, (**d**) well 4#.

4.3. Seabed Strata Stability

Currently, there is a lack of systematic evaluation criteria for instability of submarine strata. It is feasible to quantify the stability of submarine formations by evaluating the displacement components and strain components in formation in order to determine if any of the related material has reached the plastic deformation stage. Figure 9 shows only a small range of plastic deformation occurs in the hydrate fully decomposed area. It confirms that the plastic yielding remains negligible in the entire stratum. Therefore, the main risk associated with the formation is seafloor subsidence, stratum displacement due to formation strength reduction, and volume deficit, rather than large-scale submarine landslides due to plastic failure.

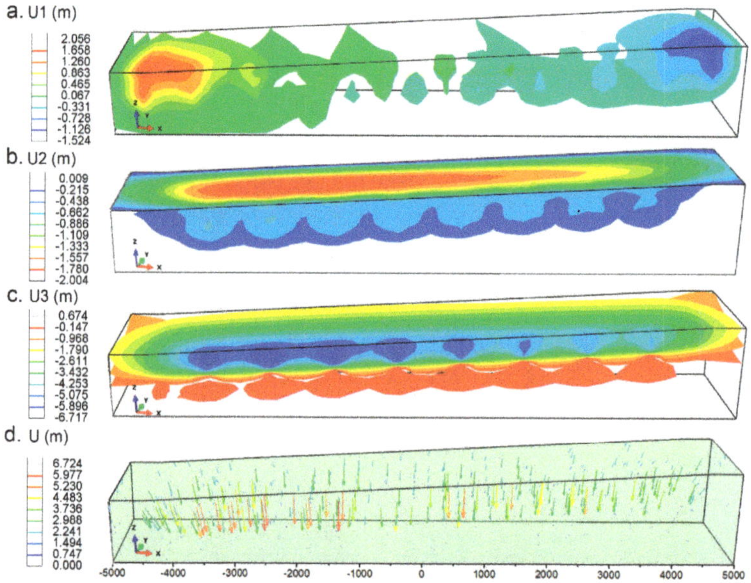

Figure 9. Distributions of the strata displacement component in different directions (**a**) x, (**b**) y, (**c**) z, and (**d**) direction and magnitude.

Figure 9 shows the displacement component in three directions. Figure 10a shows that the displacements along -xirection occurs in the slope top with a maximum of 2 m and the displacement along x direction occurs in the slope foot area with a maximum of 1.5 m. The displacements in the x direction in these two areas are caused by the seafloor subsidence. The strata exhibit minimal effective horizontal displacements due to hydrate dissociation. Model results suggest that hydrate dissociation does not trigger landslides across the entire slope in this block. A comparison between Figure 10c,d shows that displacements in the x direction, observed at the top and toe of the slope, are caused by stratum subsidence. Furthermore, hydrate dissociation results in the settling of the formation around the well, prompting surrounding formations to shift toward the center of the wellbore. Simultaneously, the slope top and toe are stress concentration areas which means that changes of formation physical parameters are more likely to produce a significant effect in these two regions.

Figure 9c shows the displacement component in the z directions at each well corresponding to settlement and rise of the stratum. Maximum subsidence in the negative z direction is 6.71 m which is almost triple that in the other two directions. The strength and density reduction in the hydrate sediment results in subsidence of the overburden strata and rising of the underburden strata. Combined with the hydrate residual saturation result in Figure 5c, the conclusion can be drawn that a higher degree of hydrate dissociation leads to a greater extent and scale of subsidence in submarine formations. Conversely, the uplift of the underlying formation is not solely dependent on hydrate saturation but is also influenced by the effective stress. With reference to Figure 9b, the distribution of the displacement in the y direction is correlated to the seafloor subsidence. The displacement vectors (arrows) and magnitudes (color) in Figure 9d indicate that the displacement directs towards the deep stratum and the middle of the slope.

In summary, the submarine strata movement is dominated by subsidence while the large-scale landslide is mainly in horizontal directions. This result indicates that the geological hazard that can be easily monitored during and after the hydrate production is submarine subsidence rather than landslides. In addition, the shear strength of cement is much lower than its compressive strength so that the horizontal displacement tends to generate a strong sheer stress on the cement ring; therefore, wellbore instability may occur. However, the large displacement in the vertical direction would influence the whole engineering and the result is hard to predict. Due to the complex mechanical behavior of hydrate sediment, it is necessary to conduct triaxial experiments on hydrate sediment constitutive model and strength dynamics to assess the geomechanical responses reliably.

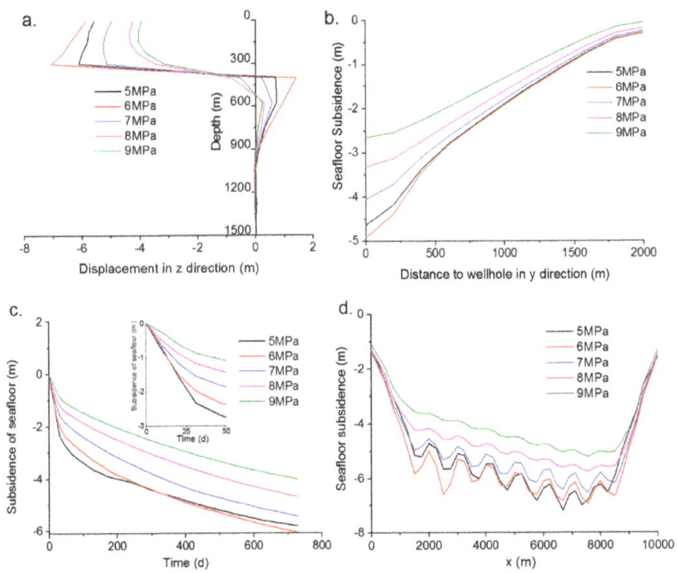

Figure 10. Evolution of displacement under different production pressure (**a**) along z direction at the well NO. 6# after 2 years' production; (**b**) along y direction at the well NO. 6# after 2 years' production; (**c**) seafloor subsidence with time at the well NO. 6#; (**d**) total seafloor subsidence on the x–z plane.

4.4. *Influence of Recovering Pressure on Formation Stability*

The strata stability response to production pressure can be evaluated through strata displacement and seafloor subsidence. Figure 10 shows the displacement of the submarine strata after using the depressurization method for two years with different production pressures. As depicted in Figure 10a, when the production pressure exceeds 6 MPa, the sedimentation of the overlying stratum and the rise of the underlying stratum increases with decreasing pressure. The settlement and uplift of the formation under a 5 MPa production pressure are less than that under a 6 MPa production pressure, which suggests that lower production pressure does not necessarily lead to greater formation displacement. This phenomenon can also be found in other subplots of Figure 10. For instance, the distribution of the seafloor subsidence in the y direction and the maximum seafloor subsidence around well 6# under the 6 MPa production pressure is greater than that under the 5 MPa production pressure, whereas the subsidence under higher production pressure is less than that under 6 MPa (Figure 10b). Excessively lower production pressure causes rapid decomposition of hydrate, which in turn leads to a rapid settlement of the formation. Rapid settlement decreases the hydrate decomposability range and compacts the surrounding formation prematurely. As a result, the hydrate decomposed range is reduced, and the settlement is reduced. Meanwhile, production pressure has a significant impact on formation settlement.

Figure 10c shows that the subsidence at 50 days under 5 MPa is equal to that at 730 days under 9 MPa and that the formation settlement for 30 days is about half of the sedimentation for two years. The seafloor level experiences a sharp decline due to an initial significant drop in pore pressure at the well, followed by a gradual decrease. Notably, the estimated seafloor subsidence in the first 50 days accounts for over half of the total subsidence observed after two years of production, as shown in Figure 10c. Specifically, seafloor subsidence reaches approximately 3.8 m after half a year and 6.2 m after two years at a production pressure of 6 MPa. The extensive range of seafloor subsidence in the horizontal direction is attributed to the low strength of unconsolidated sediments. This indicates that even a minor reduction in pressure at the production well can lead to significant seafloor subsidence, potentially causing wellbore instability and other engineering hazards and failures. Therefore, it is critical to take production pressure and related geological problems into consideration during hydrate production.

5. Conclusions

This study develops a thermal–multiphase–mechanical coupled model of hydrate sediment using ABAQUS, achieved by programming the USDFLD subroutine. It evaluates the stratum stability in response to varying production pressures applied during the depressurization recovery method. Key findings from this study include the following:

(1) The rate of hydrate dissociation is influenced by fluctuations in production pressure; an increase in pressure affects the dissociation rate. Additionally, there is a positive correlation between the distance from the hydrate temperature and pressure to the phase equilibrium curve and the rate of hydrate dissociation.

(2) The reduction of pore pressure and the corresponding increase in effective stress due to hydrate decomposition impact the overall stability of the formation. Stress redistribution leads to compaction of the surrounding strata, decreasing porosity and permeability, and trapping pore pressure, which further influences hydrate dissociation rates and patterns.

(3) Hydrate dissociation in the Shenhu area is unlikely to cause large-scale submarine landslides. However, it may result in several meters of seafloor subsidence across a broad area. Notably, the subsidence observed during the first 50 days of hydrate production accounts for approximately half of the total settlement observed after two years of production. Therefore, monitoring seafloor subsidence dynamics in the early stages of production is essential.

(4) While a higher production pressure drop can accelerate the recovery rate, it also increases the risk of severe seafloor subsidence and slippage. Therefore, it is crucial to strike a balance between the rate of hydrate production and the stability of the formation in project design and planning.

While this study initiates the discussion on the geomechanical implications of hydrate production using vertical well grids, both experimental and field tests indicate that horizontal wells are more effective for hydrate extraction but also present more serious geotechnical challenges. In future work, we aim to analyze the impact of horizontal wells on these geomechanical issues.

Author Contributions: B.S. conceptualized the study, conducted the simulations, performed data analysis, and authored the manuscript. Q.Z. provided oversight for the research, offered essential critiques, and contributed to manuscript revisions. All authors have read and agreed to the published version of the manuscript.

Funding: The research received financial support from Professor Qingping Zou through the Natural Environment Research Council of the UK (Grant No. NE/V006088/1).

Institutional Review Board Statement: Not applicable.

Informed Consent Statement: Not applicable.

Data Availability Statement: The data that support the findings of this study are available from the corresponding author upon reasonable request.

Conflicts of Interest: The authors declare no conflicts of interest.

References

1. Song, B.; Cheng, Y.; Yan, C.; Lyu, Y.; Wei, J.; Ding, J.; Li, Y. Seafloor subsidence response and submarine slope stability evaluation in response to hydrate dissociation. *J. Nat. Gas Sci. Eng.* **2019**, *65*, 197–211. [CrossRef]
2. Gajanayake, S.M.; Gamage, R.P.; Li, X.-S.; Huppert, H. Natural gas hydrates—Insights into a paradigm-shifting energy resource. *Energy Rev.* **2023**, *2*, 100013. [CrossRef]
3. Yan, C.; Chen, Y.; Tian, W.; Cheng, Y.; Li, Y. Effects of methane-carbon dioxide replacement on the mechanical properties of natural gas hydrate reservoirs. *J. Clean. Prod.* **2022**, *354*, 131703. [CrossRef]
4. Wei, N.; Pei, J.; Li, H.; Zhou, S.; Zhao, J.; Kvamme, B.; Coffin, R.B.; Zhang, L.; Zhang, Y.; Xue, J. Classification of natural gas hydrate resources: Review, application and prospect. *Gas Sci. Eng.* **2024**, *124*, 205269. [CrossRef]
5. Yan, C.; Dong, L.; Ren, X.; Cheng, Y. Stability of submarine slopes during replacement of methane in natural gas hydrates with carbon dioxide. *J. Clean. Prod.* **2023**, *383*, 135440. [CrossRef]
6. Jin, G.; Lei, H.; Xu, T.; Xin, X.; Yuan, Y.; Xia, Y.; Juo, J. Simulated geomechanical responses to marine methane hydrate recovery using horizontal wells in the Shenhu area, South China Sea. *Mar. Pet. Geol.* **2018**, *92*, 424–436. [CrossRef]
7. Zhou, S.; Zhao, J.; Li, Q.; Chen, W.; Zhou, J.; Wei, N.; Guo, P.; Sun, W. Optimal design of the engineering parameters for the first global trial production of marine natural gas hydrates through solid fluidization. *Nat. Gas Ind. B* **2018**, *5*, 118–131. [CrossRef]
8. Yamamoto, K.; Dallimore, S.R. Aurora-JOGMEC-NRCan Mallik 2006–2008 gas hydrate research project progress. Fire in the Ice. *Nat. Gas Oil* **2008**, *304*, 285–4541.
9. Yamamoto, K.; Terao, Y.; Fujii, T.; Ikawa, T.; Seki, M.; Matsuzawa, M.; Kanno, T. Operational overview of the first offshore production test of methane hydrates in the Eastern Nankai Trough. In Proceedings of the Offshore Technology Conference, Houston, TX, USA, 5–8 May 2014; pp. 1–15.
10. Chen, L.; Feng, Y.; Okajima, J.; Komiya, A.; Maruyama, S. Production behavior and numerical analysis for 2017 methane hydrate extraction test of Shenhu, South China Sea. *J. Nat. Gas Sci. Eng.* **2018**, *53*, 55–66. [CrossRef]
11. Reagan, M.T.; Moridis, G.J.; Johnson, J.N.; Pan, L.; Freeman, C.M.; Boyle, K.L.; Keen, N.D.; Husebo, J. Field-Scale Simulation of Production from Oceanic Gas Hydrate Deposits. *Transp. Porous Media* **2015**, *108*, 151–169. [CrossRef]
12. Rutqvist, J.; Moridis, G.J. Numerical studies on the geomechanical stability of hydrate-bearing sediments. *SPE J.* **2009**, *14*, 267–282. [CrossRef]
13. Ahmadi, G.; Ji, C.; Smith, D.H. Numerical solution for natural gas production from methane hydrate dissociation. *J. Pet. Sci. Eng.* **2004**, *41*, 269–285. [CrossRef]
14. Liu, X.; Flemings, P.B. Dynamic multiphase flow model of hydrate formation in marine sediments. *J. Geophys. Res.* **2007**, *112*. [CrossRef]
15. Moridis, G.J. Numerical studies of gas production from methane hydrates. *SPE J.* **2003**, *8*, 359–370. [CrossRef]
16. White, M.D.; Wurstner, S.K.; McGrail, B.P. Numerical studies of methane production from Class 1 gas hydrate accumulations enhanced with carbon dioxide injection. *Mar. Pet. Geol.* **2011**, *28*, 546–560. [CrossRef]
17. Hassanpouryouzband, A.; Joonaki, E.; Farahani, M.V.; Takeya, S.; Ruppel, C.; Yang, J.; English, N.J.; Schicks, J.M.; Edlmann, K.; Mehrabian, H.; et al. Gas hydrates in sustainable chemistry. *Chem. Soc. Rev.* **2020**, *49*, 5225–5309. [CrossRef] [PubMed]
18. Ruppel, C. Permafrost-associated gas hydrate: Is it really approximately 1% of the global system? *J. Chem. Eng. Data* **2015**, *60*, 429–436. [CrossRef]
19. Farahani, M.V.; Hassanpouryouzband, A.; Yang, J.; Tohidi, B. Insights into the climate-driven evolution of gas hydrate-bearing permafrost sediments: Implications for prediction of environmental impacts and security of energy in cold regions. *RSC Adv.* **2021**, *11*, 14334–14346. [CrossRef]
20. Yoon, H.C.; Kim, J. The impacts of scaled capillary pressure combined with coupled flow and geomechanics on gas hydrate deposits. *Géoméch. Energy Environ.* **2024**, *37*, 100529. [CrossRef]
21. Zhang, Y.; Zhang, P.; Hui, C.; Tian, S.; Zhang, B. Numerical analysis of the geomechanical responses during natural gas hydrate production by multilateral wells. *Energy* **2023**, *269*, 126810. [CrossRef]
22. Gupta, S.; Helmig, R.; Wohlmuth, B. Non-isothermal, multi-phase, multi-component flows through deformable methane hydrate reservoirs. *Comput. Geosci.* **2015**, *19*, 1063–1088. [CrossRef]
23. Gupta, S.; Deusner, C.; Haeckel, M.; Helmig, R.; Wohlmuth, B. Testing a thermo-chemo-hydro-geomechanical model for gas hydrate-bearing sediments using triaxial compression laboratory experiments. *Geochem. Geophys. Geosyst.* **2017**, *18*, 3419–3437. [CrossRef]
24. Hyodo, M.; Li, Y.; Yoneda, J.; Nakata, Y.; Yoshimoto, N.; Nishimura, A. Effects of dissociation on the shear strength and deformation behavior of methane hydrate-bearing sediments. *Mar. Pet. Geol.* **2014**, *51*, 52–62. [CrossRef]
25. He, Y.; Song, B.; Li, Q. Coupling Submarine Slope Stability and Wellbore Stability Analysis with Natural Gas Hydrate Drilling and Production in Submarine Slope Strata in the South China Sea. *J. Mar. Sci. Eng.* **2023**, *11*, 2069. [CrossRef]
26. Li, Q.; Zhao, D.; Yin, J.; Zhou, X.; Li, Y.; Chi, P.; Han, Y.; Ansari, U.; Cheng, Y. Sediment Instability Caused by Gas Production from Hydrate-bearing Sediment in Northern South China Sea by Horizontal Wellbore: Evolution and Mechanism. *Nat. Resour. Res.* **2023**, *32*, 1595–1620. [CrossRef]

27. Miramontes, E.; Sultan, N.; Garziglia, S.; Jouet, G.; Pelleter, E.; Cattaneo, A. Altered volcanic deposits as basal failure surfaces of submarine landslides. *Geology* **2018**, *46*, 663–666. [CrossRef]
28. Sun, T.; Wen, Z.; Yang, J. Research on Wellbore Stability in Deepwater Hydrate-Bearing Formations during Drilling. *Energies* **2024**, *17*, 823. [CrossRef]
29. Jamil, M.; Rahman, A.H.A.; Siddiqui, N.A.; Ibrahim, N.A.; Ahmed, N. A contemporary review of sedimentological and stratigraphic framework of the late paleogene deep marine sedimentary successions of West Sabah, North-West Borneo. *Bull. Geol. Soc. Malays.* **2020**, *69*, 53–65. [CrossRef]
30. Jamil, M.; Siddiqui, N.A.; Ahmed, N.; Usman, M.; Umar, M.; Rahim, H.U.; Imran, Q.S. Facies analysis and sedimentary architecture of hybrid event beds in submarine lobes: Insights from the crocker fan, nw borneo, malaysia. *J. Mar. Sci. Eng.* **2021**, *9*, 1133. [CrossRef]
31. Jamil, M.; Siddiqui, N.A.; Umar, M.; Usman, M.; Ahmed, N.; Rahman, A.H.A.; Zaidi, F.K. Aseismic and seismic impact on development of soft-sediment deformation structures in deep-marine sand-shaly Crocker fan in Sabah, NW Borneo. *J. King Saud Univ. Sci.* **2021**, *33*, 101522. [CrossRef]
32. Crutchley, G.J.; Mountjoy, J.J.; Pecher, I.A.; Gorman, A.R.; Henrys, S.A. Submarine slope instabilities coincident with shallow gas hydrate systems: Insights from New Zealand examples. In *Advances in Natural and Technological Hazards Research*; Springer International Publishing: Berlin/Heidelberg, Germany, 2016; Volume 41, pp. 401–409. [CrossRef]
33. Dhakal, S.; Gupta, I. Slope instability of submarine sediments due to hydrate dissociation: A case study of Northern Cascadia Margin. *Geoenergy Sci. Eng.* **2023**, *223*, 211558. [CrossRef]
34. Tan, L.; Liu, F.; Huang, Y.; Crosta, G.; Frattini, P.; Cen, X. Production-induced instability of a gentle submarine slope: Potential impact of gas hydrate exploitation with the huff-puff method. *Eng. Geol.* **2021**, *289*, 106174. [CrossRef]
35. Dong, B.-C.; Xiao, P.; Sun, Y.-F.; Kan, J.-Y.; Yang, M.-K.; Peng, X.-W.; Sun, C.-Y.; Chen, G.-J. Coupled flow and geomechanical analysis for gas production from marine heterogeneous hydrate-bearing sediments. *Energy* **2022**, *255*, 124501. [CrossRef]
36. Canals, M.; Lastras, G.; Urgeles, R.; Casamor, J.; Mienert, J.; Cattaneo, A.; De Batist, M.; Haflidason, H.; Imbo, Y.; Laberg, J.; et al. Slope failure dynamics and impacts from seafloor and shallow sub-seafloor geophysical data: Case studies from the COSTA project. *Mar. Geol.* **2004**, *213*, 9–72. [CrossRef]
37. Vanneste, M.; Sultan, N.; Garziglia, S.; Forsberg, C.F.; L'Heureux, J.-S. Seafloor instabilities and sediment deformation processes: The need for integrated, multi-disciplinary investigations. *Mar. Geol.* **2014**, *352*, 183–214. [CrossRef]
38. Wan, Z.-F.; Zhang, W.; Ma, C.; Liang, J.-Q.; Li, A.; Meng, D.-J.; Huang, W.; Yang, C.-Z.; Zhang, J.-F.; Sun, Y.-F. Dissociation of gas hydrates by hydrocarbon migration and accumulation-derived slope failures: An example from the South China Sea. *Geosci. Front.* **2022**, *13*, 101345. [CrossRef]
39. Pinkert, S.; Grozic, J.L.H.; Priest, J.A. Strain-Softening Model for Hydrate-Bearing Sands. *Int. J. Géoméch.* **2015**, *15*, 04015007. [CrossRef]
40. Hu, Z.B.; Yang, Y.X. Study on p-y curves of large-diameter steel pipe piles for offshore wind farm in sand based on in-situ tests. *J. Appl. Sci. Eng.* **2018**, *21*, 171–178. [CrossRef]
41. Kamath, V.A. A perspective on gas production from hydrates. In Proceedings of the JNOC's Methane Hydrate International Symposium, Chiba City, Japan, 20–22 October 1998; pp. 20–22.
42. Sun, X.; Mohanty, K.K. Kinetic simulation of methane hydrate formation and dissociation in porous media. *Chem. Eng. Sci.* **2006**, *61*, 3476–3495. [CrossRef]
43. Sánchez, M.; Gai, X.; Santamarina, J.C. A constitutive mechanical model for gas hydrate bearing sediments incorporating inelastic mechanisms. *Comput. Geotech.* **2017**, *84*, 28–46. [CrossRef]
44. Lin, J.; Seol, Y.; Choi, J.H. Geomechanical modeling of hydrate-bearing sediments during dissociation under shear. *Int. J. Numer. Anal. Methods Géoméch.* **2017**, *41*, 1523–1538. [CrossRef]
45. Ling, K.; Wu, X.; Zhang, H.; He, J. Improved gas resource calculation using modified material balance for overpressure gas reservoirs. *J. Nat. Gas Sci. Eng.* **2014**, *17*, 71–81. [CrossRef]
46. Miyazaki, K.; Tenma, N.; Aoki, K.; Yamaguchi, T. A nonlinear elastic model for triaxial compressive properties of artificial methane-hydrate-bearing sediment samples. *Energies* **2012**, *5*, 4057–4075. [CrossRef]
47. Uchida, S.; Soga, K.; Yamamoto, K. Critical state soil constitutive model for methane hydrate soil. *J. Geophys. Res.* **2012**, *117*. [CrossRef]
48. Chen, X.; Zhang, X.; Lu, X.; Wei, W.; Shi, Y. Numerical study on the deformation of soil stratum and vertical wells with gas hydrate dissociation. *Acta Mech. Sin.* **2016**, *32*, 905–914. [CrossRef]

MDPI AG
Grosspeteranlage 5
4052 Basel
Switzerland
Tel.: +41 61 683 77 34

Journal of Marine Science and Engineering Editorial Office
E-mail: jmse@mdpi.com
www.mdpi.com/journal/jmse

www.ingramcontent.com/pod-product-compliance
Lightning Source LLC
Chambersburg PA
CBHW040815250926
51850CB00007BA/1175
* 9 7 8 3 7 2 5 8 5 6 4 9 7 *